D0021341

Number Seven: Environmental History Series
Martin V. Melosi, *General Editor*

MEDIEVAL FRONTIER

MEDIEVAL FRONTIER

Culture and Ecology in Rijnland

by

WILLIAM H. TEBRAKE

Texas A&M University Press

COLLEGE STATION

Library of Congress Cataloging in Publication Data

TeBrake, William H. (William Henry), 1942–
Medieval frontier.

(Environmental history series ; no. 7)
Bibliography: p.
Includes index.
1. Rijnland (Netherlands)—Rural conditions.
2. Reclamation of land—Netherlands—Rijnland—History.
3. Land settlement—Netherlands—Rijnland—History.
4. Human ecology—Netherlands—Rijnland—History.
5. Hoogheemraadschap van Rijnland—History. I. Title.
II. Series.
HN510.R57T43 1984 304.2′09492′1 84-40133
ISBN 0-89096-204-9

Manufactured in the United States of America
FIRST EDITION

To my parents

Contents

List of Illustrations

FIGURE

Preface

In researching and writing this volume, I have concentrated on trying to illuminate and explore some of the essential ways in which certain human societies of the past interacted with their physical environments and how such relationships might have changed over time. My interest in doing so goes back to my days as a graduate student. For three successive semesters between September 1970 and December 1971 at the University of Texas at Austin, Paul W. English and Thomas F. Glick offered a continuing seminar on the environmental history of Western man. My initial association with it was as a more or less disinterested research associate, but after just a few sessions I became an interested participant. It was out of those weekly late-night sessions, with a dozen people arguing and discussing what environmental history is or should be, that the present investigation grew. Indeed, portions of it originally appeared, albeit in a considerably different form, as my doctoral dissertation in 1975.

Supported by funds made available to me by a Fulbright-Hays Doctoral Dissertation Fellowship, I was able to spend the academic year 1972–73 in the Netherlands collecting much of the information for my study. While there, I received tremendous assistance from D. P. Blok, the director of the place-name and settlement history section of the Dutch Royal Academy of Sciences, Amsterdam. He and his colleagues, especially R. Rentenaar, R. E. Künzel, and H. Buitenhuis, gave unstintingly of their time and knowledge. By offering me free access to their excellent collection of sources and literature and by giving me space in which to work, they provided me with the best of all possible working environments. Further, G.'t. Hart, *chartermeester* of the archives of the Hoogheemraadschap van Rijnland, in Leiden, and the staff at the Algemeen Rijksarchief, in The Hague, gave me guidance and friendly assistance whenever it was needed.

The Faculty Research Funds Committee of the University of Maine at Orono funded additional research. I was awarded a Faculty Research Grant in 1978 and a Faculty Summer Grant that allowed me to return to the Netherlands in the summer of 1980. I wish to thank the many University of Maine students who have taken my courses in medieval history and environmental history. They have been some of my toughest and yet most encouraging critics throughout. Dean Webster, meanwhile, showed a remarkable ability in transforming my rough sketches and drawings into creditable maps and figures. I owe a special debt, however, to my brother and fellow historian, Wayne TeBrake, who patiently read and criticized many versions of this work and on more than one occasion provided the necessary insight through late-night long-distance telephone conversations when I had become completely stuck.

MEDIEVAL FRONTIER

1

Introduction

BETWEEN the late tenth and early fourteenth centuries population growth, economic expansion, and urban revival laid the foundations of a uniquely European civilization. Modern medieval historiography has gone a long way toward revealing the characteristic features of this civilization and has begun to deal as well with the underlying productive forces associated with it. A series of developments in the countryside, in particular, made a significant difference. Whereas European agriculture in the early Middle Ages was geared essentially toward self-sufficiency, by the late Middle Ages it was producing a surplus of food and fiber capable of feeding and clothing growing numbers of nonagriculturalists. In between, rural Europe experienced a tremendous burst in productive capacity that helps explain more than anything else the vitality and creative energy of this formative period.

While it is easy to agree that developments in the countryside were important, perhaps even crucial, for the emergence of a new civilization during the High Middle Ages, this does not mean that such developments have received adequate treatment in the historical literature of the period. Although some studies have examined changes in field systems or agrarian technology or the introduction of urban markets in an attempt to account for changes in rural productivity, few have looked at the most obvious change of all: the tremendous expansion in the scale of agricultural activity.

During the High Middle Ages the face of Europe was transformed from a sea of wilderness with small, scattered islands of settlement and agricultural activity into a more or less continuous agrarian landscape. There is evidence that two-thirds or more of the presently occupied landscape was first brought into the world of human affairs during that period. Not only did this represent the first major change in the location of settlement and agriculture since the Bronze Age, but also it established the geographical outlines that have persisted to the present. To date, however, the effect that this incorporation of huge quantities of previously unused and thus fertile lands into the realm of culture might have had on the health and well-being of rural society, and by extension urban society, has never been properly investigated. In what follows, I attempt to treat this deficiency by examining the context, course of events, and results of land reclamation in a small, compact portion of the western Netherlands.

The Emergence of European Civilization

Modern historical research has established that the long interim between ancient and modern times was as varied, complex, and full of change as any other period of a thousand or more years. The old myth of the Dark Ages, with its visions of unrelenting sameness and monotony, has given way to the knowledge that the Middle Ages exhibited the full spectrum of human experiences. Thus there were what might be termed good times as well as bad times, phases of growth, expansion, or concentration as well as of decline, contraction, or fragmentation. Although there were times when it seemed that the great scourges of poverty, ignorance, and disease might completely overwhelm European society, there were others when, clearly, the momentum was in the opposite direction. A good example of the latter is the period known as the High Middle Ages, between 1000 and 1300, characterized by an energy and exuberance that set it off clearly from both preceding and succeeding periods.

The late tenth to the early fourteenth centuries comprised one of the most formative periods of European history. It was a time during which literacy and learning advanced greatly with the establishment of new schools and universities. Intellectual movements emerged that were more than simply extensions of Asian or African traditions, exhibiting instead a decidedly European focus. At the same time, Western Christianity, no longer requiring a united front against paganism, began developing a greater variety and complexity of forms as it sought to permeate all aspects of life. In the world of politics, meanwhile, new patterns of public authority appeared that in some respects survived in recognizable forms until modern times; indeed, some historians suggest that the prototype of the modern state was invented during this period.

These and other changes between the late 900s and the early 1300s were the harbingers of a new, uniquely European civilization,[1] and the

[1]Archibald R. Lewis, "The Closing of the Medieval Frontier, 1250–1350," *Speculum* 33 (1958): 477; P. H. Sawyer, "Baldersby, Borup and Bruges: The Rise of Northern Europe," *University of Leeds Review* 16 (1973): 75–76. Kent V. Flannery, "The Cultural Evolution of

gradual realization of this fact was a major step in the triumph of the Middle Ages in historiographic tradition. Consequently, even those who continue to insist that there was a Renaissance toward the end of the Middle Ages are required at the very least to examine it within its late-medieval context.[2]

One of the most salient features of this new European civilization was the degree to which it was an urban phenomenon. Beginning in the late tenth century, urban life made a strong reappearance for the first time since late antiquity. While most Europeans continued to live in the countryside until the Industrial Revolution of modern times, the new cities of the High Middle Ages constituted a permanent and increasingly important aspect of the European scene. It was in the urban environment, after all, that most of the new intellectual, religious, and political departures of this period originated and saw their fullest and most creative development, freed from the often depressing and energy-draining constraints imposed by the rural aristocracy that had dominated so many aspects of life in the early Middle Ages.[3]

The emergence of a new European civilization during the High Middle Ages was, however, merely a symptom of a society going through a tremendous expansionary thrust, expressed at its most elemental level by a significant surplus of births over deaths. The population of all of Europe more than doubled during the period without any apparent episodes of stagnation or decline, a truly remarkable achievement for a preindustrial society.[4] In fact, some areas saw considerably greater population growth;

Civilizations," *Annual Review of Ecology and Systematics* 3 (1972): 399–426, offers many useful and challenging perspectives on the emergence and development of "civilization." See also the interesting article by Frederic L. Cheyette, "The Invention of the State," in Bede Karl Lackner and Kenneth Roy Philp, eds., *Essays in Medieval Civilization*, pp. 143–78.

[2]This point is made in William J. Bouwsma, "The Renaissance and the Drama of Western History," *American Historical Review* 84 (1979): 9.

[3]David Sturdy, "Correlation of Evidence of Medieval Urban Communities," in Peter J. Ucko, Ruth Tringham, and G. W. Dimbleby, eds., *Man, Settlement and Urbanism*, p. 863; Jacques LeGoff, "The Town as an Agent of Civilisation, 1200–1500," in Carlo M. Cipolla, ed., *The Middle Ages*, pp. 71–106, vol. 1 in *The Fontana Economic History of Europe*, is an excellent introduction to medieval urban life in all of its facets.

[4]Josiah Cox Russell, "Population in Europe, 500–1500," in Cipolla, ed. *The Middle Ages*, p. 36, estimates 38.5 million people for the year 1000 and 73.5 million for 1340. Since population growth may have occurred earlier than most of the other changes discussed here, and since all of these changes were well under way by 1000, it may be useful to establish an earlier benchmark if possible. Norman J. G. Pounds, *An Historical Geography of Europe, 450 B.C.–A.D. 1330*, p. 184, calculates 26.5 million for the early ninth century. By this estimate the 73.5 million figure for 1340 represents a near tripling of Europe's population.

parts of Saxony reputedly experienced a tenfold increase, for example.[5] The early Middle Ages, in contrast, had been characterized by population decline and the problems of chronic underpopulation. This great burst of human vitality during the High Middle Ages was sufficient to populate the new cities, without in any way depopulating traditional areas of rural settlement.

The expansionary tendencies represented by population growth and urbanization during the High Middle Ages were expressed on another level by a considerable enlargement of Europe's area of political, military, and religious affairs. What might be termed the European sphere of influence in the early tenth century consisted roughly of central and northern Italy, France, the Low Countries, extreme western Germany, and southeastern England. By the early fourteenth century, in contrast, Europe had shifted its boundaries outward in all directions: through the conquest of Spain, southern Italy, and the major islands of the Mediterranean; the Anglo-Norman campaigns against the northern and western extremities of Great Britain and Ireland; the Christianization of Scandinavia; and the German drive into central and eastern Europe.[6] According to some observers, this external expansion exhibited many of the same features as the spread of Europeans across North America some six centuries later.[7] In any case, it showed the growing vitality and potency of the European social system.

If expansion and growth can be seen as characteristic of the High Middle Ages, clearly setting it off from the previous period, then contraction and decline can be seen as characteristic of the late Middle Ages. In fact, it has become a commonplace in late-medieval historiography to describe the fourteenth and early fifteenth centuries as times of great difficulty or crisis, when many of the great opportunities of the past had long since evapo-

[5]While total European population increase was less than threefold, France, the Low Countries, Germany, Scandinavia, and the British Isles saw more than a tripling; see Russell, "Population in Europe, 500–1500," p. 36; Pounds, *An Historical Geography of Europe*, p. 184. For Saxony see Edith Ennen and Walter Janssen, *Deutsche Agrargeschichte: vom Neolithikum bis zur Schwelle des Industriezeitalters*, p. 164.

[6]See Richard Koebner, "The Settlement and Colonization of Europe," in M. M. Postan, ed., *The Agrarian Life of the Middle Ages*, 2d ed., pp. 57–91, vol. 1 in *The Cambridge Economic History of Europe*; and the collection of documents in James Muldoon, *The Expansion of Europe: The First Phase*.

[7]This is the thesis of Lewis, "The Closing of the Medieval Frontier," pp. 475–83, but elements of it are found in the work of many other scholars of the period. See, for example, the extensive literature referred to in Bryce Lyon, "Medieval Real Estate Developments and Freedom," *American Historical Review* 63 (1957): 47–61.

rated. Most of the elements of growth and expansion had indeed come to a halt during the last years of the thirteenth century, and before long they were replaced by elements of decline and contraction.[8] For example, between the 1340s and 1440s massive depopulation and severe dislocation resulting from epidemic disease, chronic warfare, and economic depression repeatedly battered the entire European continent. Not until the sixteenth century did the population growth and relative well-being of the High Middle Ages begin to return to Europe.[9]

Severe as it was, however, this late medieval crisis was not sufficient to plunge Europe back into the circumstances it had known in the early Middle Ages. Even at its lowest point, probably in the second quarter of the fifteenth century, the total population of Europe was still far greater than it had ever been in late-ancient or early-medieval times, and it was spread out over a much greater area than ever before. In addition, although mortality rates in cities could be awesome at times, a steady flow of migrants from the countryside kept the cities alive. Urban life, in short, continued through the late Middle Ages. Finally, the forms of intellectual, religious, and political life developed during the High Middle Ages also persisted and actually came to flower during the Renaissance. While many former opportunities were gone, requiring considerable readjustment and refinement of late-medieval society, European civilization survived the period of late-medieval difficulty relatively intact.

Such themes of rise and fall, of expansion and contraction, have become the new orthodoxy of medieval historiography, replacing that older orthodoxy of a bland and featureless Middle Ages. Although there are those who think that growth may have started earlier in some places or lasted longer in others, as well as those who believe that the decline of the

[8] Lewis, "The Closing of the Medieval Frontier," pp. 477–83; B. H. Slicher van Bath, *Agrarian History of Western Europe, A.D. 500–1850*, trans. Olive Ordish, pp. 160–70.

[9] The literature on the late-medieval crisis is considerable. A good general introduction to the problems of the fourteenth and fifteenth centuries is found in Harry A. Miskimin, *The Economy of Early Renaissance Europe, 1300–1460*. Immanuel Wallerstein, *The Modern World System I: Capitalist Agriculture and the Origins of the European World Economy in the Sixteenth Century*, pp. 21–37, reviews much of the literature. One of the classic treatments of the late-medieval crisis is Léopold Genicot, "Crisis: From the Middle Ages to Modern Times," in Postan, ed., *The Agrarian Life of the Middle Ages*, pp. 660–741. Guy Bois, *Crise du féodalisme: économie rurale et démographie en Normandie orientale du début du 14e siècle au milieu du 16e siècle*, places the late-medieval rupture in a broad social, economic, and political context.

late Middle Ages was somewhat less automatic or spectacular in certain quarters, there is little reason to expect that the overall picture will see much change in the near future.

While the broad patterns of ebb and flow have become part of the modern view of the Middle Ages, the task of explaining such phenomena remains largely unattempted. Indeed, the tidal analogy suggests an important point here. Just as there is more to the movement of tides than can be explained by exclusive attention to their effects on the seashore, so too there is more to the ebb and flow of civilizations than can be understood by examining only the trappings of civilization. For example, despite their importance and usefulness in characterizing an age, new departures in intellectual, religious, and political affairs as well as the growth of cities during the High Middle Ages were merely symptoms of deeper-lying changes. Even demographic cycles, though much closer to the heart of the matter, were hardly the exogenous or totally independent variables that some scholars maintain.[10] Rather, it was a series of transformations occurring in the European countryside that lay behind or was at least inextricably intertwined with all the other great changes of the period.

The importance of changes in the European countryside becomes clearer when it is remembered that Europe as a whole remained overwhelmingly rural until the Industrial Revolution.[11] Even at the height of medieval urban development the primary production of society continued to be that of food and fiber, products of the countryside. Indeed, the lives of everyone were directly affected by rural affairs because the economy continued to be fueled by the production, distribution, and consumption of almost exclusively agricultural products. As a result, it was agriculture, that complex of activities at the juncture of human culture and natural process, that determined the health and well-being of the rest of society. European society as a whole could exhibit a greater richness and variety of

[10] Douglass C. North and Robert Paul Thomas, *The Rise of the Western World: A New Economic History*, p. 26: "Growing population was the exogenous variable that basically accounts for the growth and development of Western Europe during the high Middle Ages." Jan de Vries, *The Dutch Rural Economy in the Golden Age, 1500–1700*, p. 4, cautions that classifying population growth as exogenous is an oversimplification to fit a particular model.

[11] B. H. Slicher van Bath, "Agrarische productiviteit in het preindustriële Europa," in his *Bijdragen tot de agrarische geschiedenis*, pp. 152–53. For basic definitions and parameters of peasant societies, Eric R. Wolf, *Peasants*, especially pp. 1–17, remains a good starting point, while the essays in Jerome Blum, ed., *Our Forgotten Past: Seven Centuries of Life on the Land*, explore more fully a number of important rural themes.

forms during the High Middle Ages precisely because agriculture was becoming increasingly more productive, to the point of making large and reliable surpluses of agricultural commodities available for consumption by nonagrarian segments of society.

The productivity of European agriculture increased during the High Middle Ages because of its ability to break out of the restricted patterns of the early Middle Ages. This is seen most clearly by comparing certain agrarian features of the tenth century with various aspects of the early fourteenth century. Europe was sparsely settled in 950 by a population made up almost exclusively of agriculturalists, who produced food and fiber not only for their own consumption but also for a relatively small but powerful military and religious elite. Since it produced nothing itself, this elite was parasitic. By controlling the means of production and the patterns of distribution and consumption, however, it had been able to establish and maintain itself in a dominant position for centuries. Total productivity was low with virtually no surplus, while most commodities were consumed in the same locality in which they were produced. Although matters had improved remarkably since the difficult sixth and seventh centuries, local self-sufficiency remained the hallmark of agriculture in the early tenth century.

By the fourteenth century, in contrast, Europe was much more densely settled, with a vastly more complex economic and social picture. At the one extreme were those who continued to live as before within a system of local production, distribution, and consumption controlled by traditional rural elites. At the other extreme were growing numbers of urbanites who had withdrawn from agriculture to engage in a broad spectrum of urban activities and provisioned themselves by purchasing what they needed on urban markets that represented systems of distribution no longer solely in the hands of the traditional elites. In between, also freed from the domination of the traditional elites, were millions of agriculturalists who, in addition to maintaining themselves and other local groups, produced a surplus of agricultural products that was sufficient to fuel the entire urban market system.[12] Total productivity by the early 1300s was vastly greater than it

[12]The fundamental characteristics and early stages of this transformation are examined in Georges Duby, *The Early Growth of the European Economy: Warriors and Peasants from the Seventh to the Twelfth Centuries*, trans. Howard B. Clarke. Slicher van Bath, *The Agrarian History of Western Europe*, pp. 3–160, describes the transformation as a change from "direct" to "indirect" agricultural consumption, reaching its peak between 1150 and 1300.

had been four centuries earlier, allowing more people to live at higher levels of consumption than had ever before been possible in Europe.[13]

The tenth to early fourteenth centuries, then, stand out clearly from the rest of the Middle Ages, and one of the most important achievements of the period was a significant transformation in the European countryside. By breaking through some of the limits to earlier patterns of agricultural production, rural Europe created the wherewithal for the emergence of a new civilization.[14] Nevertheless, although the trappings of this new civilization have become well known, especially at those levels most distant from agriculture, the actual course of events in the countryside remains largely unexplored.

This book represents an attempt to uncover the true dynamics of the High Middle Ages. My starting assumption is, simply, that any complete assessment of the health and well-being of an agrarian civilization like that of preindustrial Europe must take into consideration what occurred in the countryside. By focusing attention on developments in land use and agriculture, particularly as they took place in portions of temperate Europe between the late 900s and early 1300s, I shall be able to reveal some of the primary forces lying behind the glitter and glamour of the High Middle Ages.

The Transformation of the European Landscape

The increase in agricultural productivity that lay behind the other great changes of the High Middle Ages was made possible primarily by a massive reclamation and colonization movement that drastically extended the scale of agrarian activity and irrevocably altered the physical appearance of Europe. Indeed, between the late 900s and the early 1300s, Europeans produced environmental change on a scale rarely if ever seen before or since. As late as the tenth century large portions of the Continent consisted of what might be termed wilderness, where the imprint of human culture, if it existed at all, was fleeting and slight. This was true particularly of temperate Europe, north of the Mediterranean basin, where vast areas re-

[13] Russell, "Population Growth in Europe," pp. 46–47, suggests, for example, that medieval life expectancy reached its highest point in the thirteenth century, at least according to evidence in England.

[14] John Alexander, "The Beginnings of Urban Life in Europe," in Ucko, Tringham, and Dimbleby, eds., *Man, Settlement and Urbanism*, p. 844.

mained unaffected by human activity, largely covered by deciduous forest but including enormous tracts of freshwater swamp and salt marsh as well. By the early fourteenth century, in contrast, this situation had been dramatically reversed. Only small patches of wilderness remained in a landscape that had become characterized by pastures, fields, and villages.

An example from central Germany illustrates the scale of landscape change during the High Middle Ages. In 1073, Emperor Henry IV, in danger of being captured by hostle Saxons along the northern fringe of the Harz Mountains near Harzburg, saved himself by fleeing 90 kilometers to the settlement at Eschwege. It was said that he journeyed for three days along a narrow path through what was described as primeval forest before arriving at his destination in a state of exhaustion. Two hundred years later Eschwege was entirely surrounded by a continuous expanse of fields and villages.[15]

Of course, not all regions of Europe were affected to exactly the same degree, since the proportions of wilderness to settled area varied greatly from place to place. Ever since the first appearance of sedentary farming communities Mediterranean Europe had known much more intensive settlement patterns than had areas in the north, though by the tenth century certain portions of temperate Europe had begun to catch up. In particular, the area between the Loire and Rhine rivers, the core of the early-medieval Frankish state, had become densely settled, but such densities tailed off dramatically eastward and northward from the Rhine.[16] The Harz region, in central Germany, therefore, possessed a greater potential for landscape change than did many other areas. Even so, tremendous changes of this kind were also seen in the Frankish heartland and in Mediterranean Europe, though in these regions they involved not so much vast, unbroken tracts of unused lands as the much smaller and more numerous areas of less intensively used lands on the margins of existing settlements.

While it is impossible to show exactly how much of Europe was brought into culture by reclamation and colonization during the High Middle Ages, there can be no doubt that it was extensive. One of the most eloquent testimonies of sweeping change is the large numbers of place-

[15] Wilhelm Abel, *Agricultural Fluctuations in Europe: From the Thirteenth to the Twentieth Centuries*, trans. Olive Ordish, p. 26, drawing on the *Annals* of Lampert of Hersfeld.

[16] See, for example, the calculations in B. H. Slicher van Bath, "The Economic and Social Conditions of the Frisian Districts from 900 to 1500," *A. A. G. Bijdragen*, no. 13 (1965): 101–102.

names that refer to assarts, clearings, and other aspects of the reclamation process. They are found throughout all of Europe, and most of them date from the eleventh through the thirteenth centuries.[17] Further evidence of reclamation and colonization is seen in the establishment of numerous monasteries in wooded or boggy areas, in the extension of the church's network of parishes into those portions of the countryside where it had previously been lacking, and in the increasing importance of the tithes levied on newly reclaimed lands (*novales*).[18]

The scale was most impressive in central and eastern Europe, where Saxony, Thuringia, Franconia, Pomerania, Mecklenburg, Prussia, Bohemia, Transylvania, and other areas were reclaimed and settled by colonists from the west.[19] In Saxony, for example, the density of settlement is said to have increased tenfold during the High Middle Ages.[20] In the Alps and other mountainous regions, meanwhile, the edge of human settlement crept to ever-higher elevations.[21] Even in such a long-occupied and densely settled area as the Cologne region, however, there was a noticeable in-

[17] M. Gysseling, *Toponymisch woordenboek van België, Nederland, Luxemburg, Noord-Frankrijk en West-Duitsland (vóór 1226)*, passim. See also H. C. Darby, "The Clearing of the Woodland in Europe," in William L. Thomas et al., eds., *Man's Role in Changing the Face of the Earth*, pp. 191–92, 195, 197–98; Wilhelm Abel, *Geschichte der deutschen Landwirtschaft vom frühen Mittelalter bis zum 19. Jahrhundert*, 2d ed., p. 28; Koebner, "The Settlement and Colonization of Europe," p. 82.

[18] Marc Bloch, *French Rural History: An Essay in Its Basic Characteristics*, trans. Janet Sondheimer, pp. 11–12; Roger Grand and Raymond Delatouche, *L'Agriculture au moyen âge de la fin de l'empire romaine au XVIe siècle*, pp. 250–52; Georges Duby, *Rural Economy and Country Life in the Medieval West*, trans. Cynthia Postan, pp. 69–71, 77; Duby, *The Early Growth of the European Economy*, p. 202; C. Dekker, "De vorming van aartsdiakonaten in het diocees Utrecht in de tweede helft van de 11e en het eerste kwart van de 12e eeuw," *Geografisch tijdschrift* 11 (1977): 339–60; Slicher van Bath, *The Agrarian History of Western Europe*, pp. 153–55; Ennen and Janssen, *Deutsche Agrargeschichte*, p. 164; François Louis Ganshof and Adriaan Verhulst, "Medieval Agrarian Society in Its Prime: France, the Low Countries and Western Germany," in Postan, ed., *The Agrarian Life of the Middle Ages*, pp. 292–93; Abel, *Agricultural Fluctuations in Europe*, p. 26; Koebner, "The Settlement and Colonization of Europe," pp. 68, 76–77, 81, 88.

[19] See the survey in Koebner, "The Settlement and Colonization of Europe," pp. 67–91; and the literature list in Herbert Helbig and Lorenz Weinrich, eds., *Mittel- und Norddeutschland, Ostseeküste*, 2d ed., pp. 28–36, vol. 1 in *Urkunden und erzählende Quellen zur deutschen Ostsiedlung im Mittelalter*.

[20] Ennen and Janssen, *Deutsche Agrargeschichte*, p. 164.

[21] Darby, "The Clearing of the Woodland in Europe," p. 197; Slicher van Bath, *The Agrarian History of Western Europe*, p. 133; Koebner, "The Settlement and Colonization of Europe," pp. 68, 87.

crease in *novales* collections during the eleventh and twelfth centuries.[22] Some estimates of the extent of land brought into culture between the tenth and fourteenth centuries run as high as two-thirds of the present cultivated area, even for some of the more densely settled portions of the Continent.[23] Not since the beginnings of agriculture during the Neolithic era had temperate Europe seen landscape change of such magnitude.

The transformation of the face of Europe on the scale suggested here not surprisingly had a tremendous impact on all aspects of life during the High Middle Ages. In general, it had a very stimulating effect on the rural economy in increasing agricultural productivity, which brought in its train a large measure of prosperity that eventually permeated both the agricultural and the nonagricultural sectors of society. More specifically, the relative well-being of the era was expressed through the increased tithes, rents, fees, and taxes accruing to the ruling groups; through the larger supplies of food afforded by the vastly extended area of cultivable land; and through the opportunities offered to the otherwise landless to occupy lands of their own.[24]

Some of the beneficial effects of the reclamation and colonization movement of the High Middle Ages were recognized in the early thirteenth century by Caesarius, abbot of the monastery at Prüm, about 55 kilometers north of Trier, in the Eiffel region. During the previous three centuries (893–1222), he wrote, much forested land had been cleared and many estates established, greatly increasing the tithes. Further, he added, many mills had been constructed, many vines planted, and an immeasurable amount of land brought into cultivation.[25]

The effects of bringing wilderness into the world of human culture were expressed even more clearly a century later by an observer of political affairs in the county of Holland. Philippus de Leyden reported around

[22] Ennen and Janssen, *Deutsche Agrargeschichte*, p. 164; Koebner, "The Settlement and Colonization of Europe," p. 68.

[23] H. Draye, *Landelijke cultuurvormen en kolonisatiegeschiedenis*, p. 60, produces the two-thirds figure for the Frankish heartland. Abel, *Agricultural Fluctuations in Europe*, p. 26, simply suggests that the extent of arable lands in 1300 was "many times larger" than it was around 1000.

[24] Duby, *Rural Economy and Country Life*, p. 77.

[25] Quoted in Karl Lamprecht, *Deutsches Wirtschaftsleben im Mittelalter: Untersuchungen über die Entwicklung der materielen Kultur des platten Landes auf Grund der Quellen zunächst des Mosellandes*, vol. 1, pt. 1, p. 402.

1350 that comital revenues from the new villages founded in the former peat-bog wilderness of coastal Holland constituted the most important source of income that the counts of Hainault, Holland, and Zeeland possessed in Holland.[26] In fact, his observations are supported by the fourteenth-century tithe collections recorded in the surviving comital accounts.[27]

In addition to altering drastically the physical appearance of Europe, therefore, the reclamation and settlement movement of the late tenth through early fourteenth centuries had a stimulating effect on many aspects of European society. Most important over the long term, however, was that it laid the geographical framework from which modern Europe evolved a half millennium later. The patterns of rural settlement and agriculture established during the High Middle Ages, while clearly breaking with the patterns that had persisted until then, survived with very little alteration until the early nineteenth century and remain highly visible to the present.[28]

Although the physical appearance of Europe was drastically changed by the reclamation and colonization of vast amounts of wilderness during the High Middle Ages, and although it is easy to see that such a process would have affected many aspects of medieval life, such considerations are encountered only rarely in the literature of medieval history. Even when they are found, the activities of reclamation and colonization are usually described as mere consequences of economic, political, or military growth and expansion. The possibility that the bringing of wilderness into culture might have been as much the cause as the consequence of such growth and expansion, or, indeed, that it might have set much of the tone and flavor of the entire era, remains virtually unacknowledged in medieval historiography. Ultimately only detailed study at the local and regional level will be effective in exploring these issues and eventually revealing the true significance of the reclamation and colonization of Europe's wilderness during the High Middle Ages.

[26] Philippus de Leyden, *De Cura reipublicae et sorte principantis*, ed. R. Fruin and P. C. Molhuijsen, p. 176. For Philippus de Leyden see P. Leupen, *Philip of Leyden, Fourteenth Century Jurist: A Study of His Life and Treatise "De Cura reipublicae et sorte principantis"*. From the late thirteenth century Holland and Zeeland were ruled by the counts of Hainault, or Henegouwen (Hannonia), in modern southern Belgium and northern France.

[27] See below, chap. 7.

[28] Sawyer, "Baldersby, Borup and Bruges," pp. 75–76; Clarence J. Glacken, "Man and the Earth," *Landscape* 5 (1956): 28.

The Dutch Rijnland serves very well as a case study of the physical transformation of the European continent from the late tenth to the early fourteenth centuries because it exhibited all the changes characteristic of the period. Not only was its areal extent tripled during the High Middle Ages, but it may well have been one of the first regions of temperate Europe to experience such widespread change. It also served as an important source of colonists who from the eleventh century onward undertook the reclamation and settlement of other lands in central and eastern Europe.

Rijnland as a Case Study

What happened in Rijnland between A.D. 950 and 1350 makes an interesting story that, in many respects, fits nicely into the broader, popular view of the heroic Dutch struggle against overweening water. Over a period of four centuries, a small population, initially confined to a narrow 25-kilometer-long strip of coastal sand dune just north of the The Hague, grew substantially in numbers and expanded into and across a broad band of previously uninhabited peat bogs on the east. By digging drainage ditches, they made the bogs habitable, and by constructing a network of canals, dikes, dams, and sluices, they were able to protect much of what they had gained from subsequent flooding. In the middle of the fourteenth century Rijnland measured approximately 25 kilometers by 25 kilometers, divided among about forty rural settlements, most of which survive in one form or another today. Most of the population in 1350 lived on what had been uninhabited peat bog in 950. While lacking the dramatic quality of a young boy plugging a hole in a dike with his finger and thereby sparing the entire community, this sequence of events helps show how, by dint of hard work and persistence (some might suggest typical Dutch stubbornness), residents of the western Netherlands were able to begin that long struggle with uninviting physical conditions that resulted in today's landscape, thoroughly humanized yet charming and picturesque.

Of course, there is nothing inherently wrong with scholarly research seemingly substantiating a popular perception. After all, knowledge of the processes of reclamation and colonization between the tenth and fourteenth centuries goes a long way toward explaining how, by the end of the Middle Ages, a particular piece of North Sea coastal lowland, originally among the most remote, underpopulated, and underdeveloped sections of the Continent, could have become central to one of the most populous and highly developed portions of Europe's new northwestern core. Indeed, much of what follows will show exactly that.

To see diking and draining operations in Rijnland during the High Middle Ages as nothing more than an episode in Dutch history, however, would not only be wrong but also obscure unnecessarily the reasons why they were carried out in the first place. A considerable amount of diking and draining was undertaken elsewhere, not only along the entire North Sea coast from northern France to Denmark but also in various portions of the Baltic region, in parts of eastern and southern England, and in sections of western France. Even if no activities of this sort could be found outside the Netherlands, it is important to understand exactly what they were, namely, symptoms of underlying forces that found expression in other forms of activity in different environments.

At the same time that Rijnlanders were reclaiming and colonizing peat bogs, for example, other Europeans were reclaiming and colonizing huge expanses of previously uninhabited woodland. In each endeavor, the purpose was the same: to provide more living space for a rapidly expanding social system. For this reason, one of my primary concerns is to present the reclamation and colonization of Rijnland as an important segment of a much larger movement that succeeded in transforming Europe's vast wilderness areas into cultural landscapes.

To appreciate fully the significance of the changes that were introduced into temperate Europe through reclamation and colonization during the High Middle Ages, it is important to have some knowledge of what conditions were like before they occurred. To gain such an understanding for Rijnland, one must dig deeply into the geological, archaeological, and pedological, or soil-science, literature of the western Netherlands. This search is rewarded by a wealth of information and perspectives that allows one not only to reconstruct the main features of the natural history of the western Netherlands but also to outline the ways in which human societies interacted with the natural environment before the period of reclamation and colonization. This investigation clearly shows that diking and draining set in motion a process that upset or overturned traditional patterns of settlement and land use. These patterns had begun to appear, in outline at least, with the first permanent settlement of the area during the late Neolithic era, and they survived relatively unaltered from the Bronze Age. Reclamation and colonization during the High Middle Ages apparently represented the most comprehensive change in ecological relations in Rijnland in about three and a half millennia. Similar findings will doubtless result when an extended chronology and an interdisciplinary approach are applied to the investigation of other areas of temperate Europe for the same period.

Because the reclamation and colonization of new lands during the High Middle Ages formed such an important watershed—really the first major break in agricultural and settlement patterns in some three millennia—I have organized my study around the themes of continuity and change. In the two chapters of part one I set the stage by outlining the patterns of medieval demographic and agricultural productivity and by reconstructing the natural history and the premedieval settlement patterns of the western Netherlands since the last ice age. In part two, the continuity portion, I examine the cultural and ecological parameters of early-medieval settlement in Rijnland. In part three, the section devoted to change, I first recreate the reclamation and colonization process in Rijnland during the High Middle Ages before going on to explore how it changed the cultural and ecological context of life in the region. In the final chapter I consider not only how the process of reclamation and colonization in Rijnland and adjacent areas set the stage for the Dutch golden age a few centuries later but also what the Rijnland endeavor can reveal about the process throughout temperate Europe during the High Middle Ages.

PART ONE
The Historical and Ecological Setting

2

The Frontier in Medieval Europe

THE term "frontier" was introduced around the turn of the twentieth century to describe a set of historical circumstances and experiences assumed to be unique to Europeans in North America. Since then, however, it has had many kinds of uses in a wide variety of spatial and temporal contexts.[1] Among these have been attempts to apply it to various places and periods during the European Middle Ages.[2] The major problem here is that the word "frontier" has come to mean different things to different people.[3]

In his classic essay of 1958 on the medieval frontier, Archibald R. Lewis distinguished between the external frontier and the internal frontier of the High Middle Ages. The external frontier, according to him, consisted of those areas beyond Europe's heartland (that is, outside central and northern Italy, France, extreme western Germany, the Low Countries, and southeastern England) that were incorporated into the European political, military, and religious sphere of influence during the High Middle Ages through military conquest and strategic colonization. The internal frontier, by contrast, consisted of those areas within Europe's boundaries that were transformed from wilderness into integral portions of the European cultural landscape through reclamation and essentially agricultural colonization. While the activities of the internal frontier were first put into consistent practice in Europe's heartland, once the expanded boundaries of

[1] A good introduction to the literature on frontiers is found in William W. Savage, Jr., and Stephen I. Thompson, "The Comparative Study of the Frontier: An Introduction," in William W. Savage, Jr., and Stephen I. Thompson, eds., *The Frontier: Comparative Studies*, 2:3–24.

[2] See, for example, Archibald R. Lewis, "The Closing of the Medieval Frontier, 1250 to 1350," *Speculum* 33 (1958): 477–83; Bryce Lyon, "Medieval Real Estate Developments and Freedom," *American Historical Review* 63 (1957): 47–61; Richard E. Sullivan, "The Medieval Monk as Frontiersman," in Savage and Thompson, eds., *The Frontier: Comparative Studies*, 2:25–49.

[3] A wide range of meanings is evident, for example, in the studies collected in David Harry Miller and Jerome O. Steffen, eds., *The Frontier: Comparative Studies* [vol. 1]; Savage and Thompson, eds., *The Frontier: Comparative Studies*, vol. 2.

the external frontier had been established and maintained by military conquest and colonization, the process of filling in the empty spaces by reclamation and agricultural colonization could be and often was applied there as well.[4] This was true particularly of the German drive into sparsely settled Slavic Europe, where the activities of the external frontier were followed quickly by the activities of the internal frontier.[5] Throughout this book the word "frontier" refers exclusively to the internal frontier of the High Middle Ages.

Lewis uses the word "frontier," as do many others who write on the subject, to designate various things. In addition to applying the term to areas in which such activities as conquest, reclamation, and colonization were played out, he seems to have in mind a set of advantages accruing to frontier participants (such as freedom and prosperity) as well as the period of time within which such activities and advantages occurred.[6] To avoid the confusion that such a multiplicity of uses can cause, the term "frontier" is used in this book in a geographical sense only, to designate those geographical zones that during the High Middle Ages were transformed from wilderness into a part of the cultural landscape through the activities of reclamation and colonization.

The term "colonization" is employed in this book in a very straightforward fashion, to refer simply to the movement of people into a frontier zone for the purposes of establishing residence and practicing agriculture. The word "reclamation," finally, designates the process of physically restructuring the environment of a frontier zone by simplifying or eliminating existing flora and fauna, as well as by reshaping and contouring the landscape so as to make it more amenable to a human presence. This was accomplished through a wide variety of actions, including slashing, burning, chopping, grubbing, and digging.[7] Reclamation, first of all, made possible the residence of colonists in frontier zones by preparing sites for dwellings. Most reclamation activity, however, had as its goal the creation of agricultural land where none had existed before.

[4]Lewis, "The Closing of the Medieval Frontier," pp. 475–77.

[5]Richard Koebner, "The Settlement and Colonization of Europe," in M. M. Postan, ed., *The Agrarian Life of the Middle Ages*, 2d ed., pp. 83–91, vol. 1 in *The Cambridge Economic History of Europe*.

[6]Lewis, "The Closing of the Medieval Frontier," pp. 475–83; Lyon, "Medieval Real Estate Developments and Freedom," pp. 47–61.

[7]My terminology here is similar to that used by Stanton Green, "The Agricultural Colonization of Temperate Forest Habitats: An Ecological Model," in Savage and Thompson, eds., *The Frontier: Comparative Studies*, 2:69–87.

Patterns of Medieval Productivity

Reclamation and colonization reached a fever pitch in the vast wilderness areas of Europe during the High Middle Ages. It would be wrong, however, to assume that they were never practiced earlier. Before the late tenth century reclamation and colonization were part of the normal pulses of agrarian life as were their opposites, abandonment and desertion. There is every reason to believe that a certain amount of expansion and contraction of settlement and agriculture was not out of the ordinary. Pulsations of this sort, even the lateral shifting of agricultural focus, were characteristic features of European agrarian life from the introduction of agriculture, and they continued to be well into the Middle Ages.[8]

Although reclamation and colonization as well as abandonment and desertion were common aspects of rural life in Europe, such changes had begun to occur more and more within specific geographical contexts long before the Middle Ages. Throughout Mediterranean Europe and in many of the more densely settled portions of temperate Europe as well, agriculture and settlement increasingly took place within more or less fixed territories, some of which had survived relatively intact from the early Iron Age or before.[9] Even in the less densely settled areas of Europe, where forest-fallow, or swidden, agriculture may have persisted into historical times, such shifts in agrarian focus were made according to established cycles within specific village territories.[10]

[8] See, for example, Walter Janssen, "Some Major Aspects of Frankish and Medieval Settlement in the Rhineland," in P. H. Sawyer, ed., *Medieval Settlement: Continuity and Change*, pp. 41–60; Walter Janssen, "Dorf und Dorfformen des 7. bis 12. Jahrhunderts im Lichte neuer Ausgrabungen in Mittel- und Nordeuropa," in Herbert Jankuhn, Rudolf Schützeichel, and Fred Schwind, eds., *Das Dorf der Eisenzeit und des frühen Mittelalters: Siedlungsform-wirtschaftliche Funktion-soziale Struktur*, pp. 61, 82.

[9] H. T. Waterbolk, "Siedlungskontinuität im Kustengebiet der Nordsee zwischen Rhein und Elbe," *Probleme der Küstenforschung im südlichen Nordseegebiet* 13 (1979): 1–21; Charles Thomas, "Towards the Definition of the Term 'Field' in the Light of Prehistory," in P. H. Sawyer, ed., *Medieval Settlement: Continuity and Change*, p. 151; R. W. Brandt, "De kolonisatie van West-Friesland in de bronstijd," *Westerheem* 29 (1980): 137; Herbert Jankuhn, "Rodung und Wüstung in vor- und frühgeschichtlicher Zeit," in Walter Schlesinger, ed., *Die deutsche Ostsiedlung des Mittelalters als Problem der europäischer Geschichte*, pp. 107–109, 115; Koebner, "The Settlement and Colonization of Europe," p. 10.

[10] Ester Boserup, *The Conditions of Agricultural Growth: The Economics of Agrarian Change under Population Pressure*, p. 15; Ester Boserup, *Population and Technological Change: A Study of Long-Term Trends*, p. 95; David R. Harris, "Swidden Systems and Settlement," in Peter J. Ucko, Ruth Tringham, and G. W. Dimbleby, eds., *Man, Settlement and Urbanism*, pp. 246–50; Charles Parain, "The Evolution of Agricultural Technique," in Postan, ed., *The Agrarian Life of the Middle Ages*, pp. 134, 136–37; William S. Cooter,

In many instances, the location of settlement and agriculture was not radically different in the late tenth century from what it had been a millennium and a half earlier. The intensity of land use within such long-established territories, however, could and did change a great deal. To understand the changes that occurred because of reclamation and colonization during the High Middle Ages, it is necessary to examine this territorial aspect of early-medieval productivity more closely.

Village territories represented the resource base from which residents acquired virtually everything they needed for existence. Not all portions of this resource base were exploited in exactly the same fashion, however. Residents of a particular community would employ a wide variety of activities, including hunting, fishing, collecting, livestock keeping, and crop raising, to provide themselves with food, fuel, shelter, tools, clothing, and equipment. The actual mix of these activities could vary considerably from place to place and from time to time, depending on population density, the condition of the physical environment, the efficiency of the various techniques of exploitation, and the complex of social and cultural norms that established the goals and priorities of the community.

It would be wrong to assume, however, that early-medieval food producers ever had a simple choice to make between equally effective varieties of food production. Long before the Middle Ages all of Europe had achieved full agricultural status along with a population size that was much greater than hunting, fishing, and collecting could sustain, whether singly or together. Thus the choice was between pastoral agriculture and arable agriculture (though some combination was more likely), with all other food-producing activities playing only supplementary roles of varying importance. Nevertheless, such seeming simplicity was complicated by the fact that agriculture could take a very wide variety of forms associated with both social and environmental considerations.[11]

From ancient times Europe had known two somewhat contrasting dietary patterns that stemmed from different agricultural traditions. One, having its origins in the Middle East and the Mediterranean world, was

"Ecological Dimensions of Medieval Agrarian Systems," *Agricultural History* 52 (1978): 475.

[11] B. H. Slicher van Bath, *The Agrarian History of Western Europe*, p. 244; Parain, "The Evolution of Agricultural Technique," pp. 127, 131, 136–37; David B. Grigg, *The Agricultural Systems of the World: An Evolutionary Approach*; Boserup, *Population and Technological Change*, pp. 17–23, 44–45.

centered on cereals, wine, and olive oil, supplemented by fruits, vegetables, meat, eggs, milk products, fish, and anything else that happened to be available. It was based on an agricultural tradition that devoted most attention to cereals, vines and olives, with pastoral agriculture playing only a minor role. The second dietary pattern, centered on Europe north of the Mediterranean, derived most subsistence from meat and milk products, supplemented by cereals, fruits, vegetables, and fish. This pattern was based on an essentially pastoral agricultural system with only a minor focus on crop production. Beginning with Roman expansion into temperate Europe and continuing throughout the early Middle Ages, however, Europeans became increasingly dependent on a diet derived from cereals, reinforced, no doubt, by the gradual spread northward of Christianity, which used bread consumption as a symbol of divine nourishment. By the eighth and ninth centuries the two agricultural traditions had begun to merge, forming the basis of a new, typically European system of food production that depended primarily on crop raising but contained, nevertheless, a much larger pastoral component than did the original southern agrarian tradition.[12]

Indeed, one of the major themes running through the history of medieval food production, a theme of particular importance here, was the gradual expansion of arable agriculture into all parts of temperate Europe. In the early Middle Ages this advance normally occurred at the expense of pastoral, hunting, fishing, and collecting activities within those territories that had long been inhabited. Beginning in the tenth century, however, arable agriculture was extended into the great expanses of remaining wilderness, those frontier zones beyond the realm of most human activity. In both phases, however, land reclamation, the restructuring of the physical environment, preceded the the expansion of crop agriculture. As a result we shall not go far wrong if we see the expansion of crop agriculture in tem-

[12] William H. TeBrake, "Ecology and Economy in Early Medieval Frisia," *Viator: Medieval and Renaissance Studies* 9 (1978): 16–20; Lynn White, Jr., *Medieval Technology and Social Change*, pp. 69–76; Lynn White, Jr., "The Expansion of Technology, 500–1500," in Carlo M. Cipolla, ed., *The Middle Ages*, pp. 146, 150–51, vol. 1 in *The Fontana Economic History of Europe*; Slicher van Bath, *The Agrarian History of Western Europe*, pp. 20–22, 58–62; Georges Duby, *Rural Economy and Country Life in the Medieval West*, trans. Cynthia Postan, pp. 90–92; Georges Duby, *Early Growth of the European Economy*, trans. Howard B. Clarke, pp. 17–30; Marc Bloch, *French Rural History: An Essay on Its Basic Characteristics*, trans. Janet Sondheimer, pp. 26–35; Roger Grand and Raymond Delatouche, *L'Agriculture au moyen âge de fin de l'Empire Romain au XVI siècle*, pp. 269–71.

perate Europe during the Middle Ages as a symptom of the progress of land reclamation, and vice versa. Little or no restructuring was aimed directly at providing lands for livestock that would normally feed on the fringes of the inhabited and cropped portions of newly colonized territories or on the arable land itself once the crops had been removed. Still, grazing, rooting, and browsing, in their own time, contributed to the restructuring of the natural environment as well.[13]

Land and Labor as Factors of Production

The shift in European agriculture toward a greater reliance on crop raising during the Middle Ages cannot be attributed to dietary preferences or shifts in fashions alone. It was related as well to changes in the efficiency of agricultural labor; indeed, changes in the efficiency of agricultural labor may have helped alter dietary habits. Under ideal conditions livestock keeping is almost always less labor-intensive than is crop production.[14] Still it is important to realize that there are many varieties of crop-raising possibilities that differ considerably in their labor requirements. What are considered the most primitive varieties of arable agriculture are not much more demanding of labor than are most types of pastoral agriculture, while others require very heavy labor inputs.[15]

From the point of view of medieval food producers, who possessed a simple technology that provided few if any substitutes for human labor, one of the most important considerations always had to be the return that could be expected on labor applied to a particular kind of food production. It is reasonable to assume that, within the range of their cultural preferences, they would choose a food-production system that would provide the highest return for the least amount of labor. To appreciate more fully the motives of food producers in early-medieval Europe, it is important to examine the various crop-raising possibilities a bit further. One of the best ways of distinguishing among varieties of arable agriculture is to classify them according to the length of the fallow period, the period of rest between crops.

The person who has investigated most thoroughly the role of fallow-

[13] Bloch, *French Rural History*, pp. 5–8; Koebner, "The Settlement and Colonization of Europe" p. 3.

[14] William S. Cooter, "Preindustrial Frontiers and Interaction Spheres: Aspects of the Human Ecology of Roman Frontier Regions in Northwest Europe" (Ph.D. diss., University of Oklahoma, 1976), pp. 44, 233, calculates that the return on labor from livestock raising was at least five times greater than the return from cereal production.

[15] See, for example, Boserup, *Population and Technological Change*, pp. 44–46.

ing in agricultural systems is Ester Boserup, a Danish economist, who classifies all cropping systems according to the length of the fallow.[16] The earliest form of arable agriculture in any given place, she contends, was some kind of *forest-fallow agriculture*, also known as *swidden agriculture*, featuring one or two crops followed by a fallow period of at least twenty to twenty-five years.[17] The idea behind such a long fallow was to allow sufficient time for the original vegetation to recolonize a particular plot of land before it was once again put into production. The only labor required was simple and straightforward: killing the larger trees by girdling them with a sharp instrument, burning the underbrush and ground cover, and planting by broadcasting seeds directly into the ash layer. Productivity was very high, since burning concentrated nutrients into a very fertile seedbed.[18] After the initial surge of production, however, yields quickly dropped as the flush of nutrients released by burning dwindled and as weed growth began to dominate. As a result the plot was abandoned and another designated to take its place. Such a system brought high yields for relatively light labor inputs because the extremely long fallow period, with its reestablishment of forest vegetation, restored soil nutrients through the annual leaf drop and eventually shaded out all field weeds as well. The most serious drawback to a forest-fallow, or swidden, system was the tremendous amount of potential arable land that it required, ten to fifty times as much as was cultivated at any particular time. It was ideally suited, therefore, to areas of low population density.[19]

[16]Ester Boserup, "Environnement, population et technologie dans les sociétés primitives," *Annales: économies, sociétés, civilisations* 29 (1974): 538–52; Boserup, *The Conditions of Agricultural Growth*; Boserup, *Population and Technological Change*. See also David B. Grigg, "Ester Boserup's Theory of Agrarian Change," *Progress in Human Geography* 3 (1979): pp. 64–84; B. A. Datoo, "Toward a Reformulation of Boserup's Theory of Agricultural Change," *Economic Geography* 54 (1978): 135–44.

[17]Boserup, *The Conditions of Agricultural Change*, p. 15; J. S. Otto and N. E. Anderson, "Slash-and-Burn Cultivation in the Highlands South: A Problem in Comparative Agricultural History," *Comparative Studies in Society and History* 24 (1982): 132, 136. Cooter, "Preindustrial Frontiers and Interaction Spheres," pp. 45, 233, suggests that something under 100 years was necessary for temperate deciduous-forest areas.

[18]Stanton W. Green, "The Agricultural Colonization of Temperate Forest Habitats," in Savage and Thompson, eds., *The Frontier: Comparative Studies*, 2: 83–84; Otto and Anderson, "Slash-and-Burn Cultivation in the Highlands South," p. 131. Cooter, "Ecological Dimensions of Medieval Agrarian Systems," pp. 472–73, suggests yield-to-seed ratios as high as 20 to 1.

[19]Boserup, *The Conditions of Agricultural Change*, pp. 29–30; Boserup, *Population and Technological Change*, pp. 23–26; David B. Grigg, *Population Growth and Agrarian Change: An Historical Perspective*, p. 33.

A gradual increase in population density, Boserup maintains, was the primary reason why forest-fallow agriculture eventually disappeared from many regions of the world. Initially, perhaps, the additional mouths could be fed by expanding forest-fallow agriculture into adjoining areas; indeed, migration has often been an effective response of food producers to population growth.[20] Ultimately, however, the food-producing system was altered to meet the additional needs by gradually increasing the percentage of the potential arable that was cropped at any particular time, that is, by progressively reducing the length of the fallow. Forest-fallow agriculture came to an end when the fallow period had been shortened to such a degree that forest vegetation no longer had a chance to reestablish itself.[21]

Boserup identifies four additional cropping systems besides forest-fallow, each with a shorter fallow period than the one preceding. *Bush-fallow cultivation* consists of a cropping phase of one to eight years followed by a six-to-ten-year fallow, a period long enough for bushes and small trees to reappear. *Short-fallow agriculture* is typified by one or two years of cultivation in more or less permanent fields alternating with one or two years of fallowing, during which wild grasses colonize the land. *Annual cropping*, meanwhile, though technically not a fallow system, nevertheless leaves the land uncultivated for at least part of a year, while *multi-cropping* has no fallow at all.[22] Boserup associates forest-fallow agriculture with very sparse population, bush-fallow with sparse population, short-fallow with medium population density, annual cropping with dense population, and multicropping with very dense population.[23]

As the fallow period is reduced, according to Boserup's scheme, the tools and equipment of cultivation go through a series of changes, while additional labor operations are introduced to compensate for declining natural fertility and weed control that previously had been maintained by the reappearance of forest vegetation during the longer fallow period. Under a true forest-fallow system the soil is not only rich in nutrients after burning but also loose and friable, requiring no further operations before seeding. When forest-fallow is replaced by bush-fallow agriculture, however, hoeing becomes an important additional operation in the preparation of the seedbed because the soil is less loose and friable and less free of a

[20] Boserup, *Population and Technological Change*, pp. 51–52, 57, 75.

[21] Boserup, *The Conditions of Agricultural Change*, pp. 15–18.

[22] Ibid., pp. 15–16; Boserup, *Population and Technological Change*, pp. 18–19.

[23] Boserup, *Population and Technological Change*, pp. 9, 23.

sod after the grassier vegetation produced during the shortened fallow period is burned. In addition, because of decreased ash production and, particularly when cropping phases are longer than just a year or two, declining soil fertility under bush-fallow systems must be dealt with, usually by fertilization with organic material, such as leaves and sods, collected on the bushland and hoed into the topsoil. If bush-fallow is replaced subsequently by a short-fallow system, further changes occur in tools and equipment, while additional labor inputs are necessary to maintain soil fertility. With only a year or two of fallow the characteristic vegetation to be removed before cropping consists of weeds and grasses whose root systems, while not much affected by either hoeing or burning, can be most effectively reduced by plowing. Short-fallow, however, requires not only the prior removal of tree stumps, roots, and other obstacles to plowing but also the care and feeding of the draft animals that pull the plows and the much more persistent application of fertilizers, including animal manure, to prevent total loss of soil fertility. With annual cropping, the tools of tillage are further refined, while the maintenance of soil fertility becomes an extremely important concern and involves applying animal manure and night soil, green manuring, composting, and marling. Multicropping, finally, combines intensive tillage and fertilization with irrigation.[24]

The primary benefit of adopting shorter fallow periods, according to Boserup, is considerably enhanced food production because less land is out of cultivation at any particular time or, stated differently, the land is cropped more frequently. The disadvantage of shorter fallow periods is the need to increase labor investment substantially to compensate for the forfeiture of essentially effortless fertility maintenance, soil preparation, and weed control under the forest-fallow system. Thus crop production increases per unit of land while decreasing per unit of labor. For this reason, she insists, preindustrial food producers, possessing few or no substitutes for human labor, normally did not adopt cropping systems with shorter fallow periods unless there were compelling reasons for doing so. One of the most important of such compelling reasons always has been and still is increasing population density.[25] An interesting corollary to this thesis, one that has particular relevance to medieval Europe, is the contention that declining

[24] Ibid., pp. 23–26, 43–62; Boserup, *The Conditions of Agricultural Change*, pp. 23–25; Grigg, *The Agricultural Systems of the World*, pp. 50–54, 57–58, 72–73; Otto and Anderson, "Slash-and-Burn Cultivation in the Highlands South," pp. 132, 137.

[25] Boserup, *The Conditions of Agricultural Change*.

population density can result in less intensive forms of land use, that is, in a lengthening of the fallow period.[26]

Boserup is not without her critics. Some suggest, for example, that industrial-energy inputs and improved plant breeding may actually increase the productivity of labor under shortened fallow conditions. Others maintain that changes in the market prices of agricultural products can be just as important in promoting intensification of land use as are changes in population density.[27] These and other such criticisms were made of the original formulation of her thesis as it appeared in *The Conditions of Agricultural Change* in 1965. In her more recent *Population and Technological Change* (1981), however, she has come to terms with most of her critics by refining some portions and more effectively defending other portions of her argument without in any way detracting from the original thrust of her thesis. Even so, those who originally raised objections to her thesis tended for the most part to doubt its relevance to certain modern, not to premodern, situations.[28] In fact, Boserup's thesis has great potential as a way of explaining agricultural change in the Middle Ages, when there were no industrial inputs to replace human labor and the impact of markets was virtually nonexistent.

Boserup's explanation of agricultural change is important if we are to come to a proper understanding of medieval patterns of productivity, not so much because it answers all of our questions as because it raises some extremely important ones. By forcing us to concentrate on the role of population pressure in agricultural change, it encourages us to rethink some old assumptions and explanations that now seem somehow less satisfactory than before. Her model examines the interplay of the two most essential factors of premodern food production, land and labor, and it reveals that the relationship between the two always is and always has been a dynamic one.[29] As a result, she offers an alternative to the often rigid Malthusian

[26] Ibid., pp. 62–63; Boserup, *Population and Technological Change*, pp. 97–98.

[27] See, for example, Datoo, "Toward a Reformulation of Boserup's Theory of Agricultural Change," pp. 135–44; Grigg, "Ester Boserup's Theory of Agrarian Change," pp. 64–84.

[28] So far, Boserup's ideas have received the greatest attention from anthropologists and archaeologists, especially those interested in agricultural origins and dispersals. See, for example, the collection of essays in Brian J. Spooner, ed., *Population Growth: Anthropological Perspectives*; Mark Nathan Cohen, *The Food Crisis in Prehistory: Overpopulation and the Origins of Agriculture*.

[29] Grigg, *Population Growth and Agrarian Change*, p. 21.

explanations that cast population growth solely as the result, never as the cause, of agricultural change. At the same time Boserup presents us with a challenging perspective on the relationships between agricultural change and technological development in preindustrial times. Instead of seeing technological change as the cause of agricultural development, a view that has severe limitations, especially when applied to the early Middle Ages,[30] she suggests that the kind of land use usually determines the kinds of tools and equipment that are employed.

A well-attested feature of the Middle Ages was the tremendous growth of population that occurred between roughly 650 and 1350. As noted earlier, the number of Europeans more than doubled during the High Middle Ages alone, but when the early medieval period is included as well, it amounts to more than a quadrupling. Indeed, in much of temperate Europe, population by the midfourteenth century was more than six times what it had been in the midseventh century. Although it would be incorrect to assume that such growth occurred without any interludes of stagnation or decline, the long-term population trends were nevertheless upward for about seven centuries, a real accomplishment under preindustrial conditions. This becomes all the more remarkable when it is seen in the context of the severe depopulation during the century preceding 650 in the wake of epidemics and other disasters. In fact, it was not until after 800 that the population levels known in Europe in the early 500s were achieved once again.[31]

In the light of the preceding discussion of the relations between population growth and agrarian change, it will be interesting to see whether the upward demographic spiral from 650 to 1350 was reflected to any significant degree in a shortening of fallow periods. Of course, it was not population growth by itself that would have made a difference but increasing population density, that is, a change in the ratio of labor to land. Still, to calculate the density of population for any time or place during the preindustrial era is an extremely difficult task. Such information as has come to light shows that it varied tremendously.

[30] See, for example, the criticisms offered by Cooter, "Preindustrial Frontiers and Interaction Spheres," pp. 42–43, 201–203.

[31] Josiah Cox Russell, "Population in Europe, 500–1500," in Cipolla, ed., *The Middle Ages*, p. 36; Norman J. G. Pounds, *An Historical Geography of Europe*, p. 186; Wilhelm Abel, *Geschichte der deutschen Landwirtschaft vom frühen Mittelalter bis zum 19. Jahrhundert*, pp. 25–26.

A number of population estimates have been attempted for certain portions of northern France and the Low Countries for the ninth or early tenth century. The three most densely settled sectors were the Paris region, the district of Saint-Omer in Flanders and Artois, and the Frisian district of Westergo in the northern Netherlands, at 39, 34, and 20 people per square kilometer, respectively. By all indications these were among the most densely settled portions of temperate Europe at that time. The other places for which we have information had densities ranging from 4 to 12 people per square kilometer.[32] It is important to remember, however, that all the places sampled were part of that portion of temperate Europe which saw more than a sixfold increase in total population during the Middle Ages, and considerable growth had already taken place by the ninth or early tenth century. Some very crude estimates of population density for the sixth century, for example, range from 5 or 6 people per square kilometer in Gaul to about 2 per square kilometer in the Germanic lands and England.[33]

Even after two or more centuries of population growth the Paris region, Saint-Omer, and Westergo, though among the most heavily populated sections of temperate Europe, still knew only medium population density, of the sort that Boserup has found to be associated generally with the most intensive forms of bush-fallow or short-fallow agriculture. The remaining areas exhibited densities normally seen in conjunction with the less intensive forms of bush-fallow agriculture.[34] We should not expect, therefore, to find much from the shortest end of Boserup's fallow spectrum. As a matter of fact, annual cropping is known from certain urban hinterlands only from the late Middle Ages, while multicropping has never been a common feature of European agriculture.[35] On the other hand, medieval documentation clearly reveals the existence of short-fallow cropping systems and, less clearly, of some forest- or bush-fallow cropping patterns.

[32] B. H. Slicher van Bath, "The Economic and Social Conditions in the Frisian Districts," *A. A. G. Bijdragen* 13 (1965): 101–102.

[33] Duby, *The Early Growth of the European Economy*, p. 13; Abel, *Geschichte der deutschen Landwirtschaft*, pp. 25–26.

[34] Boserup, *Population and Technological Change*, pp. 9, 23.

[35] B. H. Slicher van Bath, "The Rise of Intensive Husbandry in the Low Countries," in J. S. Bromley and E. H. Kossman, eds., *Papers Delivered to the Oxford-Netherlands Historical Conference, 1959*, pp. 130–53, vol. 1 in *Britain and the Netherlands*; Andrew M. Watson, "Toward Denser and More Continuous Settlement: New Crops and Farming Techniques in the Early Middle Ages," in J. A. Raftis, ed., *Pathways to Medieval Peasants*, pp. 67–68.

By the ninth and early tenth centuries in such places as the Paris basin, parts of Flanders, the middle Rhine Valley, and, perhaps, portions of southeastern England, areas of fairly open countryside had developed in which forest and bush vegetation had been reduced and permanent fields had been laid out.[36] These were the areas of the largest and wealthiest estates and manors, and their permanent fields were symptoms of short-fallow systems of agriculture. The so-called two-field system of crop rotation associated with the manors and estates of early-medieval sources clearly was one version of short-fallow cultivation: one crop followed by a year of fallow. The three-field system that began to appear during the eighth and ninth centuries was another version: two crops followed by a fallow of one year, though it could have been one crop followed by a two-year fallow period as well, on occasion.[37]

The kinds of land use that prevailed away from the largest and wealthiest estates and manors in the ninth and early tenth centuries are less easy to establish because there is virtually no documentation directly attributable to such areas. Nevertheless, some information can be derived indirectly. For example, Germanic law codes and written deeds from the early Middle Ages occasionally mentioned the existence of a *rothum*, a former field that had been abandoned, lying in close proximity to *nova*, land recently put back into cultivation after a rest period of a number of years.[38] This sounds very much like bush-fallow agriculture. Further, there is good reason to believe that some form of long-fallow agriculture involving clearance by burning was frequently practiced outside the most densely settled areas of temperate Europe in the early Middle Ages, particularly since such practices were known to have continued into modern times in many places where population remained sparse: near Paris as late as the twelfth century, in the Alps and other mountain areas into the seventeenth and eighteenth centuries, and in Corsica, parts of the Ardennes, and the Odenwald of Germany into the twentieth century.[39] Still, the degree to which such examples

[36] Duby, *The Early Growth of the European Economy*, pp. 23, 25–26.

[37] Bloch, *French Rural History*, pp. 26–35; Slicher van Bath, *The Agrarian History of Western Europe*, pp. 20–22, 58–62; Duby, *The Early Growth of the European Economy*, pp. 17–30; Duby, *Rural Economy and Country Life in the Medieval West*, pp. 90–93; White, *Medieval Technology and Social Change*, pp. 69–76; White, "The Expansion of Technology, 500–1500," pp. 146, 150–51; Grand and Delatouche, *L'Agriculture au moyen âge*, pp. 269–71.

[38] Duby, *The Early Growth of the European Economy*, p. 22.

[39] Bloch, *French Rural History*, pp. 26–28; Koebner, "The Settlement and Coloniza-

of shifting cultivation associated with clearance by burning in fact can be identified with one or another of Boserup's categories of long-fallow agriculture is not really clear. What is certain is that they would not have qualified as short-fallow agriculture on permanent fields.

Indeed, in much of temperate Europe during the early Middle Ages, the categories of land use we have referred to so far do not seem to apply very well except for the extremes: short-fallow cropping where population was most dense and a forest-fallow or bush-fallow system where population was most sparse. In between, however, the picture is much less clear. The evidence Boserup used to develop her scheme of agricultural classification according to the length of the fallow period came from many parts of the world, but none of it was specific to early-medieval temperate Europe. Further, some of the assumptions that she made do not fit any better. For example, she assumed a fairly even distribution of population over a given region, which then expressed itself in a pattern of crop production of a particular fallow length that was more or less evenly distributed throughout that region. As we saw above, however, population was very unevenly distributed in temperate Europe during the early Middle Ages. Patterns of land use too were far from uniform. In fact, within close proximity to the large estates and manors on which short-fallow agriculture was practiced, there were often great expanses of the countryside still given over to woodland, marsh, or swamp with no signs of permanent cultivation.[40] Even within the territory of a single community the patterns of land use were far from homogeneous, exhibiting instead a sort of hybrid pattern that included both long and short periods of fallow. Thus some portions were cropped according to one fallow length, others cropped according to another length, and still others not cropped at all. For these reasons it may be useful to see just how land was classified in the early Middle Ages and how each type was or was not exploited.

---

tion of Europe," pp. 60–61; Parain, "The Evolution of Agricultural Technique," pp. 134, 136–37; Jerome Blum, *The End of the Old Order in Rural Europe*, pp. 129–30; Slicher van Bath, *The Agrarian History of Western Europe*, p. 62; Hans-Jürgen Nitz, "The Church as Colonist: The Benedictine Abbey of Lorsch and Planned Waldhufen Colonization in the Odenwald," *Journal of Historical Geography* 9 (1983): 119; Otto and Anderson, "Slash-and-Burn Cultivation in the Highlands South," pp. 131–47; H. Uhlig, "Old Hamlets with Infield and Outfield Systems in Western and Central Europe," *Geografiska Annaler* 43 (1961): 287–88; Richard C. Hoffmann, "Medieval Origins of the Common Fields," in William N. Parker and Eric L. Jones, eds., *European Peasants and Their Markets: Essays in Agrarian Economic History*, p. 40.

[40] Duby, *The Early Growth of the European Economy*, pp. 21–30.

Ager, Saltus, and *Silva*

Since prehistoric times temperate Europe had known permanent fields, that is, plots of arable land that were cropped regularly and were not allowed to revert to bush or forest vegetation and on which plows were used routinely. In other words, such plots were exploited in a short-fallow fashion. The so-called Celtic fields of late-prehistoric times, widespread throughout northwestern Europe, were examples of such short-fallow plots.[41] Indeed, there is some evidence for permanent fields cultivated by true plows as early as the late Neolithic and Bronze ages.[42] Despite the serious decline of European population, particularly during the late sixth and early seventh centuries, resulting in the total abandonment of some arable land, permanent fields survived into the early Middle Ages.

Such a permanent plot of arable land was often referred to as an *ager* ("field") during the Middle Ages. Because the most labor-intensive forms of environmental exploitation, crop raising, took place on the *ager*, it was found near the center of a typical medieval-village territory. It lay in close proximity to the dwellings and their associated buildings, yards, and gardens.

The rest of a village territory was known as *saltus* ("woods") or *wastinae*, ("wasteland"), arrayed around the *ager*, where noncrop forms of environmental exploitation were practiced. In most of temperate Europe the *saltus* consisted of wooded or partly wooded land, though, as we shall see, it could have been swamp- or marshland in some costal areas as well. Generally, the intensity of land use in the *saltus* portion decreased as the distance from the center of the village territory increased, until the imprint of human culture faded to virtually nothing at the outer fringes of the territory.[43]

The third category of land from the point of view of early-medieval

[41] See, for example, J. A. Brongers, *Air Photography and Celtic Field Research in the Netherlands*, pp. 18–29, 56–72.

[42] See chap. 5.

[43] Duby, *The Early Growth of the European Economy*, pp. 12–13, 17; Hoffmann, "Medieval Origins of the Common Fields," pp. 39–41; Rob Rentenaar, "De Nederlandse duinen in de middeleeuwse bronnen tot omstreeks 1300," *Geografisch tijdschrift* 11 (1977): 370–71; Guy Fourquin, *Le paysan d'occident au moyen âge*, pp. 13–15; Slicher van Bath, *The Agrarian History of Western Europe*, p. 72; Michael Chisholm, *Rural Settlement and Land Use: An Essay in Location*, 2d ed., pp. 27–28; A. E. Verhulst, *De Sint-Baafsabdij te Gent en haar grondbezit (VIIe-XIVe eeuw): Bijdrage tot de kennis van de structuur en de uitbating van het grootgrondbezit in Vlaanderen in de middeleeuwen*, pp. 144, 212.

communities was the wilderness, consisting of those areas between and beyond existing community territories where the impact of human culture was slight or nonexistent. Such words as *silva* ("forest") and *nemus* ("grove") were usually used to refer to such areas, which often served as the rough boundary zones separating one settled area from another. For example, in Flanders during the early Middle Ages, the so-called Carboniferous Forest served as a border zone between Neustria and Austrasia, the two major subdivisions of Frankish territory.[44] This and other wilderness areas were the frontier zones that were reclaimed, colonized, and integrated into new or existing community territories during the High Middle Ages. In those regions in which settlements had begun to border on each other directly, of course, clear boundaries were drawn without intervening bands of *silva* or *nemus*.[45]

One of the most important factors governing the location of settlements in the early Middle Ages was soil type. The soils of temperate Europe often vary greatly over rather short distances, from light, well-drained, sandy soils to heavy, moist, clay soils. The primary instrument of cultivation on the short-fallow *ager* was the *ard*, or scratch plow, which had entered Europe from southwest Asia and Mediterranean Europe during prehistoric times. It was of light construction and ideally suited for the light soils and dry climate of its place of origin, where the object was merely to scratch the surface of the soil.[46] In temperate Europe, however, heavy, moist soils are very common. During the early stages of agriculture in this part of the Continent, when population was sparse and forest-fallow

[44] Michel Devèze, "Forêts françaises et forêts allemandes: études historique comparée," *Revue historique* 235 (1966): 365; Koebner, "The Settlement and Colonization of Europe," p. 21; W. Gordon East, *The Geography behind History*, rev. ed., p. 100; J. K. de Cock, *Bijdrage tot de historische geografie van Kennemerland in de middeleeuwen op fysisch-geografische grondslag*, pp. 22–52; Marc Bloch, "Occupation du sol et peuplement," in his *Mélanges historiques*, p. 128; Helmut Jäger, "Zur Geschichte der deutschen Kulturlandschaften," *Geographische Zeitschrift* 51 (1963): 120.

[45] Duby, *The Early Growth of the European Economy*, p. 23; Verhulst, *De Sint-Baafsabdij te Gent en haar grondbezit*, p. 144.

[46] Duby, *Rural Economy and Country Life in the Medieval West*, pp. 15, 21, 67; G. Des Marez, *Le problème de la colonisation franque et du régime agraire en Basse-Belgique*, p. 16; Herbert Jankuhn, *Vor- und Frühgeschichte vom Neolithikum bis zur Völkerwanderungszeit*, pp. 21–26; Hans Mortensen, "Probleme der Mittelalterlichen Siedlungs- und Kulturlandschaft," *Berichte zur deutschen Landeskunde* 20 (1958): 100–101; Karlheinz Paffen, "Natur- und Kulturlandschaft am deutschen Niederrhein," *Berichte zur deutschen Landeskunde* 20 (1958): 192–95; Devèze, "Forêts françaises et forêts allemandes," pp. 354, 360, 370.

cropping patterns were common, presumably most soils could be exploited, since little or no tillage or working of the soil was required. As population density increased in some areas and fallow periods were significantly shortened, however, the suitability of certain soils for crop production began to change. The *ard* of short-fallow agriculture performs poorly on the heavy, moist soils that must be dug deeply and rigorously to allow the surface to dry and to bring weed growth under control. For these reasons the short-fallow *ager* of early-medieval settlements was limited essentially to the loess, chalk, limestone, and sandy alluvial soils, as well as to the lighter loams on the flanks and crests of gently rolling hills, in short, to the soils that could be worked efficiently by the *ard*.[47]

Because in many regions light, well-drained soils had been in use as *ager* for centuries, maybe even millennia, as the fallow periods were progressively shortened to something approaching Boserup's short-fallow classification, natural soil fertility gradually disappeared. This meant that, unless substantial human labor was substituted for the benefits previously derived from the longer fallow, the yield of crops cultivated on the *ager* would drop considerably.[48] As a matter of fact, there is evidence that some yield-to-seed ratios had dropped to three to one or even lower by the early Middle Ages, a miserable performance by any standard.[49] Even with such low yields, however, the planting of crops on the *ager* may well have produced more usable calories per land unit than would grazing or any other variety of environmental exploitation.[50] The *ager*, therefore, continued to be an important focus of food-producing activity.

In general, the *ager* was not the largest feature of the agrarian landscape in the early Middle Ages except for those areas in which settlement had become dense and more or less continuous, such as the Paris region, parts of Flanders, the middle Rhine Valley, and portions of southeastern

[47] Watson, "Towards Denser and More Continuous Settlement," pp. 73, 77, 81 n. 11; Jankuhn, "Rodung und Wüstung in vor- und frühgeschichtlicher Zeit," pp. 104–105.

[48] Boserup, *Population and Technological Change*, pp. 23–26; Boserup, *The Conditions of Agricultural Change*, pp. 23–62; Cooter, "Ecological Dimensions of Medieval Agrarian Systems," pp. 475–76; Cooter, "Preindustrial Frontiers and Interaction Spheres," pp. 34–59; Otto and Anderson, "Slash-and-Burn Cultivation in the Highlands South," pp. 131–47.

[49] B. H. Slicher van Bath, "Le climat et les récoltes en haut moyen âge," in *Agricoltura e mondo rurale en Occidente nell'alto medioevo*, pp. 403–14, 423, 443–47; Duby, *Rural Economy and Country Life in the Medieval West*, pp. 26–27; Duby, *The Early Growth of the European Economy*; Watson, "Towards Denser and More Continuous Settlement," p. 69.

[50] Watson, "Towards Denser and More Continuous Settlement," p. 69.

England. In fact, most often it would have appeared as a rather small island in a sea of *saltus* and *silva*. Estimates of the amount of land incorporated into permanent fields in the sixth century, for example, suggest that it may have been less than 5 percent of the total land area in some parts of temperate Europe.[51] For this reason the *ager* simply could not have provided most of the food that was needed, particularly since yields were extremely low at times. Indeed, in parts of Germany the production of permanent fields must have amounted to less than one-third of the total caloric production.[52]

By providing what the *ager* could not, the *saltus*, as well as the streams, ponds, and lakes that it contained, was extremely important to early-medieval communities. It provided, first of all, the grazing, browsing, and rooting grounds for the livestock that still furnished the largest proportion of the calories consumed in much of temperate Europe.[53] Further, it supplied the game, fowl, fish, berries, fruits, nuts, and honey that rounded out what otherwise would have been a monotonous diet. Finally, the *saltus* was the source of cooking, heating, and industrial fuel as well as wood, the primary material for the construction of everything from tools and implements to ships and houses.[54]

There is evidence that *saltus* had yet an additional role to play by contributing both indirectly and directly to crop production. For example, the *saltus* can be seen as the outfield portion of an infield-outfield cropping system (in French, "terres froides et terres chaudes"). In such a scheme the *saltus* would function, first, as an important source of fertilizer in the

[51] Duby, *The Early Growth of the European Economy*, p. 13; Abel, *Geschichte der deutschen Landwirtschaft*, pp. 31–32. Even in the late Middle Ages, after many centuries of arable agricultural expansion, fields still comprised only about 30 to 40 percent of the countryside; see Slicher van Bath, *Agrarian History of Western Europe*, p. 22; Watson, "Towards Denser and More Continuous Settlement," p. 69.

[52] Watson, "Towards Denser and More Continuous Settlement," p. 69.

[53] TeBrake, "Ecology and Economy in Early Medieval Frisia," pp. 16–24.

[54] See, for example, Clarence J. Glacken, *Traces on the Rhodian Shore: Nature and Culture in Western Thought from Ancient Times to the End of the Eighteenth Century*, pp. 320–22; Duby, *Rural Economy and Country Life in the Medieval West*, pp. 71, 143–44; Duby, *The Early Growth of the European Economy*, p. 17; Grand and Delatouche, *L'Agriculture au moyen âge*, pp. 410–44; Fourquin, *Le paysan d'Occident au moyen âge*, pp. 9–15; Bloch, "Occupation des sol et peuplement," pp. 128–29; Bloch, *French Rural History*, pp. 6–7; B. H. Slicher van Bath, "L'Histoire des forêts dans les Pays-Bas septentrionaux," *A.A.G. Bijdragen* 14 (1967): 96–97; Slicher van Bath, *The Agrarian History of Western Europe*, pp. 72–74; Jean Birrell, "Peasant Craftsmen in the Medieval Forest," *Agricultural History Review* 17 (1969): 91–107; Devèze, "Forêts françaises et forêts allemandes," pp. 356, 372–73.

form of sods and organic matter that would be spread on the *ager* (infield). Second, and more directly related to crop production, small plots of land in the *saltus* (outfield) would be cleared by fire from time to time, planted for a few years, and allowed to revert to a bush or forest vegetation.[55] In this fashion soils not suited for *ager*, perhaps because they were too moist and heavy to be tilled with the *ard* or too lacking in fertility to be cropped more frequently, could be cultivated temporarily with good results.

While arable agriculture was an extremely important part of early-medieval alimentation, it is clear that there was much more to food production than that occurring in permanent fields. Yet most studies that have explored agriculture for this period have focused almost exclusively on fields and field systems. The reason for this is not because they produced the overwhelming majority of calories in the average diet, which they often did not, but because they were of pivotal importance in the relations between food producers and the nonfood-producing elite. Fields were the focal point of the manorial rents and labor services that provided the economic underpinnings of a particular variety of social stratification; they produced the primary taxable and storable surplus in an era noted otherwise for its essentially self-sufficient character.[56] Since most of the documentation of the period came from the largest and wealthiest estates and manors, where both crop agriculture and the control of cropland by a powerful elite had advanced the furthest, studies based on such sources of information tell us only about food production where power and wealth were most highly concentrated. They are not necessarily representative of food production elsewhere. Outside the areas of the largest estates the *ager* often paled in comparison to the *saltus*.

Even on the manors and estates of the truly mighty, however, there was a much broader food-producing base than fields alone. The set of instructions that was sent out by the Carolingian court to the managers of royal estates around 800, the *Capitulare de villis*, reveals this very clearly.

[55]Nitz, "The Church as Colonist," p. 119; Bloch, *French Rural History*, pp. 26–30; Slicher van Bath, *The Agrarian History of Western Europe*, pp. 58–59, 244, 246; Fourquin, *Le paysan d'Occident au moyen âge*, p. 14; Uhlig, "Old Hamlets with Infield and Outfield Systems," pp. 186–89, 304–305; Duby, *The Early Growth of the European Economy*, pp. 22–23; Otto and Anderson, "Slash-and-Burn Cultivation in the Highlands South," pp. 134, 139.

[56]Edward J. Nell, "The Technology of Intimidation," *Peasant Studies Newsletter* 1 (1972): 43–44; Cooter, "Ecological Dimensions of Medieval Agrarian Systems," pp. 475–76.

The most impressive characteristic of this document is the incredible variety of food sources to which it refers, ranging from duck ponds and fishponds to grazing lands and hunting reserves and especially to gardens and orchards. The message that the monarch meant to convey was the necessity of not only protecting but further enhancing such diversity of food sources as already existed, presumably to avoid the problems associated with too close a reliance on the production of fields alone in an era of chronic crop failure. In fact, Charlemagne himself had to deal with the effects of serious food shortages in 792–93 and again in 805–806, and there is evidence that these were not isolated cases.[57] Even the highest reaches of the ruling elite, therefore, whose status, wealth, and subsistence were most closely linked to the control and disposal of arable land, were well aware of the dangers of overdependence on the production of permanent fields.

Permanent fields were much less crucial to the well-being of those who knew the intricacies of food production the best in the early Middle Ages, the vast majority of the population who were the food producers, the peasant class of Europe, whose labors alone made possible the existence of a landlord class. After all, raising corps on the *ager* was not only an arduous activity but also a highly visible one that was carried out in plain view of everyone, making it very difficult to avoid paying the rents and performing the labor services associated with it. In contrast, gardening, fishing, hunting, collecting, livestock keeping, and crop raising on temporary plots in the *saltus* presumably could have been concealed more easily from the taxing ability of the landlord class. Further, we should remember that food production on permanent fields is always more demanding of labor than are the subsistence strategies normally employed in the waste. Thus, unless there were compelling reasons for doing so, early-medieval peasants most likely would not have increased their emphasis on crop production, at least in formal, fixed fields, as long as other, less visible and less labor-demanding activities could provide them with what they needed.[58]

Nevertheless, as we saw earlier, the one consistent theme that runs

[57] A. E. Verhulst, "Karolingische Agrarpolitik: Das *Capitulare de Villis* und die Hungersnöte von 792/93 und 805/06," *Zeitschrift für Agrargeschichte und Agrarsoziologie* 13 (1965): 179–89; Pierre Riché, *Daily Life in the World of Charlemagne*, trans. Jo Ann McNamara, pp. 48–49, 250–51; Michel Rouche, "La faim a l'époque carolingienne: essai sur quelques types de rations alimentaires," *Revue historique*, no. 508 (1973): 295–320.

[58] Nell, "The Technology of Intimidation," pp. 43–44; Cooter, "Ecological Dimensions of Medieval Agrarian Systems," pp. 475–76.

through the history of agriculture between 650 and 1350 was the gradual shift away from pastoral toward arable agriculture. In temperate Europe this amounted to an expansion of *ager* with its crop-raising activities at the expense of first *saltus* and eventually *silva*. This meant that an increasing proportion of the countryside came to be exploited by labor-intensive short-fallow systems of agriculture that replaced less labor-intensive forms of environmental exploitation, not only forest- or bush-fallow crop production but also such varieties of "perpetual" fallow as livestock keeping, fishing, hunting, and collecting. However, the expansion of *ager* at the expense of *saltus* and *silva* represented the same kind of change in the relationship of labor to land that a uniform shortening of the fallow period would under Boserup's scheme. In other words, the reduction of the ratio of *silva* and *saltus* to *ager* in temperate Europe during the Middle Ages was analogous to a reduction in the ratio of fallow land to cropped land elsewhere. Thus Boserup's contention that increasing population density leads to a more intensive application of labor to land seems to apply to temperate Europe after all, even though her categories of fallow length would not have been found uniformly distributed over the countryside.

Reclamation and the Expansion of Arable Agriculture

The growth of population in Europe between 650 and 1350 thus expressed itself in a gradual extension of short-fallow agricultural practices into virtually all of temperate Europe. It is important to remember, however, that short-fallow agriculture presupposes the prior restructuring of the natural environment into permanent fields. This usually meant that tree stumps, roots, and other obstacles to plowing had been removed; sometimes it involved land drainage. In all instances, however, a process of reclamation preceded the inauguration of short-fallow agriculture on permanent fields.

Reclamation and the expansion of arable agriculture went through two distinct phases during the Middle Ages. From about 650 to 950 these processes were played out for the most part within existing territories, amounting to an extension of permanent fields into former *saltus*. During the second phase, between 950 and 1350, settlement and agriculture were brought into wilderness, the *silva*, for the first time. The activities of reclamation made possible the creation of permanent fields near the centers of new settlements, while lands and waters farther from the centers were used for livestock grazing, fishing, hunting, and collecting. Both phases together were sufficient to transform temperate Europe from a vast sea of wilderness, dotted by small islands of settlement in which pastoral agriculture was more important than arable agriculture, into a more or less continuously settled landscape in which crop raising had become the predominant subsistence strategy.

Because of a series of epidemics and other disasters, the beginning of the Middle Ages was marked by a substantial decrease in European population, perhaps to the lowest level since before the expansion of the Roman Empire into Gaul and Germania.[59] Such a reduction in the number of inhabitants was responsible for a considerable shrinkage of settlement and agriculture, causing the abandonment of some areas and a tendency toward less labor-intensive subsistence strategies in others.[60] In an effort to prevent peasants from moving off into the woods to practice forest-fallow agriculture, always difficult to control and tax by ruling elites, landlords began to develop ways of keeping scarce agricultural labor focused on short-fallow agriculture carried out on permanent fields. One common solution, known to us as the "manorial system," required that peasants spend substantial portions of their time raising crops on fields that directly provisioned the households of the elites, in exchange for which they received the right to exploit other lands for their own subsistence. This was a notoriously inefficient use of labor. Peasants had little or no incentive to perform the labor operations needed to maintain productivity on the landlord's *ager*.[61] Perhaps it was this artificial bottling up of peasants when they could more easily achieve self-sufficiency by spreading out that brought on some of the miserable crop yields of the early Middle Ages. Still, in an age of declining people, money, and markets, landlords had few alternatives short of tilling the soil themselves.

[59] See Russell, "Population in Europe, 500–1500," p. 38; Duby, *The Early Growth of the European Economy*, pp. 12–13; David Herlihy, "Ecological Conditions and Demographic Change," in Richard L. DeMolen, ed., *One Thousand Years: Western Europe in the Middle Ages*, pp. 4–12.

[60] Fourquin, *Le paysan d'Occident au moyen âge*, p. 11; Duby, *The Early Growth of the European Economy*, pp. 12–13; Jankuhn, "Rodung und Wüstung in vor- und frühgeschichtlicher Zeit," pp. 111–20, 129.

[61] Duby, *The Early Growth of the European Economy*, pp. 13, 46–47, 88; Nell, "The Technology of Intimidation," pp. 43–44; Cooter, "Ecological Dimensions of Medieval Agrarian Systems," pp. 475–76; Koebner, "The Settlement and Colonization of Europe," pp. 4, 33–34.

The decline of Europe's population ended around 650 at something like two-thirds of what it had been a century earlier. Demographic growth then resumed, and by 800 the previous losses had been more or less replaced. This meant, first, that most of the places that had been abandoned earlier were occupied once again. In other areas, meanwhile, the density of settlement increased considerably. As we saw earlier, by the early ninth century some population densities approached 30 or more people per square kilometer, sufficient to encourage short-fallow agriculture on permanent fields according to Boserup's calculations. Since landlords generally continued to pursue policies designed to prevent peasants from migrating into less densely settled areas, however, the highest densities may have been somewhat forced and artificial rather than solely a result of natural increase. Indeed, the appearance of population concentrations may well have been more a measure of the ability of the elites to dominate peasant labor than an indicator of anything else, particularly since virtually empty areas often persisted within close proximity to highly populated territories.[62] Still, this is also the period in which open landscapes of more or less continuous settlement with no intervening zones of wilderness began to take shape in the Paris region, parts of Flanders, the middle Rhine Valley, and portions of southeast England.

The reoccupancy of formerly abandoned territories and the intensification of a human presence in areas that had continued to be settled were accompanied by reclamation activity designed to expand the area of arable land. This activity usually took the form of extending *ager* into the *saltus* portions of existing territories. The cumulative effects of bush-fallow agriculture and rough grazing of livestock gradually altered the character of portions of the wasteland until ultimately a parcel cleared by fire for temporary cultivation might not be allowed to revert to bush vegetation. Instead, efforts would be made to remove tree stumps, roots, and other obstacles to plowing or to dig drainage ditches. At other times the reclamation activity may simply have taken the form of adding a little each year to the margins of the permanent fields. In any case, documents from the sixth century onward contain references to *exarta*, or assarts, parcels of land reclaimed by felling trees and especially by grubbing out the stumps and

[62] Duby, *The Early Growth of the European Economy*, p. 25; Alan Mayhew, *Rural Settlement and Farming in Germany*, p. 37; Hoffmann, "Medieval Origins of the Common Fields," pp. 42–43.

roots. On occasion new settlements and clearings came into existence as well. For example, Benedictine monks had a tradition of founding monasteries in places away from existing settlements where they or their associates carried out reclamation and practiced arable agriculture on permanent fields.[63] Further, from the sixth century onward the Franks cleared and settled tracts of woodland along the Schelde and Lys rivers and some of their tributaries in Flanders and Brabant, in the Rhine Valley between Mainz and Strasbourg, and farther east near Augsburg and Würzburg. Some of these tracts had been occupied earlier, but others were virgin lands.[64] Nevertheless, the great European wilderness areas remained largely intact in the eighth and ninth centuries.[65]

Toward the end of the eighth century and the beginning of the ninth some of the most densely settled and highly developed areas of temperate Europe began to experience food shortages that look very much like symptoms of relative overpopulation.[66] The problem lay in the peasants' practice of continuing to cultivate the same soils that they always had, the lightest and best drained, which had long since lost most of their natural fertility. This situation was caused at least in part by the type of landlordism that the ruling elites continued to practice: preventing peasants from migrating into less densely settled areas and expropriating in an inefficient and stifling way a large proportion of their labor. Peasants, of course, were reluctant as they had been for centuries to do the extra plowing, weeding, and especially fertilizing required to improve production as long as the only beneficiaries were persons other than themselves. Indeed, the problems seem to have been most acute near the centers of power, where landlordism was most successful, that is, where peasants had been confined and dominated most effectively by a powerful but parasitic landlord class.[67] Away from

[63] Nitz, "The Church as Colonist," pp. 105–26.

[64] Koebner, "The Settlement and Colonization of Europe," pp. 1–43; Des Marez, *Le problème de la colonisation franque*, pp. 13–40; Grand and Delatouche, *L'Agriculture au moyen âge*, pp. 237–42; Duby, *The Early Growth of the European Economy*, p. 72; Bloch, *French Rural History*, pp. 1–9; Devèze, "Forêts françaises et forêts allemandes," pp. 360–65; Nitz, "The Church as Colonist," p. 121; Janssen, "Some Major Aspects of Frankish and Medieval Settlement in the Rhineland," pp. 51–52, 56.

[65] Bloch, *French Rural History*, p. 11; Jäger, "Zur Geschichte der deutschen Kulturlandschaften," p. 121.

[66] Verhulst, "Karolingische Agrarpolitik," pp. 179–89; Riché, *Daily Life in the World of Charlemagne*, pp. 48–49, 250–51; Rouche, "La faim à l'époque carolingienne," pp. 295–320; TeBrake, "Ecology and Economy in Early Medieval Frisia," pp. 20–21.

[67] Rouche, "La faim à l'époque carolingienne," pp. 295–320.

the centers of power, where peasants were less confined and dominated by ruling elites, food shortages may have been a less serious problem at this time.[68]

The food shortages of the late eighth and early ninth centuries were gradually resolved by a series of changes in the agricultural system. First, there were the beginnings of a move toward cultivating the *ager* more intensely by shortening the length of the fallow period. Since the introduction of permanent fields in prehistoric times, Europeans generally had divided their arable land into two fairly equal parts and, in alternate years, cultivated first one and then the other. This two-field system of crop rotation, a variety of short-fallow cultivation, consigned one-half of the *ager* to fallow each year. In the Mediterranean version planting was usually done in early winter before the rains and harvested in early summer before the drought, while the Continental tradition in the north called for planting in the spring and harvesting in late summer.

During the ninth century a hybrid of the southern and northern cropping patterns, the three-field system of crop rotation, came into use on some of the lands of the large abbeys between the Loire and Rhine rivers. In principle this version of short-fallow agriculture involved dividing the *ager* into three equal parts, one planted in early winter, one planted in the spring, and one left fallow. This pattern was rotated through the three parts of the *ager* so that each part had three uses over a three-year period. This scheme reduced the fallow to one-third of the total arable, one-sixth less than that under the two-field system. In addition it spread the most difficult and critical labor of arable farming, plowing and harvesting, through more of the year, perhaps making it possible for the same number of people to cultivate more land. Finally, combined winter and spring planting introduced new crops into regular rotation: alongside the traditional bread grains of wheat and rye on the winter fields appeared the new crops of oats and legumes on the summer fields. By this diversification of the food base the chances of famine were somewhat reduced, since each crop responded differently to variations in soil and weather, while the times of germination, growth, and ripening were spread more evenly throughout the year.[69]

[68] See, for example, TeBrake, "Ecology and Economy in Early Medieval Frisia," pp. 16–24.

[69] White, *Medieval Technology and Social Change*, pp. 69–76; White, "The Expansion of Technology, 500–1500," pp. 146, 150–51; Slicher van Bath, *The Agrarian History of Western Europe*, pp. 20–22, 58–62; Duby, *Rural Economy and Country Life in the Medieval*

The second major agricultural change introduced during the late eighth and early ninth centuries was a shift in focus away from exclusive tillage of the lightest, best-drained soils toward a greater emphasis on the cultivation of the plentiful and ultimately more productive heavy, moist soils that constituted much of the *saltus* of temperate Europe. As we saw earlier, however, the traditional instrument of cultivation, the *ard*, or scratch plow, was poorly suited to such a change. The new soils had to be dug deeply and vigorously so that their surfaces could dry out and weed growth could be controlled.[70] That this necessity was recognized early is suggested by archaeological evidence from prehistoric times revealing a modified scratch plow with a fairly broad, flared cutting end, or share, that could be maneuvered by tilting and twisting to one side and thereby made to dig and partly turn over some of the topsoil.[71] As greater attention was paid to the heavy, moister soils of the *saltus* during the early Middle Ages, a different plow, originating somewhere in eastern Europe around the beginning of our era, came to be used instead. This was a heavier, usually wheeled plow that included a colter to cut the soil vertically, a plowshare to cut it horizontally, and a moldboard to turn it over.[72]

Because the new plow had no clear advantage over the older model when it came to cultivating the traditional light, dry soils, however, it did not replace the *ard* but came into use alongside it.[73] The use of the *ard* died out when the tillage of the soils for which it was suited came to an end. For this reason the spread of the heavy, wheeled plow should be seen as a symptom of the expansion of crop agriculture into areas that previously if not totally ignored had been exploited by means of livestock keeping, fish-

West, pp. 90–93; Duby, *The Early Growth of the European Economy*, pp. 17–30; Bloch, *French Rural History*, pp. 26–35; Grand and Delatouche, *L'Agriculture au moyen âge*, pp. 269–71; Parain, "The Evolution of Agricultural Technique," pp. 137–38.

[70] Duby, *Rural Economy and Country Life in the Medieval West*, pp. 21, 67; Devèze, "Forêts françaises et forêts allemandes," p. 360.

[71] Bernard Wailes, "Plow and Population in Temperate Europe," in Brian Spooner, ed., *Population Growth: Anthropological Implications*, pp. 154–79; A. E. van Giffen, "Nederzettingen van de vroege Klokbekercultuur bij Oostwoud (N.H.)," in *In het voetspoor van A. E. van Giffen*, 2d ed., pp. 66, 68; P. J. Woltering, "Archeologische kroniek van Noord-Holland over 1978," *Holland: regionaal-historisch tijdschrift* 11 (1979): 250–51.

[72] Watson, "Towards Denser and More Continuous Settlement," pp. 73–75, 81–82 nn. 13–20.

[73] For some archaeological evidence of the continued use of the *ard* alongside the moldboard plow, see Per Kristian Madsen, "Medieval Ploughing Marks in Ribe," *Tools and Tillage* 4 (1980): 36–45.

ing, hunting, and collecting. In combination with the adoption of the harrow to pulverize clods of soil left after plowing, improvements in animal harnessing, and the slow switch to the horse as an animal of traction to meet the increased motive-power requirements, the heavy, wheeled plow provided the peasantry of temperate Europe with a technology uniquely suited to cultivating the heavy, moist soils of the *saltus* and eventually the *silva*.[74]

Both the three-field system of crop rotation and the expansion of crop agriculture onto the heavier soils of the *saltus* were adaptations to increasing population densities between the late eighth and late tenth centuries. Both, in fact, required a considerable augmentation of labor and energy inputs that only a dense population possessed. For example, the shortening of the fallow that the three-field system represented required an increased investment of labor in the forms of plowing, fertilizing, and weeding. The cultivation of the heavier soils, meanwhile, required the arduous labor of assarting, that is, removing trees, stumps, and roots, and the use of a heavy, cumbersome, and expensive new instrument of cultivation. The new plow with its greater traction demands in turn required the care and feeding of more, as well as more expensive, draft animals. Thus the first signs of both innovations appeared in the most densely settled portions of temperate Europe.[75] As late as the midtenth century Europe's population, though greater than ever, still resided in essentially the same places as it always had.[76] The introduction of the three-field system and the expansion

[74] Duby, *Rural Economy and Country Life in the Medieval West*, pp. 18–19; Slicher van Bath, *The Agrarian History of Western Europe*, pp. 62–64; White, *Medieval Technology and Social Change*, pp. 53–66, 68–69, 78–79; White, "The Expansion of Technology, 500–1500," pp. 147–48, 151–53; Bloch, *French Rural History*, pp. 50–56; Grand and Delatouche, *L'Agriculture au moyen âge*, pp. 272–75; J. F. Niermeyer, *De wording van onze volkshuishouding: hoofdlijnen uit de economische geschiedenis der nordelijke Nederlanden in de middeleeuwen*, p. 27.

[75] Duby, *Rural Economy and Country Life in the Medieval West*, pp. 18–19; Slicher van Bath, *The Agrarian History of Western Europe*, pp. 62–64; White, *Medieval Technology and Social Change*, pp. 53–66, 68–69, 78–79; White, "The Expansion of Technology, 500–1500," pp. 147–48, 151–53; Bloch, *French Rural History*, pp. 50–56; Grand and Delatouche, *L'Agriculture au moyen âge*, pp. 272–75.

[76] Duby, *Rural Economy and Country Life in the Medieval West*, pp. 15, 21, 67; Des Marez, *Le problème de la colonisation franque*, p. 16; Jankuhn, *Vor- und Frühgeschichte vom Neolithikum bis zur Völkerwanderungszeit*, pp. 21–26; Mortensen, "Probleme der Mittelalterlichen Siedlungs- und Kulturlandschaft," pp. 100–101; Devèze, "Forêts françaises et forêts allemandes," pp. 354, 360, 370. The relatively dense occupation of the German Rhineland by the Romans was also limited to the previously settled areas, though the settlement

of arable farming into the *saltus*, however, show that these same places were now being occupied and exploited much more intensely than ever before.

Toward the end of the tenth century Europeans began to spread on a large scale into the still-massive wilderness areas between and especially beyond their traditional areas of settlement and agriculture. Why this movement began then and not before is not easy to determine. A number of cultural factors may well have been involved. First, the development of an improved felling ax during the tenth century facilitated the task of clearing the hardwood forest vegetation that covered so much of the *silva*.[77] In addition, most of the groups living in temperate Europe had long traditions of valuing primeval wilderness as a frontier or border zone, generally much more effective for such purposes than a river or other easily negotiated impediment.[78] Perhaps the end of the great migrations and the establishment of a certain amount of cultural and political stability in much of temperate Europe, especially under the Carolingians in the late eighth and early ninth centuries, slowly began to undermine the need for such buffer zones. Too, for some the heart of the wilderness was the abode of the gods, an impassable and untouchable place to be treated with awe and respect; for many others the wilderness was the habitation of all manner of monsters and beasts, including the Wild Huntsman, who destroyed all in his path. Although the precise effects of such feelings on the processes of wilderness reclamation are very difficult to assess, that does not make them any less real. No doubt the activities of missionaries like Saint Boniface and others, who regularly destroyed pagan shrines, established places of Christian worship, and "Christianized" the wilderness they penetrated, went a long way toward demythologizing the forests.[79]

---

there was much more intense than it had been before: Paffen, "Natur- und Kulturlandschaft am deutschen Niederrhein," pp. 192–95.

[77] White, *Medieval Technology and Social Change*, p. 41; Georges Duby, "La révolution agricole médiévale," *Revue de géographie Lyon* 29 (1954): 363.

[78] Koebner, "The Settlement and Colonization of Europe," p. 21; East, *The Geography behind History*, p. 100; de Cock, *Bijdrage tot de historische geografie van Kennemerland*, pp. 22, 52; Bloch, "Occupation du sol et peuplement," p. 28; Devèze, "Forêts françaises et forêts allemandes," p. 365; Jäger, "Zur Geschichte der deutschen Kulturlandschaften," p. 120.

[79] Koebner, "The Settlement and Colonization of Europe," p. 21; Lynn White, Jr., "Cultural Climates and Technological Advance in the Middle Ages," *Viator: Medieval and*

Ultimately, however, it may well have been a series of political changes that made the greatest difference. The invasions of Saracens, Maygyars, and Vikings into various parts of temperate Europe during the ninth and early tenth centuries introduced something of a security crisis after several generations of relative peace and quiet. The Vikings in particular invaded some of the very areas that shortly thereafter were to undergo the large-scale transformation of wilderness, especially sections of northern France, the Low Countries, and Germany. The scattered, discontinuous settlements of these regions were vulnerable to attack, particularly since virtually all traces of former royal military power had dissipated, and there was really very little chance of receiving defensive assistance from other areas. Thus the ease with which the Vikings could sail up the great rivers of northern and western Europe and launch surprise attacks must have encouraged the contraction and retreat of settlement and agriculture in some places. Indeed, a number of villages were abandoned and fields left untilled for a time after one such attack in the Ardennes in 893.[80] When these attacks came to an end during the tenth century, however, they did so because a new set of territorial rulers organized effective defenses against them, not because there was any significant revival of centralized royal power. Many of these new power brokers, in fact, went on to assume most of the rights, privileges, and responsibilities that at least theoretically pertained to kings and emperors.

One of the privileges that the new rulers characteristically assumed was the right to the disposal of wilderness, formerly one of the *regalia*, or royal rights. Indeed, as soon as they had consolidated political control over their territories, they began to exercise this right. A large proportion of the

---

Renaissance Studies 2 (1971): 188; Lynn White, Jr., *Medieval Technology and Social Change*, p. 71; Glacken, *Traces on the Rhodian Shore*, p. 291; Russell, "Population in Europe, 500–1500," p. 39; Devèze, "Forêts françaises et forêts allemandes," p. 368; Bloch, "Occupation du sol et peuplement," p. 128. For an interesting account of the establishment of the monastery of Fulda, deep in the wilderness, and how it eventually became a civilized place "pleasing to God," with water wheels for driving machines in the workshops and the like, see the life of Saint Sturm by Eigil, abbot of Fulda, *Eigilis vita S. Sturmi abbatis Fuldensis*, in G. H. Pertz, ed., *Scriptorum*, Monumenta Germaniae Historica, 2:367–69, 375.

[80] Duby, *Rural Economy and Country Life in the Medieval West*, p. 49; Duby, *The Early Growth of the European Economy*, pp. 112–20; Glacken, *Traces on the Rhodian Shore*, pp. 291–92; Koebner, "The Settlement and Colonization of Europe," pp. 52, 55–56; White, *Medieval Technology and Social Change*, pp. 73–74; White, "The Expansion of Technology, 500–1500," pp. 146–47; Herlihy, "Ecological Conditions and Demographic Change," pp. 19, 22.

reclamation and colonization of the High Middle Ages took place with the active connivance of the territorial princes of France, the Low Countries, and Germany, who proceeded to allot the wilderness under their control to groups of colonists who agreed to bring it into the realm of human affairs. The interests of these new rulers were quite the opposite from those of the old landlords of the manorial system, who persisted in trying to keep peasants confined to the traditional areas of settlement and agriculture.

To secure the needed colonists, the new rulers often drew up agreements with representatives of prospective colonists, offering them very favorable terms that stood in sharp contrast to the restrictions placed on peasants in the old population centers. In exchange for the payment of a very small tax per homestead and about 10 percent of everything they produced, colonists in the frontier areas usually received ownership of the land they reclaimed, the rights to dispose of such land, and complete personal freedom. Further, groups of colonists were customarily organized into rural communes, or *Landgemeinden*, which had legal standing and possessed the rights to exercise local government and administer lower justice. Such rural communes were very similar to the many urban communes established during the same period.[81] What was formed in this manner was a new society of free and equal agriculturalists who decided many of their affairs themselves in the context of their communes and exploited their own property without the intervention of landlords.

The colonists who settled the frontier areas of temperate Europe during the High Middle Ages, of course, had to carry out a process of reclamation that would produce arable land where none had existed before. This process usually amounted to taking on mature hardwood forest with rather simple tools. Because of the very low population densities in the frontier zones, however, people who knew how to practice labor-intensive short-fallow agriculture characteristically changed to a less labor-intensive, pioneering version of swidden agriculture in their new environments.[82] If

[81] H. van der Linden, *De Cope: bijdrage tot de rechtsgeschiedenis van de openlegging der Hollands-Utrechtse laagvlakte*, pp. 4–16, 81–108; Janssen, "Some Major Aspects of Frankish and Medieval Settlement in the Rhineland," pp. 56–59; Karl Bosl, *Die Grundlagen der modernen Gesellschaft im Mittelalter: eine deutsche Gesellschaftsgeschichte des Mittelalters*, p. 222; Koebner, "The Settlement and Colonization of Europe," p. 35; studies collected in *Die Anfänge der Landgemeinde und Ihr Wesen*.

[82] Koebner, "The Settlement and Colonization of Europe," p. 82; Boserup, *The Conditions of Agricultural Growth*, pp. 62–63; Boserup, *Population and Technological Change*,

they had been required to remove all trees, stumps, and roots and establish permanent fields before planting the first crops, they would not have been able to survive. By employing the oldest of reclamation techniques, clearing by girdling large trees with a sharp instrument and burning underbrush and ground cover, colonists could put land into production very quickly and receive high yields before productivity dropped and weed growth choked out everything else. This should not be confused with true forest-fallow agriculture, since the object was to open up wilderness to eventual short-fallow agriculture on permanent fields rather than to establish a balanced cycle of reclamations followed by long fallow periods that allowed forest vegetation to recolonize the land. Most of the parcels allotted to colonists during the period of land reclamation were large enough that several thrusts of clearance by swidden or forest-fallow techniques could be made before their limits were reached. Only later would colonists begin the arduous task of removing trees, stumps, and roots in preparation for short-fallow agriculture using plows on permanent fields.[83] As we shall see later, a lowland variant of pioneering swidden was developed in areas where the restructuring activities of reclamation were aimed at the removal of water instead of trees.[84]

Once under way, the colonization and reclamation movement of the High Middle Ages appears to have gained a momentum all its own. It began, of course, as a response to a growing population, which provided the vitality that made it possible. However, the low population densities of the frontier zones encouraged younger marriages and larger families than those in the long-settled portions of Europe, resulting in a higher rate of population growth.[85] Thus population growth contributed to frontiering, which contributed to additional population growth, which in turn contributed to further frontiering, and so on. This process continued without any notice-

---

pp. 97–98; Green, "The Agricultural Colonization of Temperate Forest Habitats," pp. 69–87.

[83] Verhulst, *De Sint-Baafsabdij te Gent en haar grondbezit*, p. 212; Cooter, "Ecological Dimensions of Medieval Agrarian Systems," pp. 466, 471–75; Cooter, "Preindustrial Frontiers and Interaction Spheres," pp. 41–46, 233; Boserup, *The Conditions of Agricultural Growth*, pp. 62–63; Boserup, *Population and Technological Change*, p. 47; J. G. D. Clark, *Prehistoric Europe: The Economic Basis*, pp. 92–93; Harris, "Swidden Systems and Settlement," pp. 245–62; Green, "The Agricultural Colonization of Temperate Forest Habitats," pp. 69–87.

[84] See chap. 6.

[85] Green, "The Agricultural Colonization of Temperate Forest Habitats," pp. 85–86.

able interruption until virtually all the wilderness zones of western and central Europe had been incorporated into the realm of human affairs.

The expansion of settlement and agriculture into the wilderness areas of temperate Europe had profound effects on European society. First, it helped break the stranglehold that landlordism had had on peasant labor and productivity in the early Middle Ages. Even those peasants who did not migrate to frontier areas benefited since, in an effort to keep some peasants on their lands, landlords began altering some of the most stifling aspects of manorialism. For example, money payments gradually replaced labor services and payments in kind, allowing peasants a much fuller expression of productive capacity. Second, reclamation and colonization brought a tremendous burst of food production by bringing an increasing proportion of temperate Europe into crop agriculture. As a result, we know of very few food shortages in in the High Middle Ages that were more than local in scale. In fact, it was the production of agricultural surpluses during the High Middle Ages that made possible the appearance of urban life with all of its trappings. Third, the era of reclamation and colonization filled in the empty spaces so that the settlement pattern of the fourteenth century resembles the modern pattern much more closely than it does the pattern of the early Middle Ages.

Despite its obvious significance, however, there have been few attempts to chronicle and evaluate properly the reclamation and colonization movement of the High Middle Ages. The general picture that I have sketched to this point, far from being the definitive one, remains hypothetical in the sense that much of it rests on very widely scattered bits of information. In what follows, I shall explore in considerable detail the transformation of wilderness in one small corner of temperate Europe. What we learn from the course of events in Rijnland will not only explain certain developments of local importance but also answer some of the many questions that remain concerning the general picture that I have presented.

3

The Setting, Natural History, and Premedieval Settlement of Rijnland

AN examination of change in the relations between nature and human culture for portions of the western Netherlands between the midtenth and midfourteenth centuries, while significant and interesting in its own right, offers important perspectives on the general processes of growth and development characteristic of much of Europe as a whole during this period. The low-lying portions of the Netherlands, in particular, provide excellent opportunities for such inquiry because from the very beginning their inhabitants directly experienced some of the physical limits to human existence. At first it was an arctic climate, later a steadily encroaching sea. Yet in a region in which the boundary between land and water was rarely distinct or stable before A.D. 1000, human beings gradually created for themselves the very stability that long-term physical processes had always denied them. By the end of the Middle Ages they had wrested a relatively dry and secure dwelling place from sea, swamp, and river and had built on this base a new social order.[1] Nowhere was this change more evident than in Rijnland. To understand and appreciate this transformation, we must begin by examining the environmental parameters of life in the soggy western Netherlands.

Rijnland's Location

The name Rijnland may not seem immediately intelligible to anyone unfamiliar with Dutch spelling. It refers, simply, to land along the banks of the Rhine River. This was the sense of the oldest-surviving written mention of the name, a Latin form from the year 1064: "circa horas Reni."[2] Thus,

[1]H. van der Linden, *Recht en territoir: een rechtshistorisch- sociografische verkenning*, p. 5.

[2]A charter of Emperor Henry IV, in A. C. F. Koch, ed., *Oorkondenboek van Holland en Zeeland tot 1299*, vol. 1, *Eind van de 7e eeuw tot 1227*, no. 85.

the name was similar in form to those applied to areas along the Dutch coast on the south that were named after rivers as well in the early Middle Ages: Maasland (the land along the banks of the Maas, or Meuse, River) and Scheldeland (the land along the banks of the Schelde, or Scheldt, River).[3]

To locate this land along the banks of the Rhine on a modern map of the Netherlands is not as simple a matter as it may seem. In the first place, the Rhine splits into many channels soon after it enters the Netherlands and eventually mingles with the Maas River. In effect the two share one delta. The branch of the Rhine in question here, today called the Oude Rijn (Old Rhine), traces a westward course from the city of Utrecht to the North Sea by way of the city of Leiden. Originally this was the main branch of the river. The Romans recognized it as such and made it the boundary of their empire for several centuries.[4] During the Middle Ages, however, the Oude Rijn declined greatly in significance as a river, eventually silting shut where it passed through the coastal dunes. Although its mouth was dredged open again after the Middle Ages, it hardly qualifies as a river today, being completely controlled by a system of dams, sluices, and locks.

A second difficulty in trying to place the original Rijnland on a modern map stems from the fact that the phrase "circa horas Reni" is geographically imprecise. When it was so mentioned in 1064, there were few exact boundaries; the area was only partly inhabited. As we shall see, some of it was already in the process of being reclaimed, but a large proportion remained uninhabited wilderness.

The context in which Rijnland was first mentioned, however, offers some clues to establishing its location. On 30 April 1064, Emperor Henry IV gave lands to the bishop of Utrecht that were described as "comitatum omnem in Westflinge et circa horas Reni" ("the entire county in Westflinge and along the banks of the Rhine").[5] The two parts of the grant of land, Westflinge and "circa horas Reni," represented the entire coastal region

[3] J. K. de Cock, "Die Grafschaft Masalant," in *Miscellanea mediaevalia in memoriam Jan Frederik Niermeyer*, pp. 106–107.

[4] A. W. Byvanck, *Nederland in den Romeinschen tijd*, 3d ed., 1:2; D. P. Blok, "Probleme der Flussnamenforschung in den alluvialen Gebieten der Niederlande," in Rudolf Schützeichel and Matthias Zender, eds., *Namenforschung: Festschrift für Adolf Bach zum 75. Geburtstag am 31. Januar 1965*, p. 222.

[5] Koch, ed., *Oorkondenboek van Holland en Zeeland*, vol. 1, no. 85.

from the Frisian Islands in the north to the area around the Maas River (the county of Maasland or Masalant) in the south.[6] Since the southern boundary of Westflinge was the Haarlemmerhout (Woods of Haarlem) and the northern boundary of Maasland was the Woods of The Hague, or 's-Gravenhage,[7] "circa horas Reni" was the coastal region in between, extending from near Haarlem in the north to near The Hague in the south.

Rijnland was thus a naturally determined region that, as we shall see, extended to either side of the Oude Rijn as far as did the drainage area of the river.[8] The eastern border of Rijnland was clearly drawn only along the Oude Rijn near what later became known as Zwammerdam,[9] the ancient site of a Roman fort. Boundaries were marked in the surrounding peat bogs along the drainage divides later, when reclaimers from Rijnland met their

[6]D. P. Blok, "Holland und Westfriesland," *Frühmittelalterliche Studien: Jahrbuch des Instituts für Frühmittelalterforschung der Universität Münster* 3 (1969): 355; J. K. de Cock, *Bijdrage tot de historische geografie van Kennemerland in de middeleeuwen op fysisch-geografische grondslag*, p. 59.

[7]That a woods should play a role as a boundary was not unusual during the early Middle Ages; see de Cock, *Bijdrage tot de historische geografie van Kennemerland*, pp. 52, 22 n. 4; W. Gordon East, *The Geography behind History*, rev. ed., p. 100; Richard Koebner, "The Settlement and Colonization of Europe," in M. M. Postan, ed., *The Agrarian Life of the Middle Ages*, 2d ed., p. 21, vol. 1 in *The Cambridge Economic History of Europe*; H. C. Darby, "The Fenland Frontier in Anglo-Saxon England," *Antiquity* 8 (1934): 185–201. In Flanders, in the early Middle Ages, the Forêt Charbonniere served as a border zone between Neustria and Austrasia; see M. Devèze, "Forêts françaises et forêts allemandes: étude historique comparée," *Revue historique* 235 (1966): 365. Rivers, in contrast, served rather unsatisfactorily as political boundaries; see, for example, S. J. Fockema Andreae, "Stein: het ontstaan van een vrije hooge heerlijkheid op de grenzen van Holland en van hare bestuursorganen," *Tijdschrift voor geschiedenis* 47 (1932): 403–404; William S. Cooter, "Preindustrial Frontiers and Interaction Spheres: Aspects of the Human Ecology of Roman Frontier Regions in Northwest Europe" (Ph.D. diss., Univesity of Oklahoma, 1976), p. 95. For the Haarlemmerhout, see S. Jelgersma et al., "The Coastal Dunes of the Western Netherlands: Geology, Vegetational History and Archaeology," *Mededelingen van de Rijks Geologische Dienst*, n.s., no. 21 (1970): 132; Rob Rentenaar, "De Nederlandse duinen in de middeleeuwse bronnen tot omstreeks 1300," *Geografisch tijdschrift* 11 (1977): 367–70.

[8]See, for exemple, the map in Jelgersma et al., "The Coastal Dunes of the Western Netherlands," p. 131. S. J. Fockema Andreae, "De Rijnlanden," *Leids jaarboekje* 48 (1956): 45, extended the "comitatus circa horas Reni" as far south as present-day Hoek van Holland, thus including in it part of Maasland. He based this on I. H. Gosses, "De vorming van het graafschap Holland," in his *Verspreide geschriften*, ed. F. Gosses and J. F. Niermeyer, pp. 304–305. Gosses, however, distinguished between the original Rijnland and those parts south of The Hague that were added later. For convincing arguments for the existence of a county of Maasland, see de Cock, "Die Grafschaft Masalant," pp. 105–12.

[9]H. van der Linden, *De Zwammerdam*, pp. 5–6.

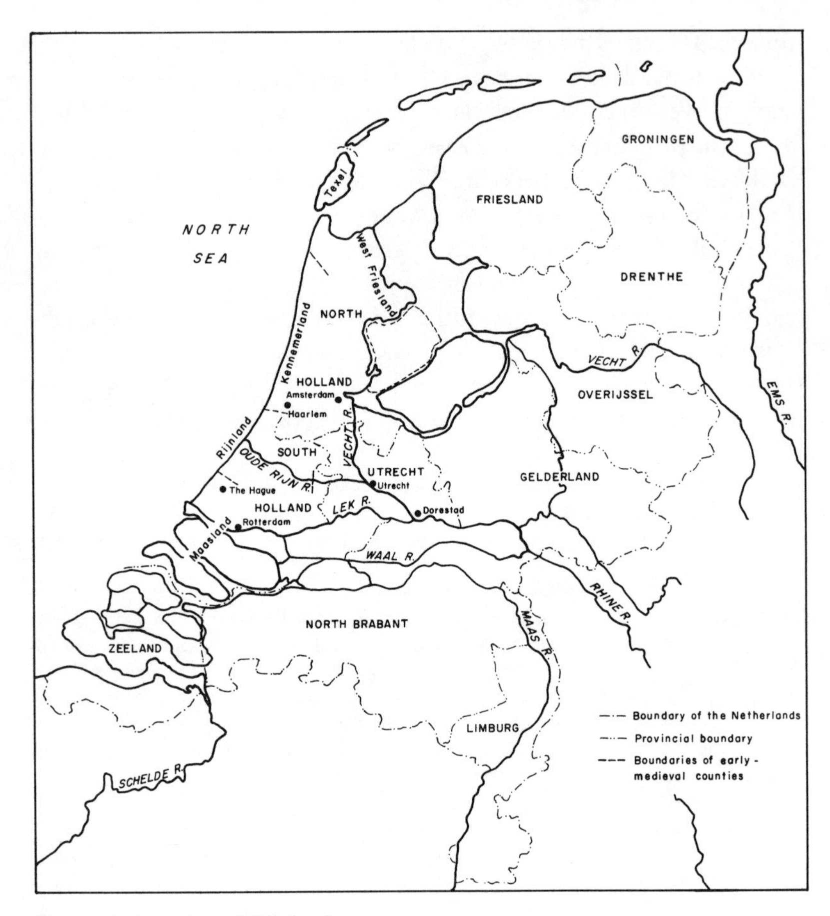

FIGURE 1. Location of Rijnland

counterparts proceeding from other directions. When the name Holland first came into use shortly after 1100, it referred specifically to the area along the banks of the Oude Rijn (see fig. 1).[10]

In the medieval county of Holland that began to assume its ultimate size and shape during the twelfth and thirteenth centuries, however, the term Rijnland was used to refer to at least three partly overlapping geographical areas (see fig. 2). It designated, first, a judicial and administrative district of the County of Holland, the Baljuwschap (related to the En-

[10]Blok, "Holland und Westfriesland," p. 358; van der Linden, *De Cope*, pp. 354–62.

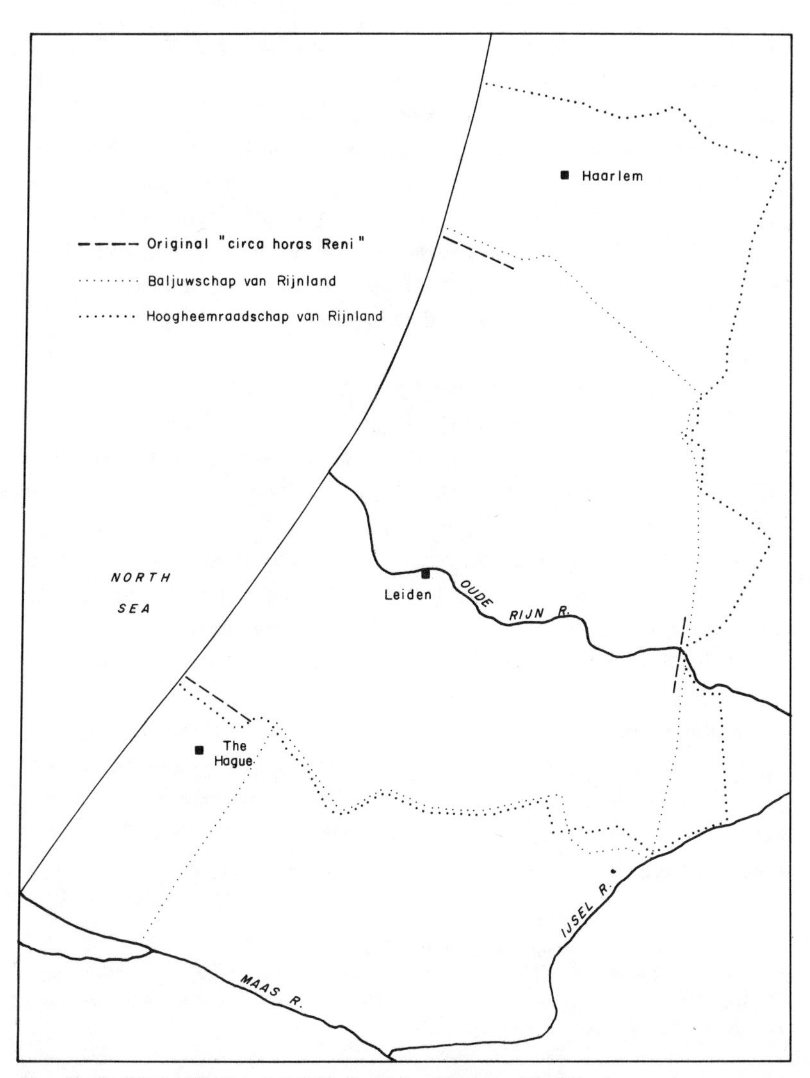

FIGURE 2. Three Medieval Rijnlands (After Foekema Andreae)

glish word *bailiwick*) van Rijnland, also known as the Baljuwschap van Noordholland. Further, it was used to delimit a subdivision of the Bishopric of Utrecht, the Dekanaat (Deanery) van Rijnland. Finally, the term Rijnland was used to describe the oldest regional drainage board, or *hoogheemraadschap*, of the Netherlands, originally known as the Hoogheemraadschap van Spaarndam but referred to from the beginning of the fourteenth century as the Hoogheemraadschap van Rijnland. The Baljuwschap van Rijnland and the Dekanaat van Rijnland comprised essentially the same area: the original land "circa horas Reni" plus a small section of the former Maasland south of The Hague. The Hoogheemraadschap van Rijnland, while not including the extension south of The Hague, was enlarged to the north by the inclusion of part of Kennemerland.[11] Although both the Baljuwschap van Rijnland and the Hoogheemraadschap van Rijnland play prominent roles in what follows, the core area of the study consists of the lands that these two Rijnlands shared, the original "comitatus circa horas Reni" and the eventually reclaimed peat lands around it.

The present appearance of Rijnland is markedly different from what it must have looked like a thousand years ago. The changes that occurred during the past millennium were due to various factors, the most important of which were geological, climatic, vegetational, and human. The geological, climatic, and vegetational forces that have shaped Rijnland were at work long before the tenth century, while the earliest human settlements also date from the dimly understood centuries of prehistory.[12]

It has been only during the last thousand years, however, that human culture has left a lasting imprint on Rijnland. Before the period under consideration settlements existed only along the coastal dunes and riverbanks, but even these had to be abandoned from time to time because of frequent periods of heightened storm activity. They became permanent only shortly before the period of reclamation began.[13] The remainder, by far the largest portion of Rijnland, consisted of an uninhabited wilderness of peat bogs.

[11] Based on Fockema Andreae, "De Rijnlanden," pp. 45–47.

[12] See, for example, J. D. van der Waals, *Prehistoric Disc Wheels in the Netherlands*, pp. 14, 34; L. P. Louwe Kooijmans, *The Rhine/Meuse Delta: Four Studies on Its Prehistoric Occupation and Holocene Geology*.

[13] J. Bennema, "De bewoonbaarheid van het Nederlandse kustgebied vóór de bedijkingen," *Westerheem* 5 (1956): 89–91; S. J. Fockema Andreae, "Rijnlandse kastelen en landhuizen in hun maatschappelijk verband," in S. J. Fockema Andreae et al., *Kastelen, ridderhofsteden en buitenplaatsen in Rijnland*, p. 1; H. Sarfatij, "Middeleeuwse mens en eeuwig water: veranderingen in landschap en bewoning aan de monden van Rijn en Maas gedurende de middeleeuwen," *Zuid-Holland* 14 (1968): 20; H. Sarfatij, "Friezen-Romeinen-Cananefaten," *Holland: regionaal-historisch tijdschrift* 3 (1971): 174–76; D. P. Blok, *De Franken in Nederland*, 3d ed., pp. 18–29; de Cock, *Bijdrage tot de historische geografie van Kennemerland*, pp. 10–12; A. J. Pannekoek et al., *Geological History of the Netherlands: Explanation to the General Geological Map of the Netherlands on the scale 1:200,000*, pp. 128–29.

The Physical Geography of the Western Netherlands during the Holocene

Rijnland is part of the western Netherlands, the low-lying coastal strip stretching from the mouth of the Schelde in the south to the island of Texel in the north. This landscape consists of a belt of sand dunes[14] at the coast with a broad, flat expanse of peat and clay[15] extending eastward about 50 kilometers. Geologically speaking, this region is very young, having been formed during the Holocene, the present geological age, which began some 10,000 years ago.[16] Only in the region east of the city of Utrecht do the underlying Pleistocene cover sands reach the surface. Westward from there these older deposits disappear under Holocene sedimentation that ranges in depth from around 10 meters along the eastern edge of Rijnland to as much as 20 meters at the coast.[17]

Because this mass of Holocene deposition lay approximately at sea level and suffered from poor drainage, it was often flooded by rivers and the sea until the first steps were taken to improve on natural conditions during the tenth century of our era. Even so, the very nature of the deposits, the fact that some would undergo subsidence or compaction after drainage, caused new problems that required new solutions in the later Middle Ages. For these reasons a brief investigation of some of the physical factors that produced the soils of the western Netherlands is necessary to understand the processes and the problems involved in their reclamation and exploitation.

FROM PERMAFROST TO INUNDATION

The late-glacial landscape bore little resemblance to the appearance of the western Netherlands of the tenth century or of today. Although the last ma-

[14] The dunes were breached and severely eroded from time to time south of the Maas River, forming the islands of South Holland and Zeeland, but northward from there they have remained essentially intact to the present.

[15] Peat is an accumulation of partly decayed plant material. Clay is composed of the silt conveyed by water.

[16] There is a difference of opinion about the use of the term Holocene. Some scholars who believe that we are living in an interstadial period between glaciations suggest the use of such terms as Late Pleistocene and Recent. I have adopted the term Holocene simply because most Dutch geologists use it, while I recognize that the other terms are as legitimate. The British equivalent of Holocene is Flandrian.

[17] See the map showing the depth of the Pleistocene surface in the western Netherlands

jor extension of snow and ice, the Würm glaciation (more or less contemporary with the Wisconsin glaciation in North America), covered most of northern and parts of central Europe, it never reached the Netherlands. Still its effects were strongly felt there as recently as 15,000 years ago.[18]

In the first place, the subsoil was permanently frozen and carried only a tundra type of vegetation, low shrubs and herbs with an occasional stunted birch or aspen tree in a sheltered spot. During the short summers the vegetation came to life and provided a surprisingly luxuriant food base for an arctic-steppe assortment of herbivores, especially reindeer, as well as some predators. Waterfowl visited the many swamps. During the long winters, however, snow and especially sand and dust storms swept the barren plain.[19]

The second important effect of the Würm glaciation was the tremendous amount of water that was locked up in snow and ice sheets, often kilometers thick, resulting in a world sea level that was perhaps as much as 100 meters lower than it is today. What is now the western coast of the Netherlands was then a great distance from the sea. The North Sea lay far to the north, beyond a broad land bridge that connected Great Britain to the European continent. The Rhine, Maas, and Schelde rivers joined the Thames River to flow through a single channel into the Atlantic, presumably by way of the Straits of Dover.[20] Groups of late Paleolithic hunters, known to archaeologists as Hamburgians, ventered into this area in pursuit

---

in L. J. Pons et al., "Evolution of the Netherlands Coastal Area during the Holocene," in *Transactions of the Jubilee Convention*, pt. 2, enclosure 2.

[18] The best general survey of past climatic change, including a thorough review of the types of evidence, is H. H. Lamb, *Climate History and the Future*, vol. 2 in *Climate Present, Past and Future*.

[19] See H. T. Waterbolk, "The Lower Rhine Basin," in Robert J. Braidwood and Gordon R. Willey, eds., *Courses Toward Urban Life: Archaeological Considerations of Some Cultural Alternates*, p. 229; H. T. Waterbolk, "The Occupation of Friesland in the Prehistoric Period," *Berichten van de Rijksdienst voor het Oudheidkundig Bodemonderzoek* 15–16 (1965–66): 13–14.

[20] S. Jelgersma, *Holocene Sea Level Changes in the Netherlands*, pp. 9, 46, 65, 71, 74 fig. 46; Pannekoek et al., *Geological History of the Netherlands*, p. 97; F. J. Faber, *Nederlandsche landschappen*, 2d ed., vol. 3 in *Geologie van Nederland*, p. 129; L. P. Louwe Kooijmans, "Archeologische ontdekkingen in het Rijnmondgebied," *Holland: regionaal-historisch tijdschrift* 5 (1973): 25–26; L. P. Louwe Kooijmans, "Mesolithic Bone and Antler Implements from the North Sea and the Netherlands," *Berichten van de Rijksdienst voor het Oudheidkundig Bodemonderzoek* 20–21 (1970–71): 66–67. The map in Lamb, *Climate History and the Future*, p. 368, is obsolete.

of reindeer during the summer, withdrawing southward again with the onset of winter.[21]

Some of the arctic characteristics of the Netherlands and surrounding areas began to disappear during an abortive warm period between 10,000 and 9000 B.C., only to return again in a last cold blast that persisted for another thousand years. Around 8000 B.C., however, the climate began to improve again, this time more permanently. As glaciers and permafrost retreated toward the north, birch forest covered more and more of the landscape, only to be replaced very quickly by pines as the dominant forest type. Because this new forest was an open, northern variety, sunlight easily penetrated to the ground, where a rich complex of grasses, herbs, and shrubs flourished. This new vegetation, together with increased precipitation, ended the sand and dust storms characteristic of the previous, colder era.[22]

As climatic improvement continued, the first warmth-loving species of deciduous trees appeared, adding a new diversity to the vegetation. Hazel, followed by elm, linden, oak, alder, and ash, began to constitute a small but important component of the still predominantly pine forests. New herbivores, including the Irish giant deer, elk, aurochs (wild cattle), red deer, roe deer, wild pig, and beaver, as well as some of their predators, became associated with the new vegetation complexes, while the numerous lakes, ponds, and swamps provided habitats for many species of fish and waterfowl. A cold climate ceased to be a serious limiting factor to human existence. Although winters remained somewhat colder than they are today, characteristic of a continental climate, summer temperatures rose to approximately modern levels.[23]

Between 8000 and 6000 B.C., therefore, the environment of what is

[21] Waterbolk, "The Occupation of Friesland in the Prehistoric Period," pp. 13–14; Waterbolk, "The Lower Rhine Basin," pp. 229–30 and the chronological table, p. 250. The Hamburgians were contemporary to the Magdalenian culture of areas in the south in what is now France.

[22] Waterbolk, "The Lower Rhine Basin," p. 232; Waterbolk, "The Occupation of Friesland in the Prehistoric Period," pp. 14–15; H. Godwin, "The Beginnings of Agriculture in North West Europe," in Joseph Hutchinson, ed., *Essays on Crop Plant Evolution*, pp. 6–7.

[23] Waterbolk, "The Lower Rhine Basin," pp. 231–34; Waterbolk, "The Occupation of Friesland in the Prehistoric Period," pp. 15–16; Godwin, "The Beginnings of Agriculture in North West Europe," p. 7; Louwe Kooijmans, "Archeologische ontdekkingen in het Rijnmondgebied," pp. 26–27.

now the Netherlands gradually became very hospitable to people depending for their subsistence on hunting, fishing, and gathering activities. It is thus not surprising that groups possessing a Mesolithic culture began to live more or less permanently in the area. Presumably they deliberately chose to live in the wetter, lower-lying areas, where they could most easily combine fishing and fowling with hunting. With canoes made from hollowed-out pine-tree trunks, they could negotiate waters, while dogs, domesticated from the European wolf, aided them in their hunting activities. Most knowledge of these people comes from Denmark, but there is reason to believe that they also lived in vast areas now submerged by the North Sea as well as in the coastal regions of the western and northern Netherlands now covered by 20 meters or more of Holocene deposition. In fact, enough artifacts have been recovered from the Rhine-Maas delta and elsewhere in recent years to lend credence to this theory.[24]

If climatic amelioration with its resultant changes in flora and fauna made the Netherlands and adjacent areas progressively more suitable for human occupancy between 8000 and 6000 B.C., further warming introduced a new kind of challenge to human existence, namely, too much water. The late Pleistocene–early Holocene era saw a rather rapid melting of the ice sheets covering the northern portions of Europe, Asia, and North America, releasing huge quantities of water and debris that produced a significant rise in global sea level (see fig. 3).

It is important to understand that geologists normally speak of *relative* sea-level movements, since the actual amount of change induced by a rise or fall in water level could be affected by tectonic shifts in the earth's crust and compaction or subsidence of sediments. It appears that the Netherlands has been rather stable during the Holocene, with a slight tectonic downwarping of no more than 3 centimeters a century.[25] On the other hand, compaction of sediments has been a very significant factor. While sand is

[24]Louwe Kooijmans, "Archeologische ontdekkingen in het Rijnmondgebied," pp. 27–30; Louwe Kooijmans, "Mesolithic Bone and Antler Implements from the North Sea and from the Netherlands," pp. 27–73; D. P. Hallewas and J. F. van Regteren Altena, "Bewoningsgeschiedenis en landschapsontwikkeling rond de Maasmond," in A. E. Verhulst and M. K. E. Gottschalk, eds., *Transgressies en occupatiegeschiedenis in de kustgebieden van Nederland en België*, pp. 161–62.

[25]Jelgersma, *Holocene Sea Level Changes in the Netherlands*, pp. 10, 15, 52.

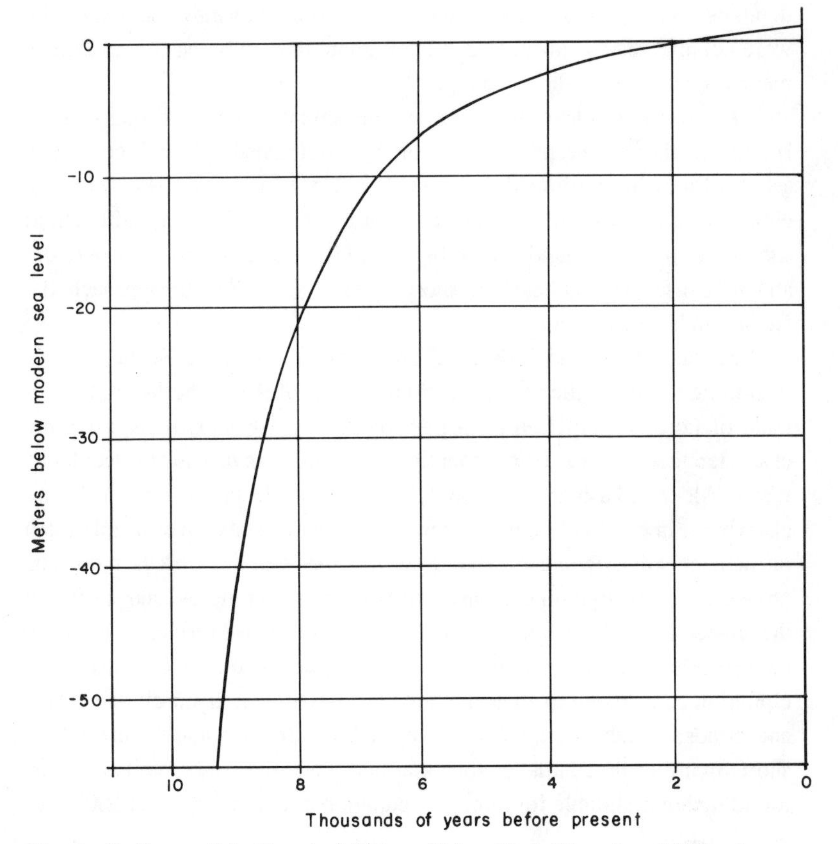

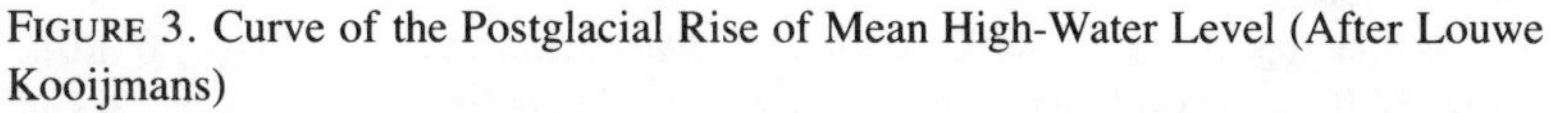

FIGURE 3. Curve of the Postglacial Rise of Mean High-Water Level (After Louwe Kooijmans)

subject to little or no subsidence, clay, depending on its organic content, can settle a fair amount. Peat, however, can compact by as much as 90 percent of its original mass. Subsidence of soils is very important in later stages of this study, since, after water was extracted from the peat bogs of the western Netherlands during the course of reclamation, the surfaces of these bogs sank considerably with respect to sea level. Nevertheless, for calculations of the rise in sea level in the western Netherlands during the Holocene, compaction is not a serious factor, since Dutch geologists normally take their samples for radiocarbon dating from the base of Holocene deposits, that is, directly above the Pleistocene cover sands that essentially were not affected by subsidence.[26] Throughout this study a sea-level change means a relative sea-level change.

The rise in sea level has been more or less continuous throughout the Holocene, though the pace has gradually diminished. At its highest rate sea level may have climbed by as much as 1.5 meters a century so that by 6000 B.C. it was within 20 meters of what it is today. This was sufficient to submerge totally the land bridge between Great Britain and the Continent and to cause the southeastern shore of the North Sea to approach the Netherlands' present coastline.[27]

Between 6000 and 5500 B.C., about the time that the North Sea was beginning to invade the western coastal regions of the Netherlands, the climate of northwestern Europe began to take on a wetter, more maritime character than before. Consequently, pine forest gave way to deciduous forest. Alder and ash covered low-lying areas, while oak became dominant elsewhere, in conjunction with birch on lighter soils or with elm and linden on heavier soils. Because the new vegetation was much denser than its predecessor, resulting in decreased growth of grasses, herbs, and shrubs on the shaded forest floor, many of the herbivores that had thrived in the open pine forests began to migrate elsewhere. At the same time increased precipitation contributed to an accelerated growth of peat in silt-clogged lakes and ponds, thereby diminishing many habitats for fish and waterfowl. In short, many of the characteristics that had made the Netherlands and surrounding areas suitable for human occupation between 8000 and 6000 B.C.

[26] Ibid., pp. 10, 16; Louwe Kooijmans, *The Rhine/Meuse Delta*, pp. 50–69; W. Roeleveld, *The Holocene Evolution of the Groningen Marine-Clay District*, pp. 26–29 and especially fig. 8; T. Edelman, *Bijdrage tot de historische geografie van de Nederlandse kuststreek*, p. 6.

[27] There have been numerous attempts to plot the Holocene sea-level rise, and many of these are represented and carefully discussed in Jelgersma, *Holocene Sea Level Changes in the Netherlands*, pp. 10–12 and chap. 4; Louwe Kooijmans, *The Rhine/Meuse Delta*, pp. 50–69. The illustration in ibid., p. 68, incorporates most recent scholarship; it is a curve corrected to solar years of the relative rise in mean high-water level since 5000 B.C. Compare this curve with those depicting mean sea level on p. 60. See also B. P. Hageman, "Development of the Western Part of the Netherlands during the Holocene," *Geologie en mijnbouw* 48 (1969): 377–79; P. C. Vos, "De relatie tussen de geologische ontwikkeling en de bewoningsgeschiedenis in de Assendelver Polders vanaf 1000 v. Chr.," *Westerheem* 32 (1983): 55–57. Perhaps the most striking representation is in Louwe Kooijmans, "Mesolithic Bone and Antler Implements from the North Sea and from the Netherlands," p. 34. Compare this Dutch material with the general picture of sea-level change summarized in Lamb, *Climate History and the Future*, pp. 111–29, 363, 370.

began to disappear thereafter. The plenty previously available to hunters, fishers, and fowlers had dwindled. It is not surprising, therefore, that many places of habitation were completely abandoned. Only where open forest, good drainage, and open water persisted did a human presence continue during the next 2,000 years.[28]

THE CYCLES OF INUNDATION

The rising and steadily encroaching sea plus the maritime climate that began making itself felt around 6000 B.C. gradually produced the conditions that literally created the western Netherlands, not only by depositing clays and sands but also by causing peat to grow on top of the older Pleistocene sands as well as on the younger clay and sand deposits from time to time. These same conditions, however, became the new primary obstacles to human occupancy. From then until they were reclaimed during the Middle Ages, the coastal regions could be inhabited only partly, and then during restricted times. Hydraulic conditions set these limits. Human occupation of the western Netherlands during the next 7,000 years or so was possible only above mean high-water level.[29]

Between 6000 and 3000 B.C. the level of the sea rose an average of approximately 50 centimeters a century.[30] Because the Pleistocene surface of the western Netherlands sloped gradually toward the west,[31] a slight rise in the level of the sea would produce a substantial shift in the shoreline toward the east, leaving behind a very well preserved record of sedimentation that reveals a number of aspects of Holocene deposition that are significant to my concerns.

[28] H. T. Waterbolk, "Food Production in Prehistoric Europe," *Science* 162 (1968): 1096; Waterbolk, "The Lower Rhine Basin," p. 234; Waterbolk, "The Occupation of Friesland in the Prehistoric Period," p. 16.

[29] Louwe Kooijmans, *The Rhine/Meuse Delta*, pp. 64, 68.

[30] Ibid., pp. 50–69; T. Edelman, *Bijdrage tot de historische geografie van de Nederlandse kuststreek*, p. 10.

[31] According to the information presented in Pons et al., "Evolution of the Netherlands Coastal Area during the Holocene," enclosure 2, the Pleistocene surface in the western Netherlands lowered approximately 20 meters over a distance of about 50 kilometers. Thus if the inclination had been uniform, the average grade would have amounted to about 0.04 percent. See also L. P. Louwe Kooijmans, "The Neolithic at the Lower Rhine: Its Structure in Chronological and Geographical Respect," in Sigfried J. de Laet, ed., *Acculturation and Continuity in Atlantic Europe Mainly during the Neolithic Period and the Bronze Age*, p. 154.

The physical record indicates that the deposition of sands and clays occurred in a series of phases or cycles that, alternating with phases or cycles of peat growth, produced a distinct succession of layers visible in cross section.[32] To explain the origins of these layers of sand or clay alternating with layers of peat, Dutch geologists have postulated a series of transgressions and regressions of the sea that alternated throughout the Holocene period.[33] In simplest terms, a transgression was a landward penetration of the sea that left behind layers of marine clay and sand. A regression, meanwhile, was a retreat of the sea that often resulted in extensive peat accumulation, though this could also occur during transgressive phases in areas not flooded. Those who put forward this explanation often point out that these alternating phases did not represent oscillations in the worldwide level of the sea. Rather, they insist, transgression-regression cycles represented fluctuations in *high*-water levels on the *regional* scale (as in the southern North Sea region).[34]

To identify and examine more closely the alternating cycles of marine sedimentation and peat growth, Dutch geologists have made numerous borings into the Holocene mass of the western Netherlands and have attempted to date with radiocarbon the layers of peat they found sandwiched between sediment layers. On the basis of these datings they developed the following chronology: there were two major periods of marine sedimentation, represented by Calais, or Old Tidal Flat, deposits, roughly dating from between 6000 and 1800 B.C., and Dunkirk, or Young Tidal Flat, deposits, put down since approximately 1500 B.C. These major periods were subdivided into individual phases and some subphases that produced the following deposits:

[32] For a cross section of Holocene deposits in the western Netherlands, see B. P. Hageman and H. Kliewe, "Neue Forschungen zur Stratigraphie mariner und perimariner Holozänsedimente in den Niederlanden," *Petermanns Geographische Mitteilungen* 93 (1969), table 10.

[33] W. Roeleveld, "De bijdrage van de aardwetenschappen tot de studie van de transgressieve activiteit langs de zuidelijke kusten van de Nordzee," in Verhulst and Gottschalk, eds., *Transgressies en occupatiegeschiedenis in de kustgebieden van Nederland en België*, pp. 292–99.

[34] Roeleveld, "De bijdrage van de aardwetenschappen," pp. 294–97; L. J. Pons, "De zeekleigronden," in *De bodem van Nederland: toelichting bij de bodemkaart van Nederland, schaal 1:200,000*, p. 45. For a discussion of various views of the Holocene sea-level rise, see Jelgersma, *Holocene Sea Level Changes in the Netherlands*, chap. 4, and especially the evaluation of storminess in sea-level oscillations, pp. 89–90; Louwe Kooijmans, *The Rhine/Meuse Delta*, pp. 8, 63; T. Edelman, *Bijdrage tot de historische geografie van de Nederlandse kuststreek*, pp. 5–21.

Deposit	Date
Dunkirk III-B	After A.D. 1100
Dunkirk III-A	A.D. 800–1000
Dunkirk II	A.D. 250–600
Dunkirk I-B	400–200 B.C.
Dunkirk I-A	600–500 B.C.
Dunkirk 0	1500–1000 B.C.
Calais IV-B	2200–1800 B.C.
Calais IV-A	2600–2300 B.C.
Calais III	3300–2800 B.C.
Calais II	4300–3400 B.C.
Calais I	6000–4500 B.C.[35]

The individual dates are intended to be no more than broad limits between which individual layers of clay or sand were deposited, based on the radiocarbon dating of the interspersed layers of peat.

This broad schema of alternating transgressions and regressions of the sea, spanning nearly eight thousand years, has become the theoretical framework within which prehistorians of the western Netherlands increasingly cast their research.[36] Thus phases of human colonization or desertion are often keyed to these underlying physical cycles. Nevertheless, as we shall see later, a number of historians and other scholars who study the western Netherlands have begun to raise a number of important questions concerning the transgression-regression model, particularly concerning the existence of transgression phases since the beginning of our era.[37]

[35] Louwe Kooijmans, *The Rhine/Meuse Delta*, pp. 8–10, 49; Hageman, "The Development of the Western Part of the Netherlands during the Holocene," p. 379; Jelgersma et al., "The Coastal Dunes of the Western Netherlands," p. 101.

[36] Louwe Kooijmans, *The Rhine/Meuse Delta*, pp. 69–76, maintains that the transgression and regression cycles established for the western Netherlands have great relevance to northwestern Germany and eastern England as well. For the Elbe region in Germany, see Gerhard Linke, "Der Ablauf der holozänen Transgression der Nordsee aufgrund von Ergebnissen aus dem Gebiet Neuwerk/Scharhörn," *Probleme der Küstenforschung im südlichen Nordseegebiet* 14 (1982): 123–57.

[37] See, for example, M. K. E. Gottschalk, *De periode 1400–1600*, vol. 1 in *Stormvloeden en rivieroverstromingen in Nederland*, pp. 818–23; and Vos, "De relatie tussen de geologische ontwikkeling en de bewoningsgeschiedenis," pp. 55–57. T. Edelman, *Bijdrage tot de historische geografie van de Nederlandse kuststreek*, pp. 6–16, thinks much of the transgression-regression terminology may be misleading and suggests an examination of sedi-

COASTAL BARRIERS AND DUNES

The physical record of Holocene sedimentation in the western Netherlands indicates not only that the material constituting it was laid down in a series of cycles but also that it was deposited behind a protective coastal barrier system. The sediments consist of tidal-flat marine clays and sands and fluvial or riverine clays and sands, interspersed with extensive layers of peat. While marine and fluvial deposition could occur in either a brackish or a freshwater milieu, peat growth could take place essentially under freshwater conditions alone, indicating the presence of lagoons protected from the direct influence of the sea by offshore coastal barriers.[38]

The coastal-barrier system began to form when the moist maritime climatic patterns that had predominated in northwestern Europe since about 6000 B.C. began to give way to drier conditions shortly before 3000 B.C. During the prolonged wet period the Rhine and Meuse rivers had transported enough silt and sand to create a substantial delta that bulged into the North Sea. This delta was ringed and flanked by a series of barrier islands that allowed raised salt marsh to develop along the somewhat sheltered coast, resembling in many respects the modern coasts of the northern Netherlands and northwestern Germany. However, the relatively drier, more Continental climatic patterns that emerged toward the end of the fourth millennium B.C. meant that the silt-and-sand transport of the rivers declined and delta building became sporadic. Eventually the sea began to erode the delta, carrying the material away laterally on both sides, generally straightening the coast and forming a series of sandy ridges offshore from the salt marsh.

The first such ridge appeared around 3000 B.C., 6 to 8 kilometers inland from the modern coast in Rijnland. Presumably, during a period of relative quiet or a sea-regressive phase the winds piled sandy material into an actual dune ridge that lay above high-water level. In a succeeding transgressive phase of the sea, dune building ended, and soil formation began, while on the west the erosion and transport of sandy deposits resumed, and

mentation and erosion phases which often could be quite independent of sea-level changes. These and many other criticisms of the transgression-regression model are found in the proceedings of a conference on the topic, published as A. E. Verhulst and M. K. E. Gottschalk, eds., *Transgressies en occupatiegeschiedenis in de kustgebieden van Nederland en België*.

[38] Jelgersma et al., "The Coastal Dunes of the Western Netherlands," p. 94; T. Edelman, *Bijdrage tot de historische geografie van de Nederlandse kuststreek*, p. 7.

a new sandy ridge developed, later becoming a dune ridge as well. By the time the coastal-barrier system was completed, around 2000 B.C., this cycle had been repeated a number of times, for it came to consist of at least four parallel barrier ridges on which dunes eventually developed.[39]

At present the coastal-barrier and dune system of the western Netherlands actually consists of two different landscapes. The original Older Dune landscape was essentially completed by the beginning of our era, though some activity continued as late as the eighth and early ninth centuries. The Younger Dune landscape began to form during the late tenth century and continued through a number of phases until the eighteenth century.[40] To this day the Older Dune landscape consists of a number of very low sandy dune ridges, never more than 4 to 5 meters above sea level,[41] separated from each other by peat- and clay-filled depressions known as *strandvlakten*. Only small portions of the Older Dunes still exist, however, for the most part along the western coast of the Netherlands. Actually they never developed in the north among the Frisian Islands, while in the south, along the modern Rhine-Maas-Schelde delta of Zeeland and south Holland, the ridges were extensively breached by storms, especially during the early Middle Ages, transforming an originally continuous coastline into a series of islands and tidal flats. Even where they remained intact, however, a large proportion of the original sand barriers with their dunes was cov-

[39]Edelman, *Bijdrage tot de historische geografie van de Nederlandse kuststreek*, pp. 29–32; Louwe Kooijmans, *The Rhine/Meuse Delta*, pp. 7, 38–42; Louwe Kooijmans, "Archeologische ontdekkingen in het Rijnmondgebied," p. 31; Pons et al., "Evolution of the Netherlands Coastal Area during the Holocene," enclosures 3–7; Hageman, "Development of the Western Part of the Netherlands during the Holocene," pp. 385–87, 386 fig. 9, pts. 1, 2; Jelgersma et al., "The Coastal Dunes of the Western Netherlands," pp. 97–100, 147, plate 1; S. Jelgersma and J. F. van Regteren Altena, "An Outline of the Geological History of the Coastal Dunes in the Western Netherlands," *Geologie en mijnbouw* 47 (1969): 237; Sarfatij, "Friezen-Romeinen-Cananefaten," p. 36; Jelgersma, *Holocene Sea Level Changes in the Netherlands*, p. 90; Hageman and Kliewe, "Neue Forschungen zur Stratigraphie mariner und perimariner Holozänsedimente in den Niederlanden," p. 129; J. C. R. M. Haans and G. C. Maarleveld, "De zandgronden," in *De bodem van Nederland: toelichting bij de bodemkaart van Nederland, schaal 1:200,000*, pp. 196–97. For the origins of coastal-barrier sands and some interesting questions that arise therefrom, see C. Kruit, "Is the Rhine Delta a Delta?" in *Transactions of the Jubilee Convention*, pp. 257–66.

[40]Rentenaar, "De Nederlandse duinen in de middeleeuwse bronnen," pp. 362–63, 373; Jelgersma et al., "The Coastal Dunes of the Western Netherlands," pp. 100–102, 145–46.

[41]Haans and Maarleveld, "De zandgronden," p. 196; G. J. Borger, "De ontwatering van het veen: een hoofdlijn in de historische nederzettingsgeografie van Nederland," *Geografisch tijdschrift* 11 (1977): 377.

ered by later Younger Dune development, while most of the remainder was excavated more recently for use in the bulb-growing industries. Those small portions still visible, as at the Keukenhof (the site of the annual floral exhibit near Lisse), provide only a rare glimpse of the character of the original coast.[42]

In contrast to the low profiles of the Older Dunes, the Younger Dunes, which dominate the picture today, attain much greater heights, in some instances up to 80 meters above sea level. In addition, instead of lying in rows atop the coastal barriers, they appear more scattered, with foredunes near the shore and parabola-shaped formations farther inland. Apparently changing currents in the North Sea during the period from 800 to 1000 gradually began to erode parts of the Older Dunes, further straightening the coastline and forming new foreshore sand drifts. The fact that the Oude Rijn silted shut during this period can be used as evidence of considerable shifting of material.[43] During relatively dry periods the winds piled the sand into high drifts. Some such activity may have begun by the late tenth century. During the twelfth and thirteenth centuries active Younger Dune formation occurred at the same time that the crests of some of the Older Dune ridges were removed by wind erosion and deposited in the *strandvlakten*, or depressions between the ridges.[44] At other times Younger Dunes eroded and changed as new ones formed elsewhere. That erosion and active dune formation took place during the Middle Ages is known from

[42] Jelgersma et al., "The Coastal Dunes of the Western Netherlands," pp. 95–96; C. H. Edelman, *Over de plaatsnamen met het bestanddeel woud en hun betrekking tot de bodemgesteldheid*, p. 18; A. M. Hulkenberg, *Keukenhof*, pp. 3–27.

[43] Actually, the Oude Rijn experienced a long and steady decline in significance with respect to the Lek and Waal, the more southerly branches of the Rhine that increasingly received more of the river's waters. The actual time of final closure of a by then relatively insignificant stream is not known for certain. That it may well have occurred between 800 and 1000 seems most reasonable. See Blok, "Probleme der Flussnamenforschung in den alluvialen Gebieten der Niederlande," pp. 222, 226; S. J. Fockema Andreae, "De Oude Rijn: eigendom van openbaar water in Nederland," in *Rechtskundige opstellen op 2 november 1935 door oud-leerlingen aangeboden aan prof. mr. E. M. Meijers*, pp. 699–700. Louwe Kooijmans, *The Rhine/Meuse Delta*, p. 120, says the Lek and Waal became the chief branches of the Rhine system during the ninth century when a stream south of Rijnland silted up. Most likely the Oude Rijn mouth closed for the same reasons at about the same time.

[44] Rentenaar, "De Nederlandse duinen in de middeleeuwse bronnen," p. 373; Hageman, "Development of the Western Part of the Netherlands during the Holocene," p. 386 fig. 9, pt. 4, p. 387; Jelgersma et al., "The Coastal Dunes of the Western Netherlands," pp. 100–101, 144–47; Jelgersma and van Regteren Altena, "An Outline of the Geological History of the Coastal Dunes," p. 340; Pons, "De zeekleigronden," p. 51; Sarfatij, "Middeleeuwse mens en eeuwig water," p. 21.

written sources, and efforts were frequently undertaken to stabilize the dunes, either by protecting their existing vegetation or by planting helm to hold the sand in place. The last phase of Younger Dune formation evidently ended in the eighteenth century.[45]

Different kinds of evidence show that during the early Middle Ages the Older Dune landscape was covered with a mature forest vegetation of oak and beech. Excavations in the dune region have uncovered large numbers of trunks and logs of these species, as well as the remains of beech mast. Pollen analysis of peat soils removed from the *strandvlakten* indicates that there was a considerable increase in the amount of beech (*Fagus*) pollen between the end of Older and the beginning of Younger Dune formation. Further, there is evidence to show that oak, birch, and alder grew in the depressions between the dune ridges. Historical sources, for example, point to the existence of a well-developed undisturbed forest north and south of present-day Haarlem, the Haarlemmerhout, which acted as the northern boundary zone of the original Rijnland. That such a forest existed was later indicated by place-names ending in *hout*, meaning "woodland," and *rode*, meaning "clearing in the woodland."[46]

The establishment of the coastal-barrier system and Older Dune landscape radically transformed the conditions of deposition between them and the higher-lying Pleistocene sands on the east. What had been a tidal-flat environment became more and more a salt marsh, bordered on the east by an extensive reed swamp. There were, of course, openings in the coastal barriers and dunes at the mouths of the four major rivers of the western Netherlands: the Schelde, northwest of Antwerpen; the Maas, west of Rotterdam; the Rhine, where the Oude Rijn is today; and the large, extinct IJ estuary near Egmond, in North Holland.[47]

The river mouths served as the inlets through which the sea could con-

[45] Rentenaar, "De Nederlandse duinen in de middeleeuwse bronnen," p. 369.

[46] Ibid., pp. 367–70; Jelgersma et al., "The Coastal Dunes of the Western Netherlands," pp. 94, 100, 131–32, 148; Gosses, "De vorming van het graafschap Holland," p. 248; Sarfatij, "Friezen-Romeinen-Cananefaten," p. 36; de Cock, *Bijdrage tot de historische geografie van Kennemerland*, pp. 52–59; C. H. Edelman, *Over de plaatsnamen met het bestanddeel woud en hun betrekking tot de bodemgesteldheid*, p. 19.

[47] Louwe Kooijmans, *The Rhine/Meuse Delta*, p. 38. For the former IJ estuary, see W. H. Zagwijn, "De ontwikkeling van het 'Oer-IJ' estuarium en zijn omgeving," *Westerheem* 20 (1971): 11–18; Vos, "De relatie tussen de geologische ontwikkeling en de bewoningsgeschiedenis," pp. 57–66. For the former situation of the Schelde, see L. P. Louwe Kooij-

tinue to exercise its influence on the salt marsh and swamp behind the coast. During presumed sea-transgressive phases, salt water entered at the inlets, eroding older sediments or peat areas and forming a network of tidal creeks. Slowly the buildup of marine sediments occurred: sands near the inlets and fine clays farther into the salt marsh. Eventually, however, particularly as the frequency of the storminess associated with a transgressive phase began to diminish, the system of tidal creeks began to silt up as well, thereby blocking drainage in areas of sedimentation and encouraging peat growth. Finally, as the influence of the sea began to decline once again, the water supply behind the coastal barriers became freshened by the rivers, causing forest or reed swamp to occur nearly everywhere. This process continued with each transgression-regression cycle, once the initial coastal-barrier system was intact.[48]

The coastal-barrier system of most of the western Netherlands, at least from Monster, in the south, to Bergen, in the north, has remained essentially intact from about 1800 B.C. The extreme northern and southern sections, however, were breached extensively during the early Middle Ages (approximately A.D. 250 to 600). Where they stood firm, these barriers effectively protected the area on the east from the direct influence of the sea except in relatively small areas near the mouths of the major rivers.[49] Yet the sea level continued to rise. Between 3000 and 1000 B.C. it rose at an average of about 15 centimeters a century. Although considerably less than the rate of 50 centimeters a century of the previous 3,000 years, it was sufficient nevertheless to affect natural conditions behind the coastal barriers, since the groundwater level rose in tandem with the level of the sea.[50]

---

mans, "Oudheidkundige boomkoorvisserij op de Oosterschelde," *Westerheem* 20 (1971): 181–83.

[48] See Louwe Kooijmans, *The Rhine/Meuse Delta*, pp. 7–8; Pons et al., "Evolution of the Netherlands Coastal Area during the Holocene," enclosures 6–9; Hageman, "Development of the Western Part of the Netherlands during the Holocene," p. 379. It is incorrect to speak of deposition or sedimentation in connection with peat. Strictly speaking, it grows in place and is not transported there by wind or water. See L. J. Pons, "De veengronden," in *De bodem van Nederland: toelichting bij de bodemkaart van Nederland schaal 1:200,000*, p. 145.

[49] Pons et al., "Evolution of the Netherlands Coastal Area during the Holocene," enclosures 6–9; and Jelgersma et al., "The Coastal Dunes of the Western Netherlands," p. 148.

[50] Waterbolk, "Food Production in Prehistoric Europe," p. 1096; Waterbolk, "The Lower Rhine Basin," p. 234; Waterbolk, "The Occupation of Friesland in the Prehistoric Period," pp. 16, 19, 22; T. Edelman, *Bijdrage tot de historische geografie van de Nederlandse kuststreek*, p. 10.

THE GROWTH OF PEAT

The rising sea level raised the groundwater table at the very time that the sea was depositing less sand and silt behind the coastal barriers. Drainage behind the coast deteriorated, causing reed swamp to cover most of the western Netherlands. Because conditions were better for growing than for decomposing vegetation, partly decomposed vegetable matter began to accumulate in the form of peat.[51] Particularly around 1800 B.C., a blanket layer of peat, called *Hollandveen*, or Holland peat, began covering almost the entire coastal region of the western Netherlands, up to the landward side of the Older Dunes as well as between the individual ridges of the dune system. During the succeeding millennia marine clays and sands were deposited near the mouths and in narrow strips along the lower courses of the major rivers with strips of fluvial clays and sands from there upstream. In the remaining portions of the western coastal regions, however, swampy conditions prevailed, and peat continued to accumulate without significant interruption until reclamation began.[52]

The growth of peat in the western Netherlands from approximately 1800 B.C.took place in a number of areas of concentration. In the eventual Rijnland there were two, one north and the other south of the Oude Rijn. In the earliest stages the primary growth was a combination of reed (*Phragmites*) and sedge (*Carex*), as well as much alder and birch. The quality of the water that supported this initial growth was mesotrophic; that is, it contained a moderate amount of nutrients.

As the groundwater level rose in step with the rise in sea level and the peat surface rose accordingly, specialized plant associations began to appear that reflected variations in the quantity and quality of nutrition in the water supply. Along the Oude Rijn and such tributaries as the Meije and

[51] Pons, "De veengronden," p. 145; Jelgersma, *Holocene Sea Level Changes in the Netherlands*, pp. 21–22; G. J. Borger, "Ontwatering en grondgebruik in de middeleeuwse veenontginningen in Nederland," *Geografisch tijdschrift* 10 (1976): 343.

[52] Hageman, "Development of the Western Part of the Netherlands during the Holocene," p. 379; W. J. van Tent, "De landschappelijke achtergronden," *Spieghel historiael* 13 (1978): 208; Pons et al., "Evolution of the Netherlands Coastal Area during the Holocene," enclosures 7–9; Pons, "De veengronden," p. 146 fig. 60. Apparently there was a period, from shortly before to shortly after the beginning of our era, during which the climate was too dry for peat growth; see H. T. Waterbolk, *De praehistorische mens en zijn milieu: een palynologisch onderzoek naar de menselijke invloed op de plantengroei van de diluviale gronden in Nederland*, pp. 16, 131.

the Aar, the milieu was eutrophic, that is, rich in nutrients, because of the presence of silt-carrying river water. Swamp forests of willow with some ash and oak appeared, from which *bosveen* or forest peat was formed. There also was a small area of eutrophic-peat development at the mouth of the Oude Rijn in a brackish combination of sea and river water. There the main constituents were the partly decomposed remains of reeds and rushes (*Scirpus*).

Bordering the area of eutrophic peat and along the smaller rivers and creeks farther into the peat landscape, the original mesotrophic environment with its growth of reed, sedge, alder, and birch persisted. Small strips of similar plant associations occurred along the dunes in the west and along the Pleistocene sands in the east, where runoff reaching the edge of the peat contained moderate amounts of nutrients.

Finally, beyond the influence of river water and seawater, a type of peat developed from a vegetation that grew in an oligotrophic milieu, that is, one depending entirely upon precipitation containing little or no nutrients. Such vegetation, consisting of peat moss (*Sphagnum*), heath (*Calluna* and *Ericacea*), and cotton grass (*Eirophorum*), began to appear once organic material had accumulated to such a depth that the water at the surface supplying such plants no longer derived nutrients from the subsoil. Pure oligotrophic peat formed the nuclei of both major concentrations of peat in Rijnland and, long before reclamation began, had spread out over most of the originally mesotrophic plant associations as well.[53]

One of the most striking characteristics of oligotrophic peat, or sphagnum peat, as it is often called, is its tendency to grow into a cushion that raises itself above the surrounding landscape.[54] Most of this type of peat in the western Netherlands was destroyed during the past ten centuries, either by human efforts or by erosion,[55] but similar peat bogs or cushions in other parts of the world provide considerable information concerning the prob-

[53] Pons, "De veengronden," pp. 145–47; C. van Wallenburg and W. C. Markus, "Toemaakdekken in het Oude Rijngebied," *Boor en spade* 17 (1971): 64–66; J. Bennema, "Het oppervlakteveen in West-Nederland," *Boor en spade* 3 (1949): 139–41, 144; C. H. Edelman, *Soils of the Netherlands*, pp. 68–71; Z. van Doorn, "Enkele waarnemingen van oorspronkelijke Indonesische veenmoerassen ter vergelijking met de Hollands-Utrechtse venen," *Boor en spade* 10 (1959): 158; Louwe Kooijmans, *The Rhine/Meuse Delta*, p. 84; Zagwijn, "De ontwikkeling van het 'Oer-IJ' estuarium en zijn omgeving," p. 14.

[54] Pons, "De veengronden," pp. 145–47.

[55] C. H. Edelman, *Soils of the Netherlands*, pp. 69–71; Edelman, *Over de plaatsnamen met het bestanddeel woud en hun betrekking tot de bodemgesteldheid*, pp. 19–20.

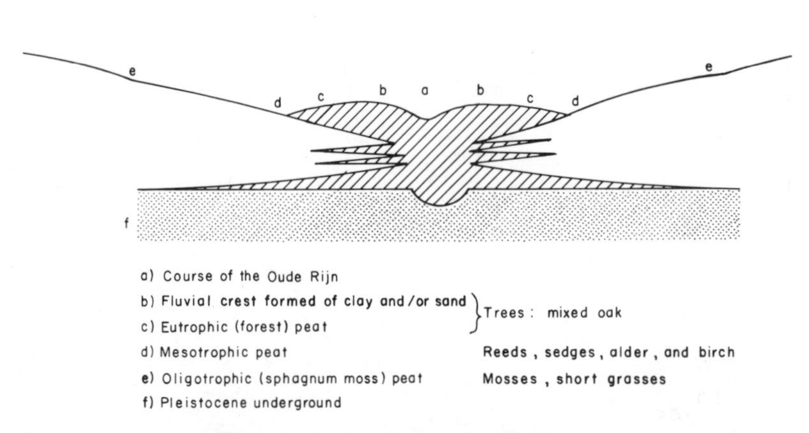

FIGURE 4. Zonation of Rijnland's Peat Bogs (After Blok)

able appearance and characteristics of large, raised masses of oligotrophic, or sphagnum, peat in Rijnland before reclamation. Untouched oligotrophic peat bogs in Indonesia, for example, reveal the following information concerning their formation: first, they grew in pace with a rising sea level that raised the groundwater table; second, because they grew in more or less bowl-shaped depressions, they were thickest at their centers and thinnest at their edges; third, the peat surface at the centers of the cushions was at least several meters higher than it was at the edges because a sort of capillary action within the mass of peat itself kept the water table higher where the peat was the thickest. As a result, the bogs achieved the form of a biconvex lens. Before the Middle Ages the main peat bogs of Rijnland, one on either side of the Oude Rijn, had come to resemble the Indonesian bogs, and they continued to do so until reclamation began (see fig. 4).[56]

That the centers of the Rijnland peat bogs were higher than their peripheries is shown in the fact that surplus water flowed away in radial fash-

[56]Z. van Doorn, "Enkele waarnemingen van oorspronkelijke Indonesische veenmoerassen," pp. 158, 161–62; van Wallenburg and Markus, "Toemaakdekken in het Oude Rijngebied," pp. 64–66; S. J. Fockema Andreae, ed., *Rechtsbronnen der vier hoofdwaterschappen van het vasteland van Zuid-Holland (Rijnland; Delfland; Schieland; Woerden)*, p. vii.

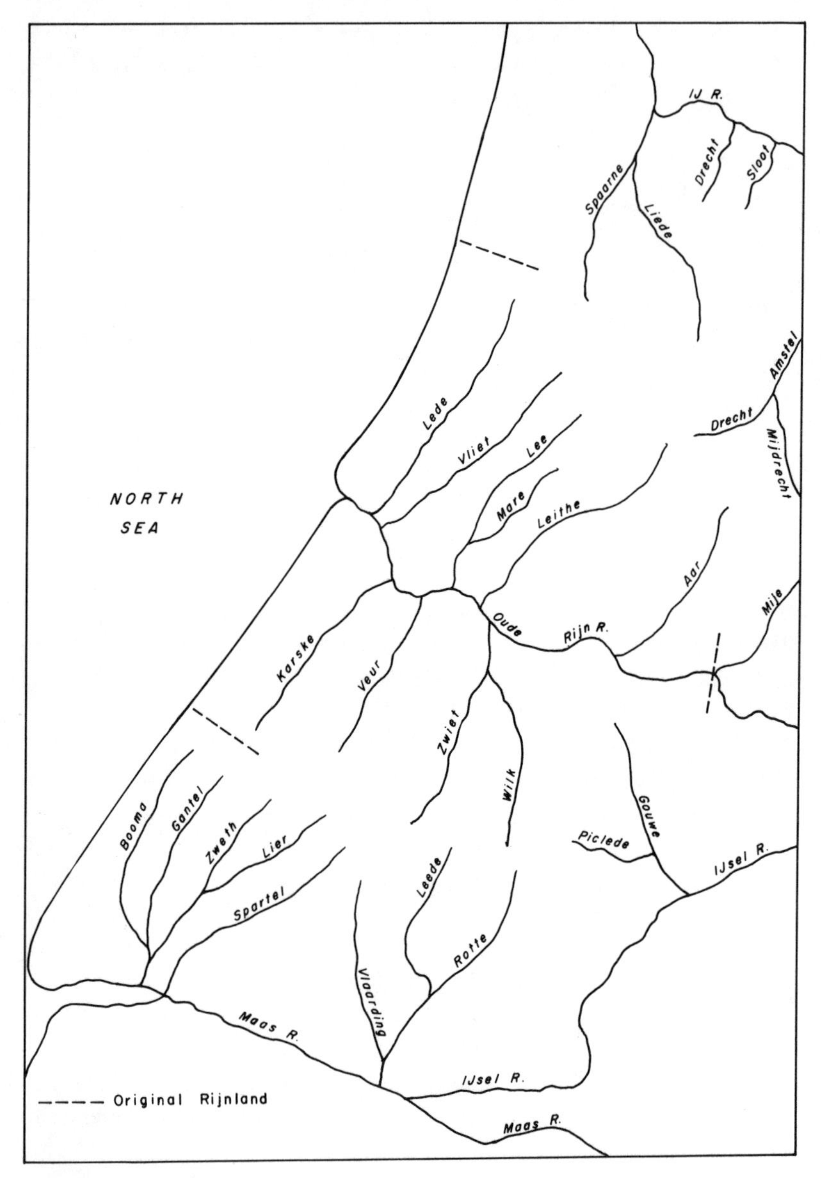

FIGURE 5. Original Radial Drainage Network

ion, fanning out from the centers in streams whose beds can still be traced (see fig. 5).[57] These streams in turn emptied into the main rivers of the western Netherlands, which flowed in courses tangential to these cushions. Thus the raised peat bog south of the Oude Rijn drained radially by way of the Booma, Gantel, Zweth, Lier, Spartel, Vlaarding, Leede, and Rotte into the Maas; by way of the Goude and Piclede into the IJsel, which joined the Maas near present-day Rotterdam; and by way of the Wilk, Zwiet, and Veur into the Oude Rijn. Similarly, the northern bog drained by way of the Spaarne, Liede, Drecht, and Sloot into the IJ; by way of the Drecht and Mijdrecht into the Amstel, which joined the IJ at present-day Amsterdam; and by way of the Mije, Aar, Aa, Leithe, Mare, Lee, and Vecht into the Oude Rijn. The Karske south and the Lede north of the Oude Rijn were old tidal creeks that drained depressions between Older Dune ridges.[58]

Although these raised cushions of peat were provided with creeks and streams to carry off surplus water, their surfaces were rarely if ever dry. They contained massive quantities of water, and capillary action within the bogs ensured a wetness throughout, even at the surface of the raised centers. With every rainfall or snowmelt, water that could not be accommodated within the peat flowed off by way of the radial networks of streams and rivers. At other times the bog streams would seem more like standing water. Even during the summers, when the rate of evapotranspiration exceeded that of precipitation, they remained soggy and virtually impossible for man or beast to traverse. The surface of the raised peat bogs was not smooth, and there were numerous small lakes or pools of water in the many depressions. A whole series of lakes and pools along the eastern side of the Older Dunes north of the Oude Rijn grew into the huge Haarlemmermeer (Haarlem Lake) during the late Middle Ages; it remained and, in fact, continued to expand until it was finally drained in 1852. In the autumn and winter, when precipitation far exceeded evapotranspiration, the entire surface of the bogs was usually under water.[59]

[57] Actually, the courses of the streams of this radial drainage network tended to shift and vary until reclamation, when they were used as reclamation bases or were canalized; see Borger, "Ontwatering en grondgebruik," p. 345.

[58] See Floris Balthasars, *Kaarten van Rijnland, 1615*, sheets 10–17; Jan Jansz. Dou and Steven Broekhuysen, *Kaartboek van Rijnland*, 3d ed., sheets 7–11; de Cock, *Bijdrage tot de historische geografie van Kennemerland*, pp. 18–20, especially p. 19 fig. 8 (for an English example of a radial drainage network, see p. 51 fig. 16); de Cock, "Die Grafschaft Masaland," pp. 109–12; Blok, "Probleme der Flussnamenforschung in den alluvialen Gebieten der Niederlande," pp. 212–27.

[59] C. H. Edelman, *Soils of the Netherlands*, p. 71; T. Edelman, *Bijdrage tot de histo-*

The streams that radiated out from the raised centers of the oligotrophic peat bogs of the western Netherlands, besides providing a modicum of drainage, served another very important function. When the level of the water in the main rivers was high, floodwaters could flow backward into the peat cushions by way of these waterways and leave behind varying amounts of fluvial or marine sediments, providing the only relatively stable or firm ground within these soggy masses. These sediments also contained the nutrients that made possible the appearance of differing types of plant associations. Where flooding occurred regularly, clay was set down on top of the peat, allowing tall gallery forest of oak, ash, and willow to develop. Farther up the stream or laterally away from it the amount of clay deposition began to taper off and peat dominated; there the vegetation shaded into a lower growth of reed, sedge, alder, and birch. Finally, in the broad treeless central portions of the bogs, beyond the reach of flooding by silt-rich water, peat moss, heath, and cotton grass prevailed.[60]

Until they were reclaimed, the raised oligotrophic peat bogs, or cushions, with their broad, treeless central portions represented the wilderness at the edge of settlement. That it was in fact an inhospitable and uninviting landscape for human settlement is indicated by certain medieval place-names applied to parts of the bogs of the western Netherlands. Names such as dat Wilt, Willens, and Wilde lant were testimony that it was "wild land." Other names, such as Rubroke, Ruvene, Rudighe Ness, Rughe Camp, Rughe Willens, Rugheka, and Ruwout (*ru*, *ruig*, *ruw*, *ruge* signify "rough," "raw," "rugged"), give a similar impression. It was, in short, a rough, wild-grown landscape of perpetual sogginess, accessible only by way of the radial network of streams.[61]

---

rische geografie van de Nederlandse kuststreek, pp. 43, 47–48; W. J. Diepeveen, *De verveening in Delfland en Schieland tot het einde der zestiende eeuw*, p.UN 4.

[60] C. H. Edelman, *Over de plaatsnamen met het bestanddeel woud en hun betrekking tot de bodemgesteldheid*, pp. 19–20; Blok, "Probleme der Flussnamenforschung in den alluvialen Gebieten der Niederlande," p. 216.

[61] D. P. Blok, De vestigingsgeschiedenis van Holland en Utrecht in het licht van de plaatsnamen," in M. Gysseling and D. P. Blok, *Studies over de oudste plaatsnamen van Holland en Utrecht*, pp. 30–31. See also Gosses, "De vorming van het graafschap Holland," p. 303; C. H. Edelman, *Over de plaatsnamen met het bestanddeel woud en hun betrekking tot de bodemgesteldheid*, p. 19; Louwe Kooijmans, *The Rhine/Meuse Delta*, pp. 84, 118; A. T. Clason, *Animal and Man in Holland's Past: An Investigation of the Animal World surrounding Man in Prehistoric and Early Historical Times in the Provinces of North and South Holland*, 1:105, 203.

THE SOILS OF EARLY MEDIEVAL RIJNLAND

The preceding discussion of the mechanics of Holocene soil formation in the coastal regions of the western Netherlands makes it possible to reconstruct, at least in general terms, the location and proportions of sand, clay, and peat at the surface in Rijnland before the start of medieval reclamation (see fig. 6).[62] A small area of marine clays and sands had settled around the mouth of the Oude Rijn and in a narrow strip along the river upstream as far as Koudekerk, about 15 kilometers from the sea. This sedimentation achieved a width of perhaps 10 to 12 kilometers between the Older Dune ridges but narrowed to about 2 to 4 kilometers along the Oude Rijn to Koudekerk. From Koudekerk eastward was a strip of fluvial deposits 2 to 3 kilometers wide extending well beyond Rijnland's eastern border, as well as some very narrow strips along the lower courses of a few tributaries, such as the Aar (north of Alphen) and the Mije (north of Zwammerdam). The Older Dune landscape, including the peat- and clay-filled *strandvlakten*, or depressions, between the ridges, was perhaps 6 to 8 kilometers wide. The rest of Rijnland, by far the largest area, consisted of varieties of peat.

Premedieval Settlement in the Western Netherlands

Until the beginning of peat-bog reclamation in the late tenth century, the western Netherlands, including Rijnland, offered only limited opportunities for permanent settlement. The primary consideration in selecting a site for occupation was the quality of drainage. Those places which dried out between floods or episodes of precipitation were most favored, while those which were permanently waterlogged were avoided.

Peat, formed under permanently waterlogged conditions, was by definition unsuitable for settlement except under the most exceptional conditions. Outside the areas of peat growth, however, the degree of wetness or dryness could vary considerably, depending on such factors as elevation,

[62] Based on information gathered from the following maps: Pons et al., "Evolution of the Netherlands Coastal Area during the Holocene," enclosure 9; Pons, "De veengronden," p. 146 fig. 60; Jelgersma, *Holocene Sea Level Changes in the Netherlands*, p. 88 fig. 50; C. H. Edelman, *Soils of the Netherlands*, p. 66; Gosses, "De vorming van het graafschap Holland," facing p. 294; Louwe Kooijmans, *The Rhine/Meuse Delta*, p. 6 fig. 1, p. 40 fig. 9; *Bodemkaart van Nederland, schaal 1:200,000*, comp. Stichting voor Bodemkaartering, sheet 6; Jelgersma et al., "The Coastal Dunes of the Western Netherlands," plate 1.

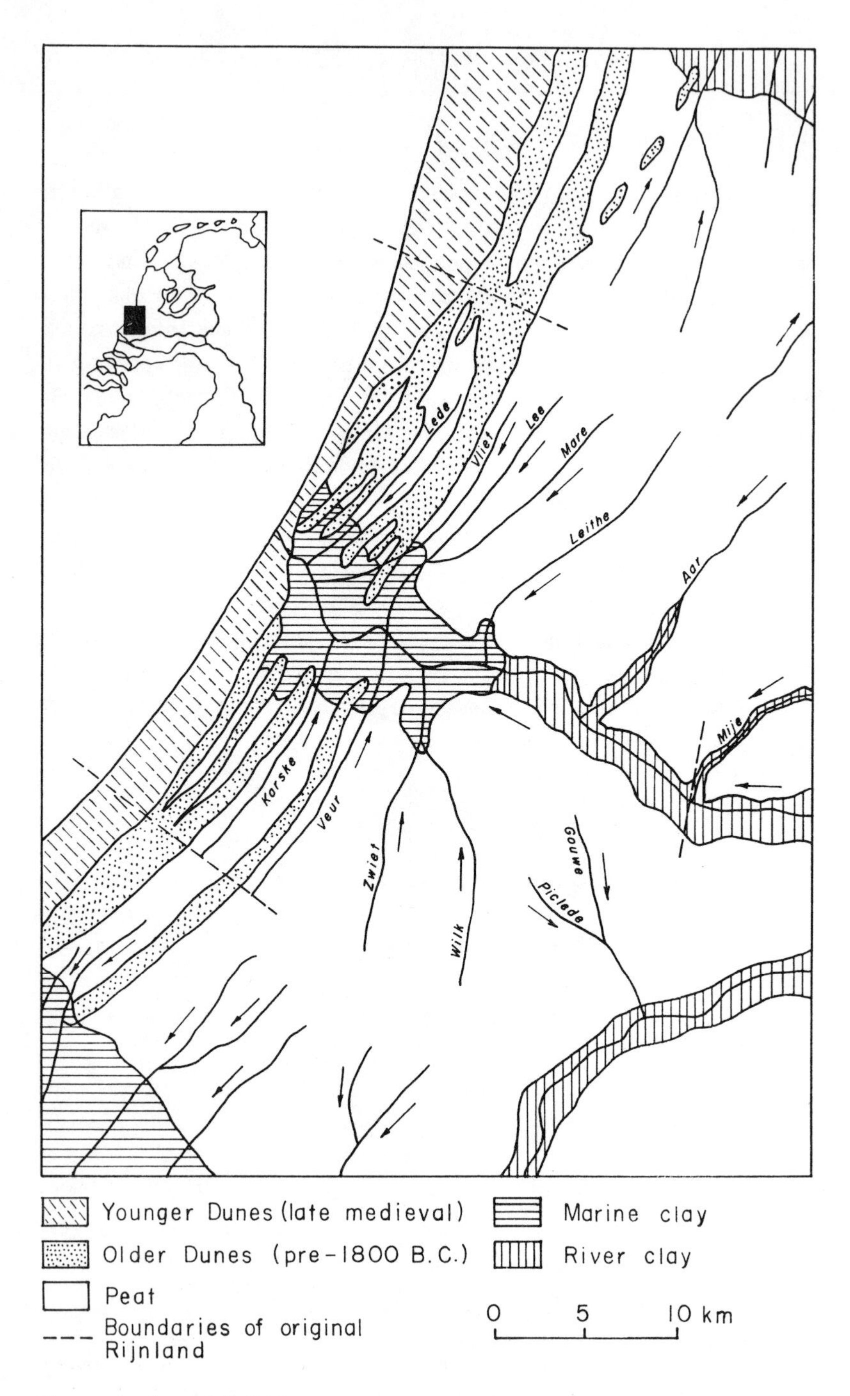

FIGURE 6. General Soil Map

slope, proximity to water courses, and types of soil. For example, a sandy ridge protruding above the surrounding landscape and sloping toward a nearby stream would dry out most quickly, while dense clay barely above groundwater level, with little or no slope and lacking easy avenues of drainage, would remain waterlogged for long periods of time. Between these two extremes were a number of possible combinations conducive to a human presence: sandy clay, clay on top of sand, clay with good slope near a stream, and others. Still, such considerations should take into account changes in river courses, in sediment transport of rivers, in tidal ranges, in precipitation amounts, and in storm-surge frequency, all of which could play a role in altering drainage patterns.[63] Thus no specific sites saw continuous settlement. Nevertheless, the conditions that determined the location of settlement remained the same from the earliest times to the tenth century. From the Neolithic period through the early Middle Ages, therefore, settlements were established exclusively on well-drained sites.

THE NEOLITHIC PERIOD

After a break of some 2,000 years, around the middle of the fourth millennium B.C., a human presence began to be felt once again in the coastal regions of the western Netherlands. In the interim Europe had been truly revolutionized by the introduction and gradual establishment of Neolithic agricultural societies through the length and breadth of the Continent. The higher-lying eastern and southern Netherlands were no exception. There as elsewhere in western and central Europe were to be found groups of agriculturalists who combined the cultivation of cereals and livestock keeping with hunting, fishing, and collecting activities.[64] The archaeological record

[63] Louwe Kooijmans, *The Rhine/Meuse Delta*, pp. 36–38; Hallewas and van Regteren Altena, "Bewoningsgeschiedenis en landschapsontwikkeling rond de Maasmond," pp. 163–64.

[64] Stuart Piggott, *Ancient Europe from the Beginnings of Agriculture to Classical Antiquity*, pp. 40–70; J. G. D. Clark, *Prehistoric Europe: The Economic Basis*, pp. 91–128; J. A. Brongers et al., "Prehistory in the Netherlands: An Economic-Technological Approach," *Berichten van de Rijksdienst voor het Oudheidkundig Bodemonderzoek* 23 (1973): 10; S. J. de Laet, *The Low Countries*, pp. 59–71; Waterbolk, "Food Production in Prehistoric Europe," p. 1100; Waterbolk, "The Lower Rhine Basin," pp. 234–41, 250; L. P. Louwe Kooijmans, "Het onderzoek van neolithische nederzettingsterreinen in Nederland anno 1979," *Westerheem* 29 (1980): 93–136; Louwe Kooijmans, "The Neolithic at the Lower Rhine,", pp. 150–73.

indicates that the people who began entering the Holocene portions of the western Netherlands had cultural affinities with groups of Neolithic farmers both on the south and on the northeast.

The earliest evidence for a human presence in the western Netherlands is both sparse and sporadic. Around 3450 B.C. a small fishing and fowling camp was established less than 10 kilometers south of Rijnland, near the modern community of Bergschenhoek, on the north side of Rotterdam. It was situated in what was then a developing peat landscape adjacent to a freshwater lake. The site was made passable by several layers of bundled reeds laid on top of the otherwise soggy surface. This was, presumably, a temporary outpost or extraction camp, occupied for short periods of time over the space of three to six years by fishers and fowlers associated with permanent settlements elsewhere. Pottery fragments found at the site seem to point to a population possessing an early Neolithic culture.[65] A short while later a sandy outcrop known as the Hazendonk, about 50 kilometers east of the coast in the central Netherlands river area, saw the first of many prehistoric occupation phases. In this instance it was a small agricultural population with a middle Neolithic culture that resided there more or less permanently.[66] Both the Bergschenhoek and the Hazendonk sites predated the establishment of the coastal-barrier system.

With the formation of the coastal-barrier system the possibilities for settlement were greatly enhanced, especially after two or more ridges had come into being. Although the barrier ridges were not habitable while they were forming, the more interior ones offered occupation sites between phases of dune building. Settlement also became possible in the salt-marsh area behind the barriers, on the natural levees of tidal creeks, after sea-transgressive maxima but before the peat growth of the sea-regressive phase began to overtake everything.[67] From around 2450 B.C., therefore,

[65] The site was investigated by L. P. Louwe Kooijmans. See H. Sartatij, "Archeologische kroniek van Zuid-Holland over 1976," *Holland: regionaal-historisch tijdschrift* 9 (1977): 245–47; H. Sarfatij, "Archeologische kroniek van Zuid-Holland over 1977," *Holland: regionaal-historisch tijdschrift* 10 (1978): 298–99; Louwe Kooijmans, "Het onderzoek van neolithische nederzettingsterreinen in Nederland," pp. 108–10, 112.

[66] Louwe Kooijmans, *The Rhine/Meuse Delta*, pp. 125–68.

[67] Jelgersma et al., "The Coastal Dunes of the Western Netherlands," pp. 137–38, 147; Louwe Kooijmans, *The Rhine/Meuse Delta*, p. 37; Hallewas and van Regteren Altena, "Bewoningsgeschiedenis en landschapsontwikkeling rond de Maasmond," p. 168; E. J. Helderman, "Enige resultaten van vijftien jaar archeologisch onderzoek in de Zaanstreek," *Westerheem* 20 (1971): 42.

the barrier system as well as the banks of some tidal creeks began to be inhabited at various times by groups of migrants from the higher-lying regions farther inland.

These new residents are known from a number of sites scattered throughout the western Netherlands from Zeeland in the south to West Friesland (the northeastern section of North Holland) in the north, as well as from the central river area on the east. Toward the center of this distribution, in and near what became Rijnland, settlements were established at Vlaardingen and Hekelingen, along the north and south sides of the Maas estuary, respectively; near Loosduinen and Voorburg, just south of Rijnland; and at Leidschendam and Voorschoten, within Rijnland. The Vlaardingen and Hekelingen sites were on tidal creek banks, and those at Loosduinen, Voorburg, Leidschendam, and Voorschoten occurred along a coastal dune ridge.[68]

None of these settlements was very large, but all of them were occupied for substantial periods of time between 2450 and 2000 B.C. The traces of a number of small rectangular houses have been uncovered, ranging in size from 3.5 by 8 meters at Haamstede, Zeeland, to 5.5 by 10 meters at Vlaardingen.[69] Pottery and other artifacts indicate that the residents of these settlements possessed a culture that was similar in many respects to contemporary late Neolithic cultures in higher-lying areas on the south and northeast. Still it was distinct enough to be classified separately and designated the Vlaardingen Culture, named for its best-known and most thoroughly investigated site.[70]

The excavation and pollen analysis of the occupation sites of Vlaardingen Culture people have yielded a considerable amount of information concerning their natural settings. For example, the Vlaardingen settlement

[68] Clason, *Animal and Man in Holland's Past*, pp. 4–7, 10–26, 105; Louwe Kooijmans, *The Rhine/Meuse Delta*, pp. 4–5, 8, 10–11, 20–45, 49, 118, 281; Helderman, "Enige resultaten van vijftien jaar archeologisch onderzoek in de Zaanstreek," p. 42; Jelgersma et al., "The Coastal Dunes of the Western Netherlands," pp. 133–38; W. Glasbergen et al., "Settlements of the Vlaardingen Culture at Voorschoten and Leidschendam (Ecology)," *Helinium* 8 (1968): 105–30; Hallewas and van Regteren Altena, "Bewoningsgeschiedenis en landschapsontwikkeling rond de Maasmond," pp. 165–67.

[69] Jelgersma et al., "The Coastal Dunes of the Western Netherlands," pp. 133–38; Louwe Kooijmans, *The Rhine/Meuse Delta*, pp. 26, 281.

[70] Louwe Kooijmans, *The Rhine/Meuse Delta*, pp. 20–23; Clason, *Animal and Man in Holland's Past*, p. 5; Waterbolk, "The Lower Rhine Basin," p. 242; B. L. van Beek, "Pottery of the Vlaardingen Culture," in B. L. van Beek, R. W. Brandt, and W. Groenman–van Waateringe, eds., *Ex horreo: I.P.P. 1951–1976*, pp. 86–100.

lay on a sandy ridge along a meandering freshwater creek whose water level was affected by the tides in the nearby Maas estuary. It was adjacent to an extensive and partly wooded swamp.[71] The Leidschendam, Voorschoten, and Voorburg sites stood on the low, sandy ridges of the oldest and easternmost set of coastal-barrier ridges at a time when a new ridge was forming farther westward. The occupied ridges were covered with an alder-elm vegetational complex. There were extensive peat on the east and a salt marsh with holophytic or salt-tolerant grasses on the west.[72] The Vlaardingen Culture settlement far north, at Zandwerven, West Friesland, was also associated with a sandy ridge, this time in a silted-up salt marsh that was coming increasingly under freshwater dominance.[73] All Vlardingen Culture settlements were focused on the best-drained portions of otherwise water-saturated environments.

Shortly after 2400 B.C. a number of Vlaardingen Culture sites began experiencing changes that have been attributed to the arrival of new groups of people from the Continental interior. Presumably the new residents, identified by Dutch prehistorians on the basis of pottery type as the Protruding Foot Beaker Culture, were coastal representatives of the Corded Ware, or Battle Ax, people, whose heartland may have been as far away as the south Russian steppe and who are thought by some to have spoken dialects belonging to or related to the Indo-European family of languages. It appears that they moved rather quickly into and through much of Europe, often occupying uninhabited lands but also settling alongside and among the older residents. In many areas the influx of newcomers was attended by violence, but in the coastal regions of the western Netherlands, they apparently merged rather peacefully with the people of the Vlaardingen Culture.[74]

Yet another group of migrants arrived at the coast toward the end of

[71] W. Glasbergen et al., "De neolithische nederzettingen te Vlaardingen (Z.H.)," in *In het voetspoor van A. E. van Giffen*, 2d ed., p. 44; Clason, *Animal and Man in Holland's Past*, pp. 10–11, 102; Louwe Kooijmans, *The Rhine/Meuse Delta*, pp. 23–26; Waterbolk, "The Lower Rhine Basin," p. 242; Brongers et al., "Prehistory in the Netherlands," pp. 12–13; Jelgersma et al., "The Coastal Dunes of the Western Netherlands," pp. 133–38.

[72] Glasbergen et al., "Settlements of the Vlaardingen Culture at Voorschoten and Leidschendam," pp. 7, 98.

[73] J. F. van Regteren Altena and J. A. Bakker, "De neolithische woonplaats te Zandwerven," in *In het voetspoor van A. E. van Giffen*, 2d ed., pp. 35, 39.

[74] Piggott, *Ancient Europe*, pp. 84–97; Brongers et al., "Prehistory in the Netherlands," pp. 10, 12; Waterbolk, "The Lower Rhine Basin," pp. 242–43; Waterbolk, "The Occupation of Friesland in the Prehistoric Period," pp. 19, 20, 22; Louwe Kooijmans, *The*

the late Neolithic period, after 2000 B.C. These newcomers are distinguishable not only by their artifacts but also by differences in their skull shape and body type.[75] Nor were these the last movements of populations. One of the characteristic features of life in the coastal regions of the western Netherlands until the tenth century A.D. was the inconstancy of settlement and the continual coming and going of people.

Despite influxes or changes of populations, as well as the much more subtle processes of acculturation evidenced by changes in artifactual assemblages and even in housing forms, it was the quality of drainage that determined whether or not a settlement was established or persisted. Indeed, until reclamation began, residence was limited to the best-drained soils, usually low, sandy ridges, in the otherwise waterlogged western Netherlands. What did change, however, was the suitability of specific sites for human habitation. For example, the late Neolithic occupation of both dune and tidal environments in the western Netherlands was nowhere continuous but appears to have shifted from one type of milieu to another. The preferred environments appear to have been successively (1) dunes (Haamstede), (2) estuary (Vlaardingen and Hekelingen), (3) dunes (Voorburg, Leidschendam, and Voorschoten), as well as the salt-marsh regions of West Friesland (Zandwerven), and (4) estuary (Vlaardingen).

According to the transgression-regression model used by most Dutch prehistorians, such shiftings of occupation sites were due to the inverse relation that apparently existed between transgressive phases of the sea and dune formation. A coastal barrier could not be occupied while it was being formed, during a sea-transgressive phase, at the same time that the possibilities for settlement in estuarine areas were at their lowest. When the maximum of the transgressive phase had passed, however, occupation became possible in the estuarine areas, but the barrier ridge began to experience dune formation that discouraged settlement. As the regressive phase of the sea progressed, access in estuarine environments gradually diminished because some tidal creeks silted shut and became clogged with peat growth. During a succeeding transgressive phase estuarine areas were subject to flooding once again, while another barrier ridge began developing west of the original ridge. At this point the interior ridge became habitable

---

Rhine/Meuse Delta, p. 26; Jelgersma et al., "The Coastal Dunes of the Western Netherlands," pp. 131, 134–35; Clason, *Animal and Man in Holland's Past*, p. 203.

[75] Louwe Kooijmans, *The Rhine/Meuse Delta*, pp. 26–36, 42–46, 319.

as dune formation ended and soil formation began.[76] While this cycle repeated itself a number of times, some sites along the interior ridges became increasingly isolated as peat growth and silting clogged drainage avenues. The result was a general movement toward the west as new barrier ridges were added.[77]

THE BRONZE AGE

As the Neolithic era began to give way to the Bronze Age, and thereafter for nearly 3,000 years, shifts in settlement occurred repeatedly as the suitability of specific sites for occupation changed. After 1950 B.C., for example, Vlaardingen saw only occasional, temporary encampments. Still, occupation apparently continued in the dunes at Voorschoten and possibly in the area of the mouth of the Maas, while a new dunal settlement appeared at Velsen, a short distance north of Rijnland. A number of sites remained habitable in West Friesland, while a new settlement was established around 1800 B.C. at Molenaarsgraaf, in the central river area on the southeast. During the early Bronze Age, however, these patterns began changing again. Settlements were established in the dunes at Monster, near the Maas, and at Veenenburg, in Rijnland, and occupation continued at Velsen, while the central river area on the east as well as West Friesland on the north were soon abandoned.[78]

During the middle Bronze Age, roughly 1500 to 1000 B.C., the West Frisian area in particular saw dense settlement at times, while a number of sites are known from the central river area as well. The dunes, in contrast, appear to have been relatively empty during this period, though settlement most likely shifted to a more recently formed dune ridge on the west as the original ones became increasingly engulfed by peat growth; we may never know, since the more westerly dune ridges were largely covered by more recent dune formation or obliterated by coastal erosion during the Middle Ages. Finally, during the late Bronze Age, between 1000 and 700 B.C.,

[76] Jelgersma et al., "The Coastal Dunes of the Western Netherlands," pp. 137–38, 147; Louwe Kooijmans, *The Rhine/Meuse Delta*, p. 37; Glasbergen et al., "Settlements of the Vlaardingen Culture at Voorschoten and Leidschendam," p. 116.

[77] Louwe Kooijmans, *The Rhine/Meuse Delta*, pp. 29–30, 37.

[78] A. T. Clason, "The Antler, Bone and Tooth Objects from Velsen: A Short Description," *Berichten van de Rijksdienst voor het Oudheidkundig Bodemonderzoek* 24 (1974): 119; Louwe Kooijmans, *The Rhine/Meuse Delta*, pp. 30–35, 107; Jelgersma et al., "The Coastal Dunes of the Western Netherlands," pp. 139–40.

human presence shrank throughout the western Netherlands. A wetter, stormier period and accelerated peat growth caused a deterioration of drainage in many places, resulting in the abandonment of many sites in the estuarine, marine-clay, and river-clay areas. Only in scattered portions of the dunes and on the most favorable sites in West Friesland did settlement continue into the Iron Age.[79] Throughout the Bronze Age as before, however, settlements were established only on the best-drained soils, usually on sandy or sandy-clay ridges, in a generally soggy environment.

Throughout the Neolithic era, the occupation phases of the coastal districts of the western Netherlands had remained relatively insignificant outliers of cultures whose heartlands lay east and south. They consisted of rather small groups of people who had made certain adaptations to the coastal environments, but they remained for the most part on the periphery of the affairs of northwestern Europe.[80] During the Bronze Age, however, the remoteness of their location gradually began to change somewhat as a result of new trading contacts between the British Isles and the Continent. The major rivers, especially the Rhine, were the only routes that adequately bridged the waterlogged Holocene portion of the western Netherlands. The people who lived on the dune ridges near the mouths of the great rivers suddenly found themselves astride important commercial routes. It is not surprising, therefore, that some depots of bronze artifacts, presumably left or lost by itinerant metalsmiths or merchants, have been discovered in the dune ridges just north of the mouth of the Oude Rijn, at Lisse and Voorhout. Among other objects these finds included bronze axes, chisels, and sickles from the British Isles as well as from both northern and southern Germany.[81]

Because the Netherlands had no native copper for making bronze,

[79]Waterbolk, "Siedlungskontinuität im Küstengebiet der Nordsee," pp. 8, 12; Waterbolk, "The Occupation of Friesland in the Prehistoric Period," p. 26; Louwe Kooijmans, *The Rhine/Meuse Delta*, pp. 42–43; Hallewas and van Regteren Altena, "Bewoningsgeschiedenis en landschapsontwikkeling rond de Maasmond," pp. 167–77; Jelgersma et al., "The Coastal Dunes of the Western Netherlands," pp. 139–40.

[80]See, for example, Waterbolk, "The Lower Rhine Basin," pp. 234–47.

[81]Brongers et al., "Prehistory in the Netherlands," pp. 25–26. For the Lisse and Voorhout finds, see J. J. Butler, "Vergeten schatvondsten uit de bronstijd," in J. E. Bogaers, W. Glasbergen, P. Glazema, and H. T. Waterbolk, eds., *Honderd eeuwen Nederland*, pp. 131–39. For the Bronze Age in the Netherlands and northwestern Europe in general, see H. T. Waterbolk, "The Bronze Age Settlement of Elp," *Helinium* 4 (1964): 97–131; J. J. Butler, *Nederland in de bronstijd*; J. J. Butler, *Bronze Age Connections across the North Sea: A Study in Prehistoric Trade and Industrial Relations between the British Isles, the Netherlands,*

something had to be traded to acquire it because, in addition to the presumed depots of traveling smiths or merchants just mentioned, bronze artifacts have been found in cultural contexts in the western Netherlands as well.[82] In some parts of the Netherlands, because of the key position along the main waterways of northwestern Europe, bronze may have come into the possession of local groups in the form of payments made by merchants to local chieftains in return for free passage or protection. Further, the inhabitants of the coastal regions could offer cattle hides, the one product of considerable commercial value that they could apparently produce in quantity.[83] Finally, amber may well have been an important collateral item in the bronze trade as it was in Scandinavia, since it occurred naturally not only in the Baltic region but also along the North Sea coast as far south as the western Netherlands.[84]

THE PRE-ROMAN IRON AGE

While the transitional period between the late Bronze Age and the early Iron Age was a difficult one for settlement in the western Netherlands, owing in large part to deteriorating drainage conditions, it was a period of significant developments in rural economy and technical capacity in the higher and drier areas on the south and east. Such changes were accompanied by considerable population growth. By the early Iron Age the sandy regions of Drenthe, in northeastern Netherlands, for example, had a population density that was similar to what was known as recently as the eighteenth century A.D.[85] As a result, the use of land there and elsewhere began to change significantly.

North-Germany and Scandinavia, c. 1700–700 B.C.; Brongers et al., "Prehistory in the Netherlands."

[82] Brongers et al., "Prehistory in the Netherlands," pp. 25–28; Louwe Kooijmans, *The Rhine/Meuse Delta*, pp. 31–36; Waterbolk, "The Lower Rhine Basin," p. 247.

[83] Brongers et al., "Prehistory in the Netherlands," pp. 13, 28.

[84] As Brongers et al., in ibid., pp. 28–29, 32–33, rightly point out, there is no reason to assume, as many have in the past, that all amber found in the Netherlands necessarily came from Scandinavia or the Baltic region. See P. Vons, "De vervaardiging van barnsteen-kralen te Velsen in de vroege bronstijd," *Westerheem* 19 (1970): 34–35, for a report on the evidence for amber working at the early Bronze Age site at Velsen, just north of Rijnland. The amber trade in general is treated in Piggott, *Ancient Europe*, pp. 137–38; and Clark, *Prehistoric Europe*, pp. 161–66.

[85] Waterbolk, "The Lower Rhine Basin," p. 247; Louwe Kooijmans, *The Rhine/Meuse Delta*, p. 38; Clason, *Animal and Man in Holland's Past*, pp. 5–6.

By the early Iron Age there were systematically laid out, permanent fields (often referred to as "Celtic fields") on the higher-lying light-soil areas of the Netherlands and surrounding regions that saw short-fallow cultivation as well as fertilization with sods.[86] The gradual transition from the use of naked barley to hulled barley, begun during the Bronze Age, was completed about the same time. Because the kernels of hulled barley were more firmly attached to the stalk, less grain was lost during harvesting, while the use of granaries, known since the middle Bronze Age, ensured fewer losses through spoilage.[87] These and many other changes represented a major advance in the agrarian capacity of northwestern Europe by the early Iron Age, 700 to 550 B.C.

In addition to advances in agrarian technology, many other economic and technological changes occurred during the Bronze Age–Iron Age transition that are considered important enough by some prehistorians to warrant the designation of a second, major prehistoric "revolution."[88] In general, the first significant moves toward the development of regional specialization with increasing economic interdependence between various geographic areas took place within the Netherlands and surrounding areas at this time. For example, a shortening of house lengths in some higher-lying areas may indicate that the sizes of cattle herds (usually stalled in one end of the farmsteads) were reduced with a concurrent increase in the reliance on cereal production. At the same time large numbers of colonists who arrived in the coastal areas at the beginning of the Iron Age specialized in livestock raising, the products of which they could trade for cereals and other commodities produced in inland areas. Salt, extracted from salt water and salt-impregnated peat, became another trade item for the inhabi-

[86]Clason, *Animal and Man in Holland's Past*, pp. 5–6. Brongers et al., "Prehistory in the Netherlands," p. 12, note, however, that the systematic laying out of permanent fields may have been anticipated during the late Bronze Age on the sandy ridges of the West Frisian salt marsh. See also P. J. Woltering, "Archeologische kroniek van Noord-Holland over 1975," *Holland: regionaal-historisch tijdschrift* 8 (1976): 237–41. For the general question of population growth and its relation to the intensification of land use, see chap. 2.

[87]J. A. Bakker et al., "Hoogkarspel-Watertoren: Towards a Reconstruction of Ecology and Archaeology of an Agrarian Settlement of 1000 B.C.," in B. L. van Beek, R. W. Brandt, and W. Groenman–van Waateringe, eds., *Ex horreo: I.P.P. 1951–1976*, p. 200; Brongers et al., "Prehistory in the Netherlands," pp. 12, 39; W. van Zeist, "Prehistoric and Early Historic Food Plants in the Netherlands," *Palaeohistoria: Acta et Communicationes Instituti Bio-Archaeologici Universitatis Groninganae* 14 (1968): 41–173.

[88]The first was, of course, the Neolithic, agricultural revolution, the symptoms of which had reached the Netherlands by about 4,000 B.C.

tants of both the western and the northern coastal regions. Further, the introduction of wood- or wattle-lined water wells allowed settlers to move for the first time into areas lacking plentiful supplies of fresh water at the surface. A large number of interrelated changes of this sort constituted the Iron Age "revolution," which was completed well before the beginning of our era.[89]

The population growth and extension of agriculture characteristic of the Bronze Age–Iron Age transitional period eventually led to serious environmental problems in some areas of the Netherlands. Particularly on the Pleistocene sands of Drenthe and North Brabant and even, perhaps, on the Pleistocene core of the island of Texel, forest clearance and more intensive land use to accommodate a growing population soon brought on extensive wind erosion. By 500 B.C. drifted sand had covered fields and settlements in Drenthe and elsewhere.[90] Because this occurred during a relatively dry period, many people migrated to the lower-lying areas along the coast: to the coastal dunes; once again to the estuarine environments of the former IJ, the Oude Rijn, and the Maas; and even for a short time to the edges of the eutrophic- and mesotrophic-peat areas near the IJ and Maas estuaries.[91]

[89] Brongers et al., "Prehistory in the Netherlands," passim. On pp. 35–39 the authors draw attention to, and outline the character and significance of, this second "revolution." Their discussions of the mechanism of cultural change interpreted from an economic-technological point of view are particularly enlightening and useful. I know of no better synthesis of the prehistoric period of the Netherlands.

[90] Clason, *Animal and Man in Holland's Past*, p. 6; Waterbolk, "The Lower Rhine Basin," pp. 247–49; Waterbolk, "The Occupation of Friesland in the Prehistoric Period," p. 33; W. J. van Tent and P. J. Woltering, "The Distribution of Archaeological Finds on the Island of Texel, Province of North-Holland," *Berichten van de Rijksdienst voor het Oudheidkundig Bodemonderzoek* 23 (1973): 56; Woltering, "Archeologische kroniek van Noord-Holland over 1975," pp. 243–44. For similar wind-erosion problems farther eastward in the Weser Valley area of Germany, see Karl W. Butzer, "Accelerated Soil Erosion: A Problem of Man-Land Relationships," in Ian R. Manners and Marvin Mikesell, eds., *Perspectives on Environment*, p. 69.

[91] Clason, *Animal and Man in Holland's Past*, pp. 17–18; Brongers et al., "Prehistory in the Netherlands," pp. 29, 34–35; Jelgersma et al., "The Coastal Dunes of the Western Netherlands," p. 140; Louwe Kooijmans, *The Rhine/Meuse Delta*, pp. 42–44; D. P. Hallewas, "Een huis uit de vroege ijzertijd te Assendelft (N.H.)," *Westerheem* 20 (1971): 19–35; L. Havelaar, "Een huisplattegrond uit de vroege-ijzertijd te Vlaardingen," *Westerheem* 19 (1970): 120–27; E. J. Helderman, "Een nederzetting van de Zeijener cultuur te Assendelft," *Westerheem* 16 (1967): 183–90; Helderman, "Enige resultaten van vijftien jaar archeologisch onderzoek in de Zaanstreek," pp. 43–44; H. Sarfatij, "Archeologische kroniek van Zuid-Holland over 1975," *Holland: regionaal-historisch tijdschrift* 8 (1976): 266; Woltering, "Archeologische kroniek van Noord-Holland over 1975," p. 248; A. M. Dumon Tak and J. van den Berg, "Een pottenbakkersoven uit de ijzertijd te Serooskerke (Walcheren)," *Westerheem* 22 (1973): 242–47.

Finally, migrants also arrived for the first time on the coastal marshes of Friesland and Groningen in the north. Most of them came from Drenthe and parts farther eastward, though there may have been a slightly earlier movement of late Bronze Age peoples from West Friesland who could have provided the nuclei of settlement to which the Iron Age settlers adhered.[92]

The return of wet conditions around 400 B.C. quickly brought an end to estuarine occupation of the western Netherlands as well as along the Ems River, in northwestern Germany, but settlement continued in the western coastal dunes and on the salt marshes of Friesland and Groningen.[93] In the latter region, in particular, continued occupancy had an important impact on the future settlement of the western Netherlands. For the first time a truly maritime culture developed that was more than an extension of neighboring inland cultures and was capable of withstanding worsening drainage conditions.

Whereas the western Netherlands consisted of a low-lying landscape of peat, sand, and clay soils, largely protected from the direct influence of the sea by coastal barriers and dunes, the northern coastal regions had no such protection. There was, however, a series of low, sandy offshore barrier islands in the north that provided some protection to the coast, similar in many respects to the situation in the western Netherlands before the formation of the continuous coastal-barrier system. The coast of Friesland and Groningen was a flat, treeless salt marsh of marine clays, deposited at the end of the Bronze Age and by the early Iron Age covered with a lush vegetation of various halophytic, or salt-loving, grasses. The entire landscape was intersected by a network of tidal creeks and gullies. Twice a day the tides forced salt water up these creeks and gullies and into the salt marsh; at low tide the waters flowed out to the sea once again. Between the salt marsh and higher-lying Pleistocene sands on the south was a broad band of

[92] Waterbolk, "The Occupation of Friesland in the Prehistoric Period," p. 26; until the post-Roman period West Friesland and Friesland were part of the same landscape, separated only by a large river. On the origin and culture of the Iron Age colonists on the northern marshes, see also A. Russchen, "Origin and Earliest Expansion of the Frisian Tribe," *It Beaken: tydskrift fan de Fryske Akademy* 30 (1968): 143, 149 map 7; Russchen, "Tussen Aller en Somme," *It Beaken: tydskrift fan de Fryske Akademy* 29 (1967): 95.

[93] See, for example, Waterbolk, "The Occupation of Friesland in the Prehistoric Period," p. 29; W. Haarnagel, "Die Prähistorischen Siedlungsformen in Küstengebiet der Nordsee," in *Beiträge zur Genese der Siedlungs- und Agrarlandschaft in Europa*, pp. 70–71; Clason, *Animal and Man in Holland's Past*, p. 19; Louwe Kooijmans, *The Rhine/Meuse Delta*, pp. 42–43; Roeleveld, *The Holocene Evolution of the Groningen Marine-Clay District*, p. 110.

lagoons and peat growth, traversable only by boat along the natural waterways.[94] While the lagoon-peat area remained essentially uninhabited until the Middle Ages, the Iron Age colonists settled on the salt marsh along the coast.

The earliest settlements in the coastal marsh of Friesland and Groningen were established on low, sandy ridges that occurred here and there or on the lighter clays of the marsh itself. The sea was not a constant threat. The fact that the inhabitants could at times raise cereals on the marsh shows that a certain amount of desalination of the marine clays occurred.[95] The return of wetter conditions shortly before 400 B.C,. however, began causing some serious flooding, and, for the first time, the reaction to rising mean high-water levels was to stay instead of flee. The settlers responded by constructing low mounds of sods upon which they rebuilt their farmsteads, granaries, storage pits, and workshops. At first these dwelling mounds, or *terpen*, as they were called, provided space for one or more farm units, but gradually they fused into larger, compact *terp* villages with as many as fifteen farmsteads arranged in a circle around an open space in the center.[96]

The residents of the *terpen* continued to keep large herds of cattle, stalled at one end of their farmsteads, and raised vegetables and cereals

[94] Roeleveld, *The Holocene Evolution of the Groningen Marine-Clay District*, sec. 3 and figs. 57–64; W. A. van Es, "Friesland in Roman Times," *Berichten van de Rijksdienst voor het Oudheidkundig Bodemonderzoek* 15–16 (1965–66): 37–39; Pons, "De zeekleigronden," pp. 26–34; *Bodemkaart van Nederland*, sheets 1–2; D. P. Blok, "Histoire et toponymie: l'example des Pays-Bas dans le haut moyen âge," *Annales: économies, sociétés, civilisations* 14 (1969): 919, 922; P. C. J. A. Boeles, *Friesland tot de elfde eeuw: zijn vóór- en vroege geschiedenis*, 2d ed., pp. 80–82; J. M. G. van der Poel, "De landbouw in het verste verleden," *Berichten van de Rijksdienst voor het Oudhiedkundig Bodemonderzoek* 10–11 (1960–61): 174–75; T. Edelman, "Oude ontginningen van de veengebieden in the Nederlandse kuststrook," *Tijdschrift voor economische en sociale geografie* 49 (1958): 239–40; T. Edelman, *Bijdrage tot de historische geografie van de Nederlandse kuststreek*, pp. 35–36.

[95] TeBrake, "Ecology and Economy in Early Medieval Frisia," pp. 6–7 n. 14; Roeleveld, *The Holocene Evolution of the Groningen Marine-Clay District*, pp. 106–109.

[96] Roeleveld, *The Evolution of the Groningen Marine-Clay District*, pp. 110–11; Waterbolk, "The Occupation of Friesland in Roman Times," pp. 42, 44–46; Boeles, *Friesland tot de elfde eeuw*, pp. 85–96; TeBrake, "Ecology and Economy in Early Medieval Frisia," p. 7. *Terp* (plural, *terpen*) is a Frisian word, originally meaning "village," that has come into common usage to describe the elevated dwelling mounds of the coastal-clay regions of the modern province of Friesland. In the province of Groningen the word most used is *wierde*, while across the Ems in Germany the most common word is *Wurt*. Other forms, such as *werd*, *wird*, or *ward*, appear widely in both the Netherlands and Germany as place-name endings. See H. T. Waterbolk, "Terpen, milieu en bewoning," in J. W. Boersma, ed., *Terpen-mens en milieu*, p. 71.

wherever possible. This they did on the edges of the *terpen* and also on the highest portions of the salt marsh itself. They obtained further supplies of cereals from inland areas in exchange for cattle hides as well as wool, woolen products, and other locally available commodities. In short, between 400 and 200 B.C., during a period of deteriorating hydraulic conditions, the inhabitants of the Friesland-Groningen coastal marshes made the necessary adaptations to an environment dominated by the sea that constituted the proto-Frisian Culture. It, and the early Frisian Culture that followed after 200 B,.C. achieved a remarkable success in dealing with a marshy environment and a cultural independence that served as the focus of a maritime way of life for some 1,500 years, gradually spreading its influence along the entire North Sea coast from Flanders to southern Denmark.[97]

THE PROTOHISTORIC PERIOD

The return of relatively dry conditions around 200 B.C. ushered the Netherlands into the late pre-Roman Iron Age. For the first time the evidence supplied by archaeology can be associated with groups of people who a little later were observed and described by Roman writers. Thus, instead of merely talking about groups of Bronze or Iron Age peoples known only by their artifactual remains, it becomes possible to speak about peoples who begin to take on historical validity. The Frisians were just such a group, and they remained an important cultural entity throughout the Middle Ages and after. This is not to suggest that settlement in the coastal regions of the Netherlands suddenly took on a static and unchanging character. On the contrary, there was a great deal of fluidity and movement among and by cultural groups, as there had been for thousands of years throughout northwestern Europe and continued to be for another thousand years or more. Yet, despite the flux, certain forms began to take dim shape around the beginning of our era.

The period between 200 B.C. and A.D. 250 was one of unusually dry conditions in the coastal regions of the Netherlands. The repopulation that subsequently occurred was on an unprecedented scale. In the northern Netherlands, especially in Friesland, new settlements were once again es-

[97] See, for example, Waterbolk, "The Occupation of Friesland in the Prehistoric Period," pp. 29–34. The maritime way of life pioneered by the proto-Frisian and Frisian cultures will be discussed in greater detail in chap. 5. See also TeBrake, "Ecology and Economy in Early Medieval Frisia," pp. 1–29.

tablished directly on the salt-marsh surface, alongside and among the still-occupied *terpen*.[98]

In the western Netherlands both estuarine and dunal environments were resettled by immmigrants from inland areas and coastal areas both north and south. Between the IJ and Maas estuaries, thus including all of Rijnland, two streams of cultural influence overlapped. From the north came representatives of the early Frisian Culture with their large, three-aisled houses and *streepband* ("stripe-decorated") pottery. They developed a large concentration of settlement along the former IJ, with traces as far south as The Hague or the mouth of the Maas. Settlement was particularly dense north of Rijnland, on the old dune ridges and stream banks between Velsen and Uitgeest, and even on the eutrophic and mesotrophic peat on the east between Assendelft and Krommenie.[99]

In the Maas River estuary a similar concentration of settlement occurred by members of a group associated with the coast on the south, perhaps the Menapii of Roman sources. They are represented archaeologically by a type of two-aisled house and De Panne pottery (named for modern-day De Panne, Belgium, on the coast near the French border). Outliers of this southern tradition also extended through Rijnland. Like the houses of the early Frisian Culture on the north, some houses were built on the edges of peat bogs, particularly south of Rijnland in the vicinity of Vlaardingen, near the Maas estuary.[100]

[98] J. H. F. Bloemers, "Rijswijk (Z.H.) 'De Bult,' een nederzetting van de Cananefaten," *Hermeneus: tijdschrift voor antieke cultuur* 52 (1980): 96; van Es, "Friesland in Roman Times," pp. 40, 44; Haarnagel, "Die Prähistorischen Siedlungsformen im Küstengebiet der Nordsee," pp. 7–8. On the relative dryness of the period, see Waterbolk, *De praehistorische mens en zijn milieu*, pp. 16, 131.

[99] Between 80 and 90 farmsteads have been found on the edge of the peat, some of which have been excavated. There were a northern nucleus and a southern nucleus with a 12-kilometer occupied strip between them; Sarfatij, "Friezen-Romeinen-Cananefaten," pp. 38–39, 96–105: R. W. Brandt, "De archeologie van de Zaanstreek," *Westerheem* 32 (1983): 120–37; Helderman, "Enige resultaten van vijftien jaar archeologisch onderzoek in de Zaanstreek," p. 54; P. Stuurman, "Archeologie van het jaar nul," *Westerheem* 18 (1969): 70–73; H. T. Waterbolk, "Evidence of Cattle Stalling in Excavated Pre- and Protohistoric Houses," in A. T. Clason, ed., *Archaeozoological Studies*, pp. 383–94; Waterbolk, "The Occupation of Friesland in the Prehistoric Period," pp. 29, 33; W. A. van Es, *De Romeinen in Nederland*, pp. 131, 135; van Es, "Friesland in Roman Times," pp. 46–49, 135; J. K. de Cock, "Veenontginningen in West-Friesland," *West-Frieslands oud en nieuw* 36 (1969): 156.

[100] Stuurman, "Archeologie van het jaar nul," pp. 66–70; Hallewas and van Regteren Altena, "Bewoningsgeschiedenis en landschapsontwikkeling rond de Maasmond," pp. 168–83; S. J. de Laet, "Les limites des cités des Ménapiens et des Morins," *Helinium* 1 (1961):

This mixing and overlapping of two coastal cultural traditions[101] between the Maas and IJ was further complicated during the last half-century B.C. by the arrival of the Cananefates. They and the related Batavians apparently broke away from a confederation with the Chatti[102] on the upper reaches of the Weser River in Germany and established themselves between the Maas and Rhine rivers. The Cananefates settled on the coast, including the southern half of Rijnland, while the Batavians stayed east of the peat bogs, in the central Netherlands river region.[103]

Although the Cananefates were mentioned by classical authors, attempts to associate them with specific archaeological material, such as pottery, have been unsuccessful.[104] Perhaps they and the Batavians farther inland were simply new ruling groups who replaced the existing power holders without much affecting the material culture of those they ruled.[105] The possibility exists, however, that the settlement excavated at Rijswijk, just south of Rijnland, may have been a Cananefate settlement.[106] In any case, their presence serves to underline once again the fluid and mixed nature of settlement and culture in the western Netherlands during the late pre-Roman Iron Age.[107]

20–34; Waterbolk, "The Occupation of Friesland in the Prehistoric Period," pp. 33–34; J. E. Bogaers, "Waarnemingen in Westerheem," *Westerheem* 17 (1968): 217–23; van Es, *De Romeinen in Nederland*, p. 136; Clason, *Animal and Man in Holland's Past*, pp. 19–20, 203; C. Wind, "Een nederzetting uit de voor-Romeinse ijzertijd te Rockanje," *Westerheem* 19 (1970): 243. The Schelde area, too, saw extensive settlement, including some on peat soils, during both the late pre-Roman Iron Age and the Roman period itself; C. Dekker, *Zuid-Beveland: de historische geografie en de instellingen van een Zeeuws eiland in de middeleeuwen*, pp. 10–16.

[101] Whether this mixing and overlapping resulted primarily from processes of acculturation or those of migration is impossible to determine at this time; see Stuurman, "Archeologie van het jaar nul," p. 73; Helderman, "Enige resultaten van vijftien jaar archeologische onderzoek in de Zaanstreek," pp. 50–51, 56–57.

[102] See, for example, the accounts of Tacitus, *Histories*, 4:12, 15.

[103] J. E. Bogaers, *Civitas en stad van de Bataven en Caninefaten*, pp. 3–4; Stuurman, "Archeologie van het jaar nul," pp. 73–74; van Es, *De Romeinen in Nederland*, pp. 28, 195.

[104] See Stuurman, "Archeologie van het jaar nul," pp. 73–75, where he discusses the general problem of trying to pinpoint the existence of distinct groups mentioned by classical authors in the archaeological record. The best attempt to do so to date is R. Wenskus, *Stammesbildung und Verfassung: Das Werden der frühmittelalterlichen gentes*, especially pp. 113–42.

[105] At least van Es, *De Romeinen in Nederland*, pp. 170–71, suggests this.

[106] Bloemers, "Rijswijk (Z.H.) 'De Bult,' een nederzetting van de Cananefaten," pp. 95–106.

[107] For the mixed ethnic and cultural heritage of the late pre-Roman Netherlands, see van

During the last two decades B.C., the Romans began entering this region of mixed population and culture in the western Netherlands.[108] Julius Caesar, during his conquest of Gaul between 57 and 49 B.C., had penetrated into what is now Belgium and perhaps into extreme southeastern Netherlands on several occasions, but he never reached the western Netherlands. In 12 B.C., however, Drusus arrived in the Rhine delta on orders from the emperor Augustus. Official policy apparently called for pushing the Roman Empire as far northeastward as the Elbe River, and the Rhine delta came to be used as a staging area for such a campaign.

The insula Batavorum (the Island of the Batavians, east of the coastal peat bogs between the Waal branch of the Rhine and the Maas), Fectio (a military installation at Vechten, just east of Utrecht, where the Kromme Rijn branched off into the Oude Rijn and the Vecht, which in turn was connected to the former IJ estuary), and perhaps some temporary encampments near the mouth of the Oude Rijn and along the IJ estuary were used for repeated military thrusts into and through the land of the Frisians, in the northern Netherlands, and the land of the Chauci, in northwestern Germany, over a period of about sixty years. In A.D. 47, however, the emperor Claudius abandoned the Elbe policy and ordered General Gnaius Corbulo to pull back to the Rhine delta. From then until the middle of the third

---

Es, *De Romeinen in Nederland*, pp. 195, 197; Sarfatij, "Friezen-Romeinen-Cananefaten," pp. 33–47, 89–105, 153–79; Russchen, "Tussen Aller en Somme," pp. 90–96. Cooter, "Pre-Industrial Frontiers and Interaction Spheres," pp. 73–86, 95–108, 141–59, develops a very strong case for the existence of an important "interaction sphere" in the Lower Rhine watershed (including the western coastal regions of the Netherlands) during the late pre-Roman Iron Age. The mixed nature of the population and material culture of the western Netherlands at that time certainly helps support such a notion. An interaction sphere could have facilitated the movement of goods, people, and ideas that would have been responsible for the overlapping of the two cultures in the western Netherlands suggested by archaeological research. Van Es, *De Romeinen in Nederland*, p. 170, hints at common social and political developments in the Lower Rhine watershed during the last century B.C. that lend further credence to Cooter's fascinating interaction-sphere argument. For example, aristocracies generally began to wrest power and influence from traditional, semicultic kings. See also Wenskus, *Stammesbildung und Verfassung*, pp. 409–28; R. S. Hulst, "Bewoning op het oostelijke rivierengebied in the Romeinse tijd," *Westerheem* 23 (1974): 233–34.

[108] A. W. Byvanck, ed., *Teksten*, vol. 1 in *Excerpta Romana: de bronnen der Romeinsche geschiedenis van Nederland*, is an important collection of the relevant Roman literary references to the Netherlands. It formed the basis of Byvanck, *Nederland in den Romeischen tijd*. The best recent survey is van Es, *De Romeinen in Nederland*, which incorporates the latest archaeological and historical research in a very readable and richly illustrated volume. See also A. van Doorselaer, "De Romeinen in België en Nederland," *Hermeneus: tijdschrift voor antieke cultuur* 52 (1980): 74–86.

century the Oude Rijn was considered the official boundary of the Roman Empire, though some Roman military presence may have continued along the IJ at Velsen until about A.D. 69 or 70.[109] There is some evidence to suggest, however, that, just before he was ordered back to the Rhine, Corbulo was busy establishing the boundaries of a Roman territory of Frisia in the northern Netherlands, preparatory to introducing a senate, magistrates, and laws that would have created a civitas Frisionum as a permanent part of the Roman Empire.[110] During the same period there was considerable Roman activity in the area of the Ems estuary as well.[111]

During the first century A.D., a number of Roman fortifications were built along the naturally raised levees of the Oude Rijn as part of the larger Rhine-Danube *limes*, or fortified frontier. In Rijnland, *castella*, forts large enough for a cohort of 480 men, were built around the middle of the first century at a number of places: at the mouth of the Oude Rijn (perhaps the legendary Brittenburg, destroyed by coastal erosion in the Middle Ages), Katwijk (Lugdunum), Valkenburg (Praetorium Agrippinae), Roomburg-Leiden (Matilone), Alphen aan den Rijn (Albaniana), and Zwammerdam (Nigrum Pullum), on the eastern edge of Rijnland.[112]

All of the Oude Rijn fortifications were repeatedly rebuilt. At first they were made of earth and wood, but by the end of the second century they were constructed for the most part of tufa, stone, or brick.[113] On a number of occasions the rebuilding was necessitated by the destruction associated with anti-Roman uprisings. For example, all installations were apparently burned in 69–70 in the revolt of the Batavians and their allies un-

[109] Van Es, *De Romeinen in Nederland*, pp. 22–37, 76–82; P. Vons, "Op zoek naar een castellum," *Westerheem* 23 (1974): 59–69; P. Vons, "The Identification of Heavily Corroded Roman Coins Found at Velsen," *Berichten van de Rijksdienst voor het Oudheidkundig Bodemonderzoek* 27 (1977): 139–63; Woltering, "Archeologische kroniek van Noord-Holland over 1975," pp. 250–52; Woltering, "Archeologische kroniek van Noord-Holland over 1978," p. 260.

[110] Van Es, *De Romeinen in Nederland*, p. 37.

[111] Günter Ulbert, "Die römischen Funde von Bentumersiel," *Probleme der Küstenforschung im südlichen Nordseegebiet* 12 (1977): 33–65.

[112] The discovery and identification of Roman military sites along the Oude Rijn are far from complete; see, for example, P. C. Beunder, "Tussen Laurum (Woerden) en Nigrum Pullum (Zwammerdam?) lag nog een castellum," *Westerheem* 29 (1980): 2–23; A. Wassink, "Het Romeinse castellum te Alphen aan den Rijn," *Westerheem* 32 (1983): 296–302.

[113] Van Es, *De Romeinen in Nederland*, pp. 82–86, 93–96; A. E. van Giffen, "Three Roman Frontier Forts in Holland at Utrecht, Valkenburg and Vechten," in Eric Birley, ed., *The Congress of Roman Frontier Studies 1949*, pp. 31–40.

der Julius Civilis,[114] toward the end of the second century, during the early third century, and again around 270 in conjunction with the massive attacks of the Franks and their allies. None of the *castella* associated with the Oude Rijn were rebuilt after the last-mentioned attacks.[115]

The various building and rebuilding phases are best known from the *castellum* at Valkenburg, the most thoroughly examined Roman fortification in western Europe. The 3-meter-high artificial mound on which Valkenburg village is situated has been dug into repeatedly since 1941, revealing seven consecutive *castella* built between the middle of the first century and the third quarter of the third century.[116] The Valkenburg site seems to have been particularly important as a storage and transshipment point for grain from the dune and riverbank areas and especially from England.[117]

At first Roman presence in the western Netherlands was almost exclusively military, especially along the border. In addition to building and manning the frontier forts, the military controlled the land between them in a strip along the left, or southern, bank of the Oude Rijn that may have been as much as 15 kilometers wide. Between A.D. 47 and 70 native settlement was prohibited in this frontier zone. The Roman army also controlled a belt of land across the Oude Rijn, which was perhaps used for the grazing of livestock and in which native settlement also may have been forbidden from time to time.[118]

An integral part of the *limes* system was a network of roads. A single road ran along the silted-up left bank of the Rhine, Kromme Rijn, and Oude Rijn, all the way to the North Sea, though its location in many areas has yet to be determined.[119] Another road ran from Katwijk (Lugdunum),

[114] P. G. van Soesbergen, "The Phases of the Batavian Revolt," *Helinium* 11 (1971): 238–56, especially p. 241; van Es, *De Romeinen in Nederland*, pp. 39–44, 93; H. Sarfatij, "Archeologische kroniek van Zuid-Holland over 1978," *Holland: regionaal-historisch tijdschrift* 11 (1979): 313.

[115] Van Es, *De Romeinen in Nederland*, pp. 93–96.

[116] Ibid., pp. 65–69; van Giffen, "Three Roman Frontier Forts in Holland," pp. 31–40; W. Glasbergen et al., *De Romeinse castella te Valkenburg Z.H.: de opgravingen in het dorpsheuvel in 1962*; W. Glasbergen and W. Groenman-van Waateringe, *The Pre-Flavian Garrisons of Valkenburg, Z.H.:Fabriculae and Bipartite Barracks*; J. K. Haalebos, *Zwammerdam-Nigrum Pullum: Ein Auxiliarskastell am niedergermanischen Limes*.

[117] W. Groenman–van Waateringe, "Grain Storage and Supply in the Valkenburg Castella and Pretorium Aggrippinae," in B. L. van Beek, R. W. Brandt, and W. Groenman–van Waateringe, eds., *Ex horreo: I.P.P. 1951–1976*, pp. 226–40.

[118] Van Es, *De Romeinen in Nederland*, pp. 85–86, 177; Bogaers, "Waarnemingen in Westerheem," pp. 173–79.

[119] J. K. Haalebos, "Het einde van de weg," *Westerheem* 25 (1976): 24–29; P. C.

near the mouth of the Oude Rijn, along the dune ridges southward to the Maas estuary. This road was paralleled by another road that connected Roomburg-Leiden (Matilone) with the Maas. The latter road ran along the easternmost dune ridge in association with a canal that linked the Oude Rijn and the Maas; the canal, completed around A.D. 47, was formed by channeling and connecting the Veur to a peat-bog stream in the Maas watershed. Finally, the Maas estuary, known at that time as the Helinium, was connected to the east by a road that ran along the right bank of the Maas and the Waal branch of the Rhine to the insula Batavorum.[120]

Around A.D. 100 the strict military character of the Roman presence began changing somewhat as the number of troops declined and the number of civilians increased. Settlements of service people, merchants, crafts people, and others grew up beside the frontier forts. Meanwhile, excavations at Rijswijk, just south of Rijnland, have indicated that rural settlement too went through considerable development, no doubt by producing surplus people and products for the nearby army. By 200, in fact, there was a highly developed and complex society inside the *limes*, with many types of settlements, from small to large, and considerable population density.[121]

Nor did the no-man's-zone across the *limes* remain unchanged. During the second and third centuries, small native settlements grew up directly across the Oude Rijn from a number of *castella*. At Koudekerk, for example, traces of a Roman-age, native three-aisled house have been unearthed.[122] In addition, about 45 kilometers north of the Oude Rijn, about 10 kilometers northeast of Velsen, was a small settlement in the vicinity of modern Krommenie, occupied by people who combined agriculture with weaving and possibly iron smelting. On the basis of pottery and other imports found in the excavations of this area, it is evident that these people had a brisk trade with the Roman military as early as A.D. 40 to 70.[123]

Beunder, "De Romeinse legerweg tussen Zwammerdam en Bodegraven," *Westerheem* 23 (1974): 216–25.

[120] Van Es, *De Romeinen in Nederland*, pp. 85–90, 96. The Maas estuary, or Helinium, actually was broad enough to resemble a small inland sea during the Roman period. It was the home of a Roman fleet and was ringed by a number of *castella*, some of which were manned by cavalry. See J. E. Bogaers, "Romeinse militaren aan het Helinium," *Westerheem* 23 (1974): 70–78.

[121] Bloemers, "Rijswijk (Z.H.) 'De Bult,' een nederzetting van de Cananefaten," pp. 96–100; Sarfatij, "Archeologische kroniek van Zuid-Holland over 1975," p. 267.

[122] Sarfatij, "Archeologische kroniek van Zuid-Holland over 1978," pp. 315–34.

[123] W. Groenman–van Waateringe et al., "Een boerderij uit de eerste eeuw na Chr. te

About halfway along the road and canal linking Roomburg-Leiden with the Maas River, atop the easternmost ridge of the Older Dunes, an important civilian settlement appeared at the site of present-day Voorburg, just across the southern boundary of Rijnland. This settlement, known as Forum Hadriani and eventually Municipium Aelium (or Aurelium) Cananefatium, was one of the few settlements in the Netherlands to achieve true urban status during the Roman period, while in the southern Netherlands a number of places achieved de facto though not de jure urban status. This city served, among other things, as the administrative center of the civitas, or land, of the Cananefates. The other city, near Nijmegen, in the central river region of the insula Batavorum east of Rijnland, was Municipium Batavodurum, the administrative center of the civitas, or land, of the Batavians. Virtually nothing is known from written sources about these two municipia, though archaeological evidence suggests that some stone structures, presumably public buildings, were built on their sites.[124]

Roman occupation of the western Netherlands ended abruptly toward the end of the third century A.D. During the course of the century the general good times of the Roman period gradually diminished as a consequence of political unrest, money devaluations, cash shortages, and in the western Netherlands a rising water table that began to tax the agrarian sector in particular. These pressures were felt, for example, in the agricultural settlement at Rijswijk, which began shrinking and finally was abandoned.[125] Finally, around 260, the entire *limes* system was overrun by groups attacking from beyond the empire, the Franks and the Frisians and many others. While much of the imperial boundary was reestablished by the early fourth century, the Romans never reoccupied the Rhine delta or much of the central river area on the east. The failure to reestablish a military presence in the western Netherlands may have been related at least in part to the onset of a new transgressive phase of the sea, a phenomenon for which the Romans apparently had no response.[126] In any event, archaeologists have

---

Krommenie (N.H.)," in *In het voetspoor van A. E. van Giffen*, 2d ed., pp. 81–92, 158, 176–77; Brandt, "De archeologie van de Zaanstreek," p. 136.

[124] Van Es, *De Romeinen in Nederland*, pp. 103–28.

[125] Bloemers, "Rijswijk (Z.H.) 'De Bult,' een nederzetting van de Cananefaten," p. 106; G. de Boe, "De Romeinse villa te Haccourt en de landelijke bowoning," *Hermeneus: tijdschrift voor antieke cultuur* 52 (1980): 112–13.

[126] Van Es, *De Romeinen in Nederland*, pp. 96–97; Blok, *De Franken in Nederland*, pp.

found the remains of Roman occupation along the Oude Rijn covered by Dunkirk II deposits. Native population also appears to have dwindled or disappeared by the end of the third century retreating from increased flooding and a rising ground water table.[127]

The almost complete lack of archaeological or written evidence of a human presence in the coastal regions of the western Netherlands between the third and sixth centuries A.D. indicates the degree to which deteriorating hydraulic conditions served as a deterrent to settlement and agriculture, regardless of whether or not this situation was caused by a true transgression of the sea.[128] South of Rijnland flooding was severe enough to breach the coastal barriers and dunes, creating the broken coastline that still exists in the province of Zeeland. Further, a large part of Maasland, especially the so-called Westland portion, on the north side of the Maas mouth, received a substantial layer of marine clay at this time. North of the Maas the barrier and dune system survived intact, but a rising groundwater level produced flooding at the Rijn mouth and contributed substantially to the final silting shut of the IJ mouth in North Holland. In general, soggy conditions prevailed behind the protective dunes, encouraging the continued growth of peat and thereby further aggravating drainage problems. In Rijnland, there is a slight possibility of more or less continuous occupation from the pre-Roman Iron Age only in the neighborhood of Katwijk, near the mouth of the Oude Rijn.[129]

17–19; Sarfatij, "Friezen-Romeinen-Cananefaten," p. 175; Sarfatij, "Middeleeuwse mens en eeuwig water," p. 20; W. J. van Tent, "De landschappelijke actergronden," *Spieghel historiael* 13 (1978): 212; R. G. den Uyl, "Dorpen in het rivierkleigebied," *Bulletin van de Koninklijke Nederlandse Oudheidkundig Bond*, 6th ser., 11 (1958), col. 101; W. Jappe Alberts and H. P. H. Jansen, *Welvaart in wording: sociaal-economische geschiedenis van Nederland van de vroegste tijden tot het einde van de middeleeuwen*, p. 16.

[127] Pannekoek et al., *Geological History of the Netherlands*, pp. 119, 125–26; Sarfatij, "Middeleeuwse mens en eeuwig water," p. 20; Sarfatij, "Friezen-Romeinen-Cananefaten," pp. 175–76; Fockema Andreae, "De Rijnlandse kastelen en landhuizen in hun maatschappelijk verband," p. 1; Jelgersma et al., "The Coastal Dunes of the Western Netherlands," p. 144.

[128] See, for example, Roeleveld, "De bijdrage van de aardwetenschappen," pp. 292–99; T. Edelman, *Bijdrage tot de historische geografie van de Nederlandse kuststreek*, pp. 6–16; Verhulst and Gottschalk, eds., *Transgressies en occupatiegeschiedenis in de kustgebieden van Nederland en België*.

[129] Fockema Andreae, "De Rijnlandse kastelen en landhuizen in hun maatschappelijk verband," p. 1; Pons et al., "Evolution of the Netherlands Coastal Area during the Holocene," p. 206; Pannekoek et al., *Geological History of the Netherlands*, pp. 119, 125–26; Sarfatij, "Middeleeuwse mens en eeuwig water," p. 20; Sarfatij, "Friezen-Romeinen-

Although the almost complete lack of direct evidence of continuing settlement and agriculture suggests that the coastal regions of the western Netherlands were essentially vacated for several centuries, certain varieties of indirect evidence suggest otherwise. A number of place-names that display characteristics of name giving from before the great migrations, and thus before the fourth century, have survived in the coastal regions. In Rijnland water names such as Aar, Vennep, and Wilk and the place-name Alphen are of this type. These old names most likely would have fallen into disuse had there been a total discontinuity of human presence.[130] Thus even though settlements from Roman times were covered with layers of clay, there must have been some people who continued to use these names: perhaps fishermen or hunters, perhaps people who lived in settlements in the dunes that archaeologists have not yet discovered, perhaps simply boatmen who regularly traveled the Oude Rijn.[131]

In the northern Netherlands, however, coastal settlement persisted. The response to deteriorating hydraulic conditions there was the creation of a second generation of *terpen*, or dwelling mounds.[132] When, during the sixth century A.D., drier conditions returned to the western Netherlands, it was, not surprisingly, *terp*-dwelling Frisians from the northern Netherlands, experienced at living in a soggy coastal lowland, who substantially repopulated the dune ridges and riverbanks of Rijnland and adjacent areas.

Throughout the premedieval period, including the Roman era, the human occupation of the western Netherlands continued to be limited by natural conditions. Only the best-drained soils could be settled. Even so, no specific sites saw a continuous presence. Changes in the sediment transport of rivers, in tidal ranges, in precipitation patterns and amounts, and in storm-surge frequency all played a role in altering local drainage patterns. The largest proportion of the western Netherlands, however, remained

Cananefaten," pp. 175–76; J. C. Besteman and A. J. Guiran, "Het middeleeuws-archeologisch onderzoek in Assendelft, een vroege veenontginning in middeleeuwse Kennemerland," *Westerheem* 32 (1983): 148.

[130]Blok, "De vestigingsgeschiedenis van Holland en Utrecht in het licht van de plaatsnamen," pp. 13–15; de Cock, *Bijdrage tot de historische geografie van Kennemerland*, p. 252; de Cock, "Veenontginningen in West-Friesland," p. 156.

[131]Sarfatij, "Friezen-Romeinen-Cananefaten," p. 170, maintains that the archaeological picture for Roman and early medieval times is far from complete. See also Blok, "Probleme der Flussnamenforschung in den alluvialen Gebieten der Niederlande," p. 218.

[132]TeBrake, "Ecology and Economy in Early Medieval Frisia," p. 8.

uninhabited. Until they were reclaimed, beginning in the late tenth century A.D., the peat bogs constituted the wilderness at the edge of civilization, occasionally visited, perhaps, for the purposes of hunting, fishing, or fowling.[133]

[133] See Louwe Kooijmans, *The Rhine/Meuse Delta*, p. 42: ". . . no fundamental changes in the relation men-natural environment occurred before c. A.D. 1000." See further, ibid., p. 118; Sarfatij, "Friezen-Romeinen-Cananefaten," pp. 167, 170. The prehistoric stone ax found near Leimuiden, deep within the peat bog north of the Oude Rijn, doubtless came from the Pleistocene underground or was left by someone traveling along one of the bog streams; see Louwe Kooijmans, *The Rhine/Meuse Delta*, p. 36.

PART TWO
Continuity in Rural Society

4

Early-Medieval Settlement in Rijnland: The Cultural Context

WHEN Rijnland was once again repopulated during the sixth and seventh centuries A.D., natural conditions determined the location of settlements just as they had for the previous three thousand years. Occupation continued to be limited to the best drained sites, essentially to the coastal dune ridges and the raised levees of the Oude Rijn. Even so, it is difficult to find any direct evidence of continuous residence at any specific location before the tenth century. The peat bogs, meanwhile, persisted as wilderness, beyond the realm of normal human affairs. Such settlement patterns were not unique to Rijnland, however. They applied to the entire western Netherlands, from the Belgian border in the south to West Friesland in the northeast.

For the early Middle Ages it becomes possible for the first time to go beyond simply indicating where people lived in the western Netherlands and why and to begin examining coastal settlement in the context of larger social systems. It would be a mistake, however, to think that life in the waterlogged western Netherlands ever occurred in a cultural vacuum. Even in the late Neolithic period coastal residents had contacts and connections to higher-lying inland areas that went beyond pottery-making and tool-using traditions. Still, it is only in the early Middle Ages that the political, social, and economic parameters of life in Rijnland and surrounding areas first become visible. It appears, in fact, that the western Netherlands was deeply involved in the affairs of northwestern Europe from the very moment that repopulation began.

The Location of Early-Medieval Settlement

During the course of the sixth century changes in storm-surge frequency or precipitation patterns or both began to diminish somewhat the soggy character of the western Netherlands. As a result the dune ridges and riverbanks in the coastal regions became generally more suitable for settlement once again. The earliest medieval archaeological evidence for a human presence comes from areas north and south of Rijnland. For example, digging operations for the North Sea Canal near Velsen, Kennemerland, in 1866 uncovered a hoard of coins, presumably buried sometime between A.D. 570 and 575. Other archaeological remains from the same period have been uncovered in northern Kennemerland, among them the traces of settlements north of Alkmaar and an arable field containing pottery fragments under the Younger Dunes near Castricum. Remains of human habitation datable to approximately A.D. 500 have also been detected in Zeeland.[1]

During the course of the seventh century, however, settlement began to appear along the entire coast of the western Netherlands. In Rijnland an important new concentration of population came into existence at the mouth of the Oude Rijn, near Katwijk, Valkenburg, and Rijnsburg. Archaeological investigations have been most fruitful so far within the present municipality of Rijnsburg, near the juncture of the Vliet and Oude Rijn, turning up many signs of human occupation at a number of locations dating from the seventh to the tenth centuries.[2] These finds, all closely associated with the Vliet, include sherds of both Merovingian and Carolingian wheel-thrown pottery, a cemetery, holes made by the supporting posts of large oblong buildings and related smaller buildings, and some plow traces in the soil.

One of the most important discoveries was made at Rijnsburg town

[1] J. Bennema, "De bewoonbaarheid van het Nederlandse kustgebied vóór de bedijkingen," *Westerheem* 5 (1956): 90; S. Jelgersma et al., "The Coastal Dunes of the Western Netherlands: Geology, Vegetational History and Archaeology," *Mededelingen van de Rijks Geologische Dienst*, n.s., 21 (1970): 144; J. Ypey, "De verspreiding van vroeg-middeleeuwse vondsten in Nederland," *Berichten van de Rijksdienst voor het Oudheidkundig Bodemonderzoek* 9 (1959): 98; J. K. de Cock, *Bijdrage tot de historische geografie van Kennemerland op fysisch-geografisch grondslag*, pp. 10, 12; D. van Deelen and A. Schermer, "Middeleeuwse akkerland onder de Castricummer duinen," *Westerheem* 12 (1963): 136–44; H. Sarfatij, "Middeleeuwse mens en eeuwig water: veranderingen in landschap en bewoning aan de monden van Rijn en Maas gedurende de middeleeuwen," *Zuid Holland* 14 (1968): 20; and L. P. Louwe Kooijmans, *The Rhine/Meuse Delta: Four Studies on Its Prehistoric Occupation and Holocene Geology*, p. 45.

[2] See the map in H. Sarfatij, "Die Frühgeschichte von Rijnsburg (8.-12. Jahrhundert): ein historisch-archäologischer Bericht," in B. L. van Beek, R. W. Brandt, and W. Groenman-van Waateringe, eds., *Ex horreo: I.P.P. 1951–1976*, p. 294; as well as the map in D. P. Blok, *De Franken in Nederland*, 3d ed., p. 84.

center, under the clay layer on which the former Benedictine convent was established in 1133. The excavations revealed portions of a settlement situated on a particularly high area of sandy sedimentation in the Oude Rijn estuary. The large tidal gully that ran along the north edge of the settlement can be identified with the Vliet, a stream that originated in the large, raised peat bog north of the Oude Rijn. A side creek of this gully, about 6 meters wide, dissected the complex and was bridged over or dammed in at least two places by small structures resting on pilings driven into the ground. Both the Vliet and the side creek were shored with timber. The early-medieval Rijnsburg settlement was apparently occupied in two distinct phases. The first phase, associated with the Merovingian pottery, was a rather intensive settlement of a largely agrarian character, datable to the seventh and eighth centuries. The second occupation phase, associated with Carolingian pottery from the ninth to tenth centuries, was less intensively inhabited but contained some signs of possible industrial activities, among them iron smelting.[3]

More recently another early-medieval site has been investigated along the silted-up riverbank at Koudekerk, a short distance upstream from Rijnsburg. Along a 15-meter-wide tributary or side creek of the Oude Rijn lay a small settlement that was apparently inhabited in various stages over a number of centuries. The many postholes and ditches point to houses and farmyards. A nearby early-medieval cemetery containing both urns and skeletal inhumations is associated with this settlement. The finds include considerable quantities of pottery that can be broken down into both Merovingian and Carolingian types, suggesting two settlement phases here as well. In addition, there was a high percentage of contemporary imported pottery, suggesting some trade no doubt associated with the Oude Rijn.[4] The Carolingian pottery, particularly the Badorf type, can in fact be com-

[3]W. A. van Es, "Early Medieval Settlements," *Berichten van de Rijksdienst voor het Oudheidkundig Bodemonderzoek* 23 (1973): 281–85; Sarfatij, "Die Frühgeschichte van Rijnsburg," pp. 295–97; H. Sarfatij, "Middeleeuwse mens en eeuwig water," p. 20; W. Glasbergen, "De abdijkerk van Rijnsburg: opgravingen in 1949," *Leids jaarboekje* 47 (1950): 100–102; W. Glasbergen, "Sporen van Rothulfuashem, het vroeg-middeleeuwsche Rijnsburg," *Leids jaarboekje* 36 (1944): 101–109; A. T. Clason, *Animal and Man in Holland's Past: An Investigation of the Animal World surrounding Man in Prehistoric and Early Historical Times in the Provinces of North and South Holland*, pp. 23–24.

[4]H. Sarfatij, "Archeologische kroniek van Zuid-Holland over 1977," *Holland: regionaal-historisch tijdschrift* 10 (1978): 308–309; H. Sarfatij, "Archeologische kroniek van Zuid-Holland over 1978," *Holland: regionaal-historisch tijdschrift* 11 (1979): 331–35.

pared to other pottery finds along the Oude Rijn, ranging from Rijnsburg and Valkenburg through Leiden, Leiderdorp, Koudekerk, and Alphen aan den Rijn and finally all the way to Dorestad, about 50 kilometers east of Rijnland, in the central river area of the Netherlands. All the finds were associated with the raised, natural levees of the Oude Rijn.[5]

That the early-medieval settlement at Rijnsburg, and possibly the one at Koudekerk as well, was actually inhabited in two distinct phases raises once again the problem of the discontinuity of human presence at specific sites in the western coastal regions before reclamation. Unfortunately the archaeological record does not explain why the site was temporarily abandoned. Dutch geologists, soil scientists, and physical geographers have postulated the Dunkirk III-A transgression of the sea, beginning shortly after 800, that would have adversely affected settlement and agriculture.[6] Indeed, marine sediments from this period are known from the coastal area north of Alkmaar as well as from the Oude Rijn estuary. Whatever the cause, flooding could have been serious enough along the Oude Rijn to cause some people to leave, without necessarily disrupting settlement in the nearby dunes.[7]

For the first time documentary evidence seems to support the physical evidence of deteriorating living conditions. On 26 December 838, for example, a great storm surge caused serious marine flooding along the entire Netherlands coast. Further, the frequency of river flooding seems to have increased considerably during the ninth century.[8] The question remains,

[5] A. Wassink, "Ligt de oorsprong van de stad Leiden bij het Romeinse castellum Matilo?" *Westerheem* 27 (1978): 294–98; H. Sarfatij, "Archeologische kroniek van Zuid-Holland over 1975," *Holland: regionaal-historisch tijdschrift* 8 (1976): 267; H. Sarfatij, "Archeologische kroniek van Zuid-Holland over 1976," *Holland: regionaal-historisch tijdschrift* 9 (1977): 257–58; Sarfatij, "Archeologische kroniek van Zuid-Holland over 1977," p. 303; P. C. Beunder, "Waarnemingen langs de Romeinse Rijnoever te Alphen aan den Rijn," *Westerheem* 26 (1977): 275.

[6] See chap. 3. De Cock, *Bijdrage tot de historische geografie van Kennemerland*, pp. 10, 12, identified a number of eighth-century peat settlements in northern Kennemerland and on the island of Texel that disappeared after A.D. 800 because of flooding. See also Besteman and Guiran, "Het middeleeuws-archeologisch onderzoek in Assendelft, een vroege veenontginning in middeleeuws Kennemerland," *Westerheem* 32 (1983): 146.

[7] Sarfatij, "Middeleeuwse mens en eeuwig water," p. 21; de Cock, *Bijdrage tot de historische geografie van Kennemerland*, pp. 10, 12; S. J. Fockema Andreae, "De Rijnlandse kastelen en landhuizen in hun maatschappelijk verband," in S. J. Fockema et al., *Kastelen, ridderhofsteden en buitenplaatsen in Rijnland*, p. 1.

[8] M. K. E. Gottschalk, *Stormvloeden en rivieroverstromingen in Nederland*, 1:18, 27–29, 40–41.

however, whether such episodes constituted an actual transgression of the sea with consistently higher mean high water levels.[9] Presumably a true transgression would consist of more than a single storm surge. Nevertheless, settlement was once again interrupted at Rijnsburg and possibly elsewhere, and to attribute it to a worsened hydraulic situation, though not necessarily a true sea transgression, remains a reasonable hypothesis. Settlements along the Oude Rijn presumably would have been most susceptible to a rising ground-water table, attributable not only to true transgressions but also to changes in currents or tidal ranges or frequency of storm surges.

By the late ninth or early tenth century the possibilities for human settlement in the western coastal regions had improved once again. Areas that a short time earlier had been too wet because of poor drainage began to dry out and were quickly reoccupied.[10] This time, however, the resettlement became permanent. During the late tenth century the residents of the western Netherlands began taking control of hydraulic conditions and managing them to suit their own purposes. Although they were not completely successful in warding off the damage of floods during the late Middle Ages, never again did the western coastal regions become an uninhabited wilderness.

For the period beginning in the seventh century, written evidence can be used for the first time to fill in the rather vague picture of settlement in the western Netherlands drawn by archaeological research. In the last quarter of that century Anglo-Saxon Christian missionaries began the conversion of the Frisians, who controlled the entire North Sea coast from northern Flanders to northwestern Germany. The earliest churches in the western coastal regions were founded on the coastal ridges and riverbanks. Saint Willibrord, the first bishop of Utrecht (690–739), preached there himself and founded five churches on lands given to him by a number of landowners, including Charles Martel, the ruler of Austrasia or East Francia.

[9]For a discussion of this question, see A. E. Verhulst and M. K. E. Gottschalk, eds., *Transgressies en occupatiegeschiedenis in de kustgebieden van Nederland en België*; Gottschalk, *Stormvloeden en rivieroverstromingen in Nederland*, 1:34, 40–41; 2:818–23; T. Edelman, *Bijdrage tot de historische geografie van de Nederlandse kuststreek*, pp. 11–16; P. C. Vos, "De relatie tussen geologische ontwikkeling en de bewoningsgeschiedenis in de Assendelver Polders vanaf 1000 v. Chr," *Westerheem* 32 (1983): 55–57.

[10]De Cock, *Bijdrage tot de historische geografie van Kennemerland*, p. 10.

The churches were in Vlaardingen, in Maasland, south of Rijnland; Velsen, Heilo, and Petten, in Kennemerland, north of Rijnland; and Kerkwerve, an early name for Oegstgeest, adjacent to Rijnsburg.[11] Some centuries later these same churches, which Willibrord turned over to the monastery at Echternach, in modern-day Luxemburg, served as the mother churches of many chapels and churches founded in the reclaimed peat bogs.[12]

Churches like those founded by Willibrord in the early eighth century were, of course, established in settled places, not in the wilderness areas. In fact, the sites chosen for the new churches often had a prior religious or public character.[13] In any event, the establishment of the church at Oegstgeest suggests the preexistence of a settlement there. Early churches also were established by others at Alphen aan den Rijn, Zoeterwoude, Leiderdorp, and Valkenburg, along the Oude Rijn, as well as at Noordwijk and Voorhout, in the dunes.[14]

Further, in a charter granting some property to the monastery at Lorsch, Germany, in 772 or 776, a certain Gericus and his wife mentioned a number of places in Frisia, the coastal regions from Flanders to northwestern Germany. One of these, Elisholz, may have been in what later became Rijnland.[15] The text merely says that Elisholz was somewhere between the Rhine and Maas rivers. The place in question may have been

[11] The churches were mentioned in a charter of Willem I, bishop of Utrecht, 28 December 1063, in A. C. F. Koch, ed., *Oorkondenboek van Holland en Zeeland tot 1299*, 1:84, col. 2: "Flardinge, Kiericwerve, Velsereburg, Heiligelo, Pethem, aliquando a Karolo et orthodoxis patribus aliis beato Willibrordo, patrono nostro ac primae sedis nostrae archiepiscopo."

[12] D. P. Blok, "De Hollandse en Friese kerken van Echternach," *Naamkunde* 6 (1974): 167–84; Blok, *De Franken in Nederland*, pp. 49–55; and S. J. Fockema Andreae, "Middeleeuwsch Oegstgeest," *Tijdschrift voor geschiedenis* 50 (1935): 257.

[13] S. J. Fockema Andreae, "Warmond," *Leids jaarboekje* 41 (1949): 71, suggests this possibility for Warmond; see also Fockema Andreae, "Middeleeuwsch Oegstgeest," p. 259. J. M. Wallace-Hadrill, "Early Medieval History," in his *Early Medieval History*, pp. 3–4, points to further religious continuity between local pagan dieties and saints of the new Christian order.

[14] S. J. Fockema Andreae, "Een verdwenen dorp? Zwieten bij Leiden," in *Varia historica aangeboden aan professor doctor A. W. Byvanck ter gelegenheid van zijn zeventigste verjaardag door de Historische Kring te Leiden*, pp. 121–28; S. J. Fockema Andreae, ed., *Rechtsbronnen der vier hoofdwaterschappen van het vasteland van Zuid-Holland (Rijnland; Delfland; Schieland; Woerden)*, p. ix; Blok, *De Franken in Nederland*, pp. 49–55.

[15] See the reconstructed text in Koch, ed., *Oorkondenboek van Holland en Zeeland tot 1299*, vol. 1, no. 6: ". . . in loco qui dicitur Forismarische et in Engilbrechtes ambehte in Thesla, et in Elisholz inter Renum et Masam, et ad Masamuda, et mancipia VI, stipulatione subnixa."

upstream of Rijnland and Maasland, though not necessarily so. Elisholz is a compound of the Germanic *aliso-* ("alder") and *hulta* ("woods").[16] Alder was a common vegetational element in mesotrophic-peat associations, particularly where dune ridges and peat met. The Dutch word for "alder" (*els*) still exists in such names as Elsgeest and Elsgeesterpolder, along the dunes just north of the Oude Rijn, in Rijnland.[17]

Finally, the Merovingian and Carolingian Franks had extensive royal domains scattered throughout the southern, middle, and western Netherlands. In particular there seems to have been a large concentration near Noordwijk in the Rijnland dunes north of the mouth of the Oude Rijn.[18] In 889, King Arnulf gave some of this royal property to a certain Gerulf, the founder of a dynasty of West Frisian counts, later known as the counts of Holland. The grant consisted of, among other things, forest and arable land near Nordcha and Osprehtashem, between the Oude Rijn and Suuithardeshaga.[19] The identification of Nordcha with Noordwijk and the placing of Osprehtashem nearby offer no linguistic, toponymic, or geographical problems.[20] The location of Suuithardeshaga, however, is more difficult to determine. The name comes from the Germanic words *swintha* ("strong"), *hardu* ("brave"), and *hagō* ("woods"), thus Woods of Swinthahard.[21] Although some earlier scholars had thought it necessary to place Suuithardeshaga somewhere on the north side of Kennemerland,[22] there

[16] Maurits Gysseling, *Toponymisch Woordenboek van België, Nederland, Luxemburg, Noord-Frankrijk en West-Duitsland (vóór 1226)*, 1:310.

[17] S. J. Fockema Andreae, *Poldernamen in Rijnland*, Bijdragen en mededelingen der Naamkunde-Commissie van de Koninklijke Nederlandse Akademie van Wetenschappen te Amsterdam, no. 4 (Amsterdam, 1952), nos. 40–41.

[18] Blok, *De Franken in Nederland*, pp. 74–85.

[19] Koch, ed., *Oorkondenboek van Holland en Zeeland tot 1299*, vol. 1, no. 21: "Dedimus namque ei inter Renum et Suuithardeshaga in comitatu ipsius in locis Nordcha et Osprehtashem nominatis silvam unam et terram arabilem unam . . . et quicquid ad easdem hobas et mansas jure legitimeque pertinere videtur, cum curtilibus et edificiis, mancipiis, campis, agris, pascuis, pratis, silvas, acquis aquarumque decursibus, molinis, piscationibus, viis et inviis, accessibus et regressibus, quesitis et inquisitis, cultis et incultis, et cum universis appendenciis et adiacentiis finibus ad prefatas hobas juste aspicientibus."

[20] D. P. Blok, "De vestigingsgeschiedenis van Holland en Utrecht in het licht van de plaatsnamen," in M. Gysseling and D. P. Blok, *Studies over de oudste plaatsnamen van Holland en Utrecht*, p. 18; de Cock, *Bijdrage tot de historische geografie van Kennemerland*, p. 55.

[21] Gysseling, *Toponymisch woordenboek* 2:949.

[22] See, for example, I. H. Gosses, "Vorming van het graafschap Holland," in his *Verspreide geschriften*, ed. F. Gosses and J. R. Niermeyer, pp. 247–51.

seems to be general agreement at present that it is an unidentified place near Noordwijk, in the vicinity of or in association with the Haarlemmerhout, the so-called Woods of Haarlem, which served as the border zone between Rijnland and Kennemerland.[23]

In 922, King Charles II gave additional royal possessions in the Noorwijk-Haarlemmerhout area to Gerulf's successor, Diederik I. This time they were designated as lying in the vicinity of Suuithardeshaga and extending to Fortrapa and Kinnem.[24] Some historians in the past were inclined to look for Fortrapa far south in Zeeland and for Kinnem far north on Terschelling, in the Frisian Islands.[25] Recent scholarship, however, places them in the dunes north of the Oude Rijn. The *apa* suffix to Fortrapa indicates an old water name of pre-Germanic origins, perhaps associated with the settlement at Voorhout. Kinnem must have been related in some way to Kennemerland.[26]

[23]De Cock, *Bijdrage tot de historische geografie van Kennemerland*, pp. 52–59, thought that *haga* could also designate a thorn hedge on top of a low earthen wall associated with a stream or ditch. For this reason he tried to show that Suuithardeshaga was a hedge plus ditch or stream marking the southern boundary of Haarlemmerhout. See also Jelgersma et al., "The Coastal Dunes of the Western Netherlands," pp. 130–32. This may be similar to the Anglo-Saxon *haga*, which apparently designated a hedge plus ditch designed to prevent livestock and other grazing animals from damaging a protected forest; see Oliver Rackham, *Ancient Woodland: Its History, Vegetation and Uses in England*, pp. 156, 188, 191. While D. P. Blok agrees that Suuithardeshaga lay in the Noordwijk-Haarlemmerhout vicinity, he considers de Cock too arbitrary in trying to identify it with a sharply drawn border; personal communication, 6 March 1974. See also the critique of de Cock by Rob Rentenaar, "De Nederlandse duinen in de middeleeuwse bronnen tot omstreeks 1300," *Geografisch tijdschrift* 11 (1977): 367. It is well to remember that boundaries tended to be only vaguely drawn in this area during the early Middle Ages; de Cock himself, in *Bijdrage tot de historische geografie van Kennemerland*, pp. 22, 52, argued for boundary zones instead of clearly drawn lines in the Haarlemmerhout area.

[24]Koch, ed., *Oorkondenboek van Holland en Zeeland tot 1299*, vol. 1, no. 28: ". . . ecclesiam videlicet Ekmunde cum omnibus ad eam jure pertinentibus a loco qui dicitur Suuithardeshaga usque Fortrapa et Kinnem. Cuius petitionem binigne, uti decebat, suscipientes concedimus prefato fideli nostro hec omnia ex integro cum manicipiis, quesitis et inquirendis, pratis, silvis, pascuis, aquis sive aquarum decursibus, precipientesque jubemus, ut sicut reliquis possessionibus quibus jure hereditario videtur uti, ita et hiis nostri muneris largitate rebus impensis valeat secure omni tempore vite sue frui ipse et omnis eius posteritas."

[25]Gosses, "Vorming van het graafschap Holland," pp. 267–69; O. Oppermann, ed., *Fontes Egmundenses*, intro., p. 105*. On the reliability of Oppermann's introduction, see Rentenaar's advice, "De Nederlandse duinen in de middeleeuwse bronnen," p. 364, to tear it out and throw it away.

[26]De Cock, *Bijdrage tot de historische geografie van Kennemerland*, pp. 52–59; and Gysseling, *Toponymisch woordenboek* 1:370, 562; M. Schönfeld, *Nederlandse Waternamen*, pp. 113–23, 214.

Both royal grants to the West Frisian–Holland counts had to do with the transfer of property that was already within the purview of human activities, not with wilderness. They referred, in short, to settled lands containing woods, pastures, water, and many other appurtenances, and to the peasants or serfs (*mancipiis*) who used them. All the places mentioned were associated with the ridges of the Older Dune landscape.

The summary of the possessions of the Church of Saint Martin in Utrecht, drawn up during the first half of the tenth century, gives a clearer picture of prereclamation settlement in the coastal regions of the western Netherlands.[27] During the entire second half of the ninth century and the first decades of the tenth, the Utrecht bishops sought refuge in the eastern Netherlands from the frequent incursions of the Vikings, who occupied much of the western coastal region during this time. When Bishop Balderik was able to reestablish his see in Utrecht in or about 918, he immediately set out to reassert the church's control along the coast. The list of possessions was an inventory of the property that presumably belonged to the Church of Saint Martin before the Viking presence, as well as a few that may have been granted during the exile. This list apparently was presented to King Otto I, who in a charter of 948 restored these former possessions to the church.[28] In the list were mentioned forty-three place-names that with reasonable certainty can be placed in what later became Rijnland. Once again, all identifiable places without exception were associated with the dune ridges and the riverbanks (see fig. 7).[29]

The preceding investigation of prereclamation settlement does not present a complete picture. Archaeologists have made only a beginning at a systematic reconnaissance of Rijnland. No doubt future research will turn up

[27] See the text in M. Gysseling and A. C. F. Koch, eds., *Diplomata Belgica ante annum millesimum centesimum scripta*, pt. 1, *Teksten*, no. 195.

[28] Koch, ed., *Oorkondenboek van Holland en Zeeland tot 1299*, vol. 1, no. 34. D. P. Blok, "Het goederenregister van de St.-Maartenskerk te Utrecht," *Mededelingen van de Vereniging voor Naamkunde te Leuven en de Commissie voor Naamkunde te Amsterdam* 33 (1957): 89–104, localizes many of the places mentioned, discusses the circumstances under which the list was drawn up, and dates the document to the period between the return of the bishop to Utrecht (918) and the date of the charter of Otto I restoring the possessions to the church (948).

[29] The map is based on app. A, in which I have attempted to assemble those parts of the Saint Martin's list that are relevant to Rijnland. For the locations of places outside Rijnland, see the map in Blok, "Het goederenregister van de St.-Maartenskerk te Utrecht," facing p. 90.

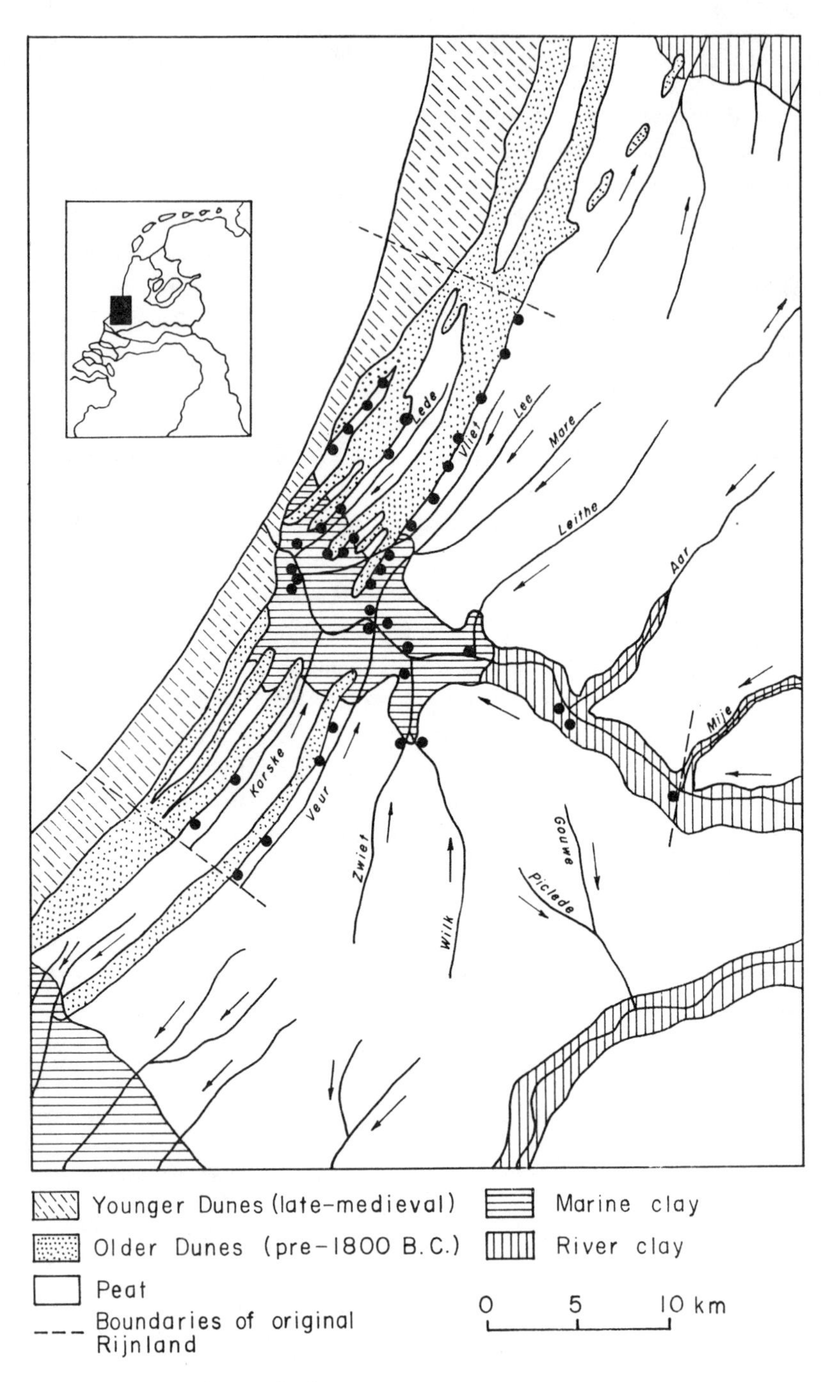

FIGURE 7. Tenth-Century Settlement

important new evidence. Some of the unidentifiable places mentioned in the Saint Martin's list may have been situated in Rijnland as well. No doubt many documents from the period before midcentury have not survived. Almost all of the written sources that still exist have to do with large landowners, such as kings, counts, and religious establishments. One can only assume that there were many more small landowners who left no written records of their possessions.

The available information, however, does allow some tentative conclusions. First, the evidence points to an occupancy that continued to be associated exclusively with the dune ridges and riverbanks. There was a definite concentration of population at the mouth of the Oude Rijn. From there the settled area extended north and south along the dune ridges and east along the banks of the Oude Rijn. Natural conditions determined the boundaries. Dune settlements were bounded on the north and south by forests, on the east by the peat wilderness, and on the west by the North Sea. The settled areas along the Oude Rijn were sandwiched between the peat bogs on the north and south. Only along the river toward the east was there any possibility of continuous settlement, ultimately touching perhaps the nucleus of population at Utrecht.

Second, the settlement of Rijnland in the middle of the tenth century was surprisingly dense. The Saint Martin's list alone mentioned forty-three inhabited places, which together accounted for 128½ manors, a villa, a church, three entire villages and one-third of another, and the inheritances of several individuals. There must have been much more that did not belong to Saint Martin's Church.[30] No doubt these were small settlements, but then the total area made up of the dune ridges and riverbanks was also very small, since most of what became Rijnland still consisted of uninhabited peat bogs.

Frisians and Franks in the Western Netherlands

From a wide variety of evidence it becomes clear that the people who occupied the western coastal regions of the Netherlands just before the era of

[30]For example, the list mentioned 13 *mansa* in Rijnsburg belonging to Saint Martin's Church, but this presumably was only half of the total associated with the villa there; see Sarfatij, "Die Frühgeschichte von Rijnsburg," p. 291. Blok, *De Franken in Nederland*, pp. 50, 79, says that many of the items of property mentioned in the Saint Martin's list were the tenth part of royal possessions previously given to the church. Theoretically, therefore, every manor on the list could have represented ten in that locality.

reclamation were part of a larger group known as the Frisians. They had moved there from their heartland in the northern part of the Netherlands during the unusually dry period associated with the late pre-Roman Iron Age, but deteriorating drainage conditions toward the end of the Roman period had presumably forced them to retreat somewhat from the western coast. They returned, however, with the resumption of drier conditions from the fifth century onward.[31]

The Lex Frisionum, the earliest compilation of Frisian law carried out at the command of Charlemagne at the beginning of the ninth century, recognized three main divisions of Frisian territory. West Frisia, or Inter Fli et Sincfalam, included all of the western Netherlands from the Vlie in the northern Netherlands to as far south as the Sincfal, or Zwin, in northern Flanders.[32] This is not to say that Frisians had settled the entire coast as far south as Flanders; rather, they exercised strong political and economic influence in the area. The southern limit of concentrated Frisian settlement should be placed somewhere in the vicinity of The Hague or the area of the mouth of the Maas river, including, therefore, all of Rijnland. Place-names betraying peculiarly Frisian characteristics have been found as far south as the mouth of the Maas, while this was also the southern limit of the diffusion of the *aasdom*, the typically Frisian manner of administering justice and regulating inheritance.[33] Consequently, the counts of the coastal regions of the western Netherlands were referred to almost exclusively as the counts of Frisia or West Frisia until the beginning of the twelfth century, while their subjects were known as Frisians or West Frisians.[34]

[31] See, for example, Gosses, "De vorming van het graafschap Holland," pp. 239–60; Otto Oppermann, *Die Grafschaft Holland und das Reich bis 1256*, vol. 1 in *Untersuchungen zur nordniederländischen Geschichte des 10. bis 13. Jahrhunderts*, pp. 1–21; P. C. J. Boeles, *Friesland tot de elfde eeuw: zijn vóór- en vroege geschiedenis*, 2d ed., pp. 269–93; D. P. Blok, "Holland und Westfriesland," *Frühmittelalterliche Studien: Jahrbuch des Instituts für Frühmittelalterforschung der Universität Münster* 3 (1969): 347–61.

[32] See, for example, Karl von Richthofen, ed., *Lex Frisionum*, in Monumenta Germaniae Historica, *Legum*, sec. 1, *Legum nationem Germanicarum*, 3:656–82.

[33] Based on the findings of D. P. Blok, especially his "Plaatsnamen in Westfriesland," *Philologia Frisica anno 1966*, no. 319 (1968): 15–18; "De vestigingsgeschiedenis van Holland en Utrecht in het licht van de plaatsnamen," pp. 15, 24; *De Franken in Nederland*, p. 38; "Holland und Westfriesland," pp. 348–50; "Opmerkingen over het assdom," *Tijdschrift voor rechtsgeschiedenis* 31 (1963): 243–74. See also S. J. Fockema Andreae, "Friesland van de vijfde tot de tiende eeuw," in *Algemene geschiedenis der Nederlanden* 1:394–95; S. van Leeuwen, *Costumen, Keuren ende ordinatien van het baljuschap ende lande van Rijnland*, pp. 13, 231.

[34] See some representative designations, such as "comites Fresonum" and "Frisiones qui vocantur occidentales," in Blok, "Holland und Westfriesland," pp. 348–50.

Besides a West Frisia or Frisia Inter Fli et Sincfalam, the ninth-century Lex Frisionum also referred to two other major areas of Frisian settlement or influence. Frisia Inter Laubachi et Wiseron included the entire North Sea coast from the Lauwers Sea, the approximate location of the Friesland-Groningen provincial boundary today, all the way to the Weser River, in northwestern Germany. This territory was often referred to as East Frisia, or Ostfriesland in German.

The third and final portion of Frisia, Inter Laubachi et Flehi, lay between the Lauwers Sea and the Vlie, roughly the modern Netherlands province of Friesland. It constituted the Frisian heartland,[35] while the first two areas were those into which Frisians expanded during the early Middle Ages. It is not really known whether this expansion occurred through the agency of outright conquest, colonization, or simple merger of a number of discrete groups into a sort of confederation. Most likely it was a combination of all three processes. In any event, by the early Middle Ages the word Frisian had become a collective designation which, in much the same manner as did names like Frank or Saxon, reflected the commonality of political, economic, social, and linguistic interests of many groups formerly considered separate.[36]

The precise reasons why the Frisians began moving or extending their political and economic influence southward along the western coast of the Netherlands in the early Middle Ages are not clear. No doubt this movement should be brought into connection with the widespread migrations that took place generally in the fourth to sixth centuries.[37] It should be remembered, for example, that groups known as Franks and others were on the move for some time in northwestern Europe. By around 260 they had already penetrated the Roman frontier along the Rhine east of Rijnland, especially in the area of the Insula Batavorum.[38] The Frisians underwent a process of expansion at about the same time. By 290, for example, they had more or less gained dominance over the other inhabitants of the west-

[35] See von Richthofen, ed., *Lex Frisionum*; Blok, "Holland und Westfriesland," p. 347; H. Halbertsma, "The Frisian Kingdom," *Berichten van de Rijksdienst voor het Oudheidkundig Bodemonderzoek* 15–16 (1965–66): 73–74.

[36] Blok, *De Franken in Nederland*, pp. 11–17; I. H. Gosses, *Handboek tot de staatkundige geschiedenis der Nederlanden*, rev. ed., ed. R. R. Post, pp. 23–24, 28; Lucien Musset, *The Germanic Invasions: The Making of Europe A.D. 400–600*, trans. Edward James and Columba James, pp. 11–12, 69.

[37] Musset, *The Germanic Invasions*, passim.

[38] See Blok, *De Franken in Nederland*, p. 17; R. Wenskus, *Stammesbildung und Verfassung: Das Werden der frühmittelalterlichen Gentes*, pp. 512–41.

ern coastal regions, including such groups as the Cananefates of classical sources.[39]

In one very important respect the Frisian expansion in the early Middle Ages differed from most of the other movements of peoples in this period. Groups like the Franks, for example, literally migrated, that is, they left old places or residence for new ones.[40] The Frisians, in contrast, never abandoned their heartland. Apparently they were the only group beyond the old Roman frontier that survived the early-medieval period of the great migrations in the same locality, that is, the modern province of Friesland.[41] Even though they moved into both West and East Frisia, this central portion, Inter Laubachi et Flehi, always remained their homeland as well as the focus of their densest settlement.

When the Frisians reoccupied the western Netherlands during the sixth century, they did so in considerable strength. They were responsible for the appearance of a number of important new population centers at this time, one of which was established in the area around the mouth of the Oude Rijn. For example, the early settlements visible in the archaeological record at Rijnsburg, Koudekerk, and elsewhere were part of this complex.[42]

In the intervening period many changes had taken place in northwestern Europe. One of the most important of these was the emergence of a vital new cultural realm around the North Sea, with the Frisians constituting one of its most significant components. In short, the new inhabitants of the coastal regions of the western Netherlands were part of a larger interaction sphere that included linguistic, cultural, and economic similarities and linkages not only to the Frisian heartland in the northern Netherlands but also to groups living across the North Sea in England and Scandinavia.

This new North Sea interaction sphere[43] came into existence at least

[39] Gosses, *Handboek tot de staatkundige geschiedenis der Nederlanden*, p. 23; Boeles, *Friesland tot de elfde eeuw*, pp. 71–72.

[40] Blok, *De Franken in Nederland*, p. 15, suggests that they may have been associated with the Chauci of Roman sources, originating from somewhere along the North Sea coast in what had become East Frisia by the early Middle Ages.

[41] J. F. Niermeyer, *De wording van onze volkshuishouding: hoofdlijnen uit de economische geschiedenis der noordelijke Nederlanden in de middeleeuwen*, p. 13.

[42] See the discussion above, this chapter.

[43] William H. TeBrake, "Ecology and Economy in Early Medieval Frisia," *Viator: Medieval and Renaissance Studies* 9 (1978): 12–14. A number of Dutch historians use the term "North Sea Culture" in this connection: Boeles, *Friesland tot de elfde eeuw*, pp. 241–49;

in part because of the widespread movements of peoples in the early Middle Ages and the cultural contacts that they afforded. During the fifth century, for example, Anglo-Saxon groups from coastal areas on the northeast moved through the Frisian heartland in the northern Netherlands. Although there were some hostilities, suggested by ash layers excavated in a number of the Frisian *terpen*, or dwelling mounds, there occurred a general mixing and a partial merging of two already similar groups of people. While many of these Anglo-Saxons and similar groups from elsewhere in northwestern Europe, including some Frisians, took part in the better-known Anglo-Saxon migration to England a little later, many others stayed in Frisia and helped create a revitalized Frisian Culture.[44] One should re-

---

B. H. Slicher van Bath, "Problemen rond de Friese middeleeuwse geschiedenis," in his *Herschreven historie: schetsen en studiën op het gebied der middeleeuwse geschiedenis*, pp. 265–69; A. N. Zadoks-Josephus Jitta, "Looking Back on 'Frisians, Franks and Saxons,'" *Bulletin van de Vereniging tot Bevordering der Kennis van de Antieke Beschaving te 's-Gravenhage* 36 (1961): 41–59. A. Russchen, *New Light on Dark-Age Frisia*, p. 62, talks instead of a North Sea "Koine" and "Commonwealth." This entire question was the subject of a history and archaeology project at the University of Amsterdam from 1972 to 1974, and the findings were published in *De "Noordzeecultuur": een onderzoek naar culturele relaties van de landen rond de Nordzee in the vroege middeleeuwen*. The participants in this project found many linguistic, cultural, and economic similarities and connections in the lands around the North Sea, but they also found differences. In fact, some constituents of this "culture" seemed closer at times to neighboring groups not part of the North Sea world. In short, they concluded that the term "culture" suggests a far greater degree of cohesiveness than actually existed. The findings of this project are summarized in H. H. van Regtern Altena and H. A. Heidinga, "The North Sea Region in the Early Medieval Period (400–950)," in B. L. van Beek, R. W. Brandt, and W. Groenman–van Waateringe, eds., *Ex horreo: I.P.P. 1951–1976*, pp. 47–52. The third edition of Blok, *De Franken in Nederland*, pp. 27–33, the edition I used, was greatly influenced by the findings of the Amsterdam project; compare this with the two earlier editions, in which he leaned heavily in favor of a "North Sea Culture." It seems to me that the term "interaction sphere" might be more appropriate. While it allows contacts, linkages, and perhaps forms of acculturation, it does not require the unity or cohesiveness implied in the term "culture"; see the discussion in William S. Cooter, "Preindustrial Frontiers and Interaction Spheres: Aspects of the Human Ecology of Roman Frontier Regions in Northwest Europe" (Ph.D. diss., University of Oklahoma, 1976), pp. 73–86, 95–108, 144–59.

[44] Boeles, *Friesland tot de elfde eeuw*, pp. 207–58; Gosses, *Handboek tot de staatkundige geschiedenis der Nederlanden*, p. 28; W. Jappe Alberts and H. P. H. Jansen, *Welvaart in wording: sociaal-economische geschiedenis van Nederland van de vroegste tijden tot het einde van de middeleeuwen*, pp. 17, 22; H. Halbertsma, *Terpen tussen Vlie en Eems: een geografisch-historische benadering*, 2:67–69, 71; Dirk Jellema, "Frisian Trade in the Dark Ages," *Speculum* 30 (1955): 16, 24; Musset, *The Germanic Invasions*, pp. 97–100; A. Russchen, "Keramiek en ritueel in de vijfde eeuw," *It Beaken: tydskrift fan de Fryske Akademy* 32 (1970): 129.

member, however, that a body of water like the North Sea, despite its legendary storminess, was more a unifying influence in human affairs than a dividing one in preindustrial times. Therefore, those Anglo-Saxons and others who went on to England could continue to maintain close ties to their places of origin across the North Sea.[45]

The early-medieval North Sea interaction sphere was expressed, first, in a number of linguistic similarities. Languages like Old Frisian, Old English, and to a lesser extent Old Saxon, as well as some of the dialects of Holland, Zeeland, and Flanders, shared many characteristics, and all belonged to the North Sea or Coastal Germanic group of languages. Since some of the similarities date from after the migration period, continued contact and interaction must have been maintained between the British Isles and the Continent.[46]

Further, archaeological research has shown that Frisia, England, and Scandinavia maintained wide-ranging commercial contacts in the early Middle Ages based on extensive waterborne trade, ultimately expanded into southern Europe. The most important connection between the Mediterranean region and the North Sea world was through the Alpine passes and down the Rhine River. When the trading routes between the Baltic and the Mediterranean that passed through central Europe began falling victim to attacks by Avar and Slavic groups around 550, the Rhine connection took on even greater importance, carrying not only much of the North Sea trade but also a good share of the Baltic trade with southern Europe. Lan-

[45]Blok, *De Franken in Nederland*, pp. 27–28; Musset, *The Germanic Invasions*, pp. 97, 108–109, 208; David Wilson, *The Anglo-Saxons*, 2d ed., p. 27; J. M. Wallace-Hadrill, "A Background to St. Boniface's Mission," in his *Early Medieval History*, p. 138. A body of water such as the North Sea offered no real difficulties for communications. Once they were in possession of boats and navigational skills, it was much easier for groups such as the Anglo-Saxons or Frisians to travel by water than by land; see *De "Noordzeecultuur"*, p. 5. Close connections existed between the British Isles and the Continent at least from the Bronze Age; see J. J. Butler, *Bronze Age Connections across the North Sea: A Study in Prehistoric Trade and Industrial Relations between the British Isles, the Netherlands, North Germany and Scandinavia, c. 1700–700 B.C.*

[46]Blok, *De Franken in Nederland*, p. 28; *De "Noordzeecultuur"*, pp. 20–23; van Regteren Altena and Heidinga, "The North Sea Region in the Early Medieval Period," p. 49; Slicher van Bath, "Problemen rond de Friese middeleeuwse geschiedenis," p. 266; Wenskus, *Stammesbildung und Verfassung*, p. 238; A. Russchen, "Jutes and Frisians," *It Beaken: tydskrift fan de Fryske Akademy* 26 (1964): 30; Musset, *The Germanic Invasions*, p. 7. There were additional similarities between Old Frisian and Old Norse that were reinforced by extensive commercial contacts; see D. P. Blok, "De Wikingen in Friesland," *Naamkunde* 10 (1978): 45–47.

gobard disturbances in northern Italy from 568 onward, threatening the security of the Alpine bottlenecks, caused much of this traffic to shift westward to a route along the Maas, Saône, and Rhone rivers to Marseilles for a time. By 630, however, renewed peace in northern Italy and Arab pressures on Marseilles had helped restore the Rhine-Alps link, making possible the emergence of the Frisian settlement at Dorestad, in the central Netherlands river area, as one of the preeminent trading centers of northwestern Europe. From the sixth century Anglo-Saxons, Scandinavians, and especially Frisians had dominated the northern end of this trade, especially along the lower Rhine and through the Baltic and North seas.[47]

Finally, an interaction sphere incorporating the lands around the North Sea is indicated by a significant number of cultural similarities that appeared in the early Middle Ages. Archaeological research in England,

[47]Blok, *De Franken in Nederland*, pp. 30–33; A. C. F. Koch, "Phasen in der Entstehung von Kaufmannsniederlassungen zwischen Maas und Nordsee in der Karolingerzeit," in G. Droege et al., eds., *Landschaft und Geschichte: Festschrift für Franz Petri zu seinem 65. Geburtstag am 22. Februari 1968*, pp. 314–18; Slicher van Bath, "Problemen rond de Friese middeleeuwse geschiedenis," pp. 260, 269–70; Zadoks-Josephus Jitta, "Looking Back on 'Frisians, Franks and Saxons,'" 52; Richard Hodges, "Trade and Urban Origins in Dark Age England: An Archaeological Critique of the Evidence," *Berichten van de Rijksdienst voor het Oudheidkundig Bodemonderzoek* 27 (1977): 197–206; Edith Ennen, *The Medieval Town*, trans. Natalie Fryde, p. 37; H. Jankuhn, "Der fränkisch-friesische Handel zur Ostsee im frühen Mittelalter," *Vierteljahrschrift für Sozial- und Wirtschaftsgeschichte* 40 (1953): 202–209; *De "Noordzeecultuur"*, p. 138; Archibald R. Lewis, *The Northern Seas: Shipping and Commerce in Northern Europe, A.D. 300-1100*, pp. 110–78; J. Ypey, "Frankisch goud in Beuningen (Gld.)," in W. A. van Es et al., eds., *Archeologie en historie: opgedragen aan H. Brunsting bij zijn zeventigste verjaardag*, p. 45; Kazimiers Slaski, "North-Western Slavs in Baltic Sea Trade from the VIIIth to the XIIIth Century," *Journal of European Economic History* 8 (1979): 84–85; Jellema, "Frisian Trade in the Dark Ages," pp. 16–24; Georges Duby, *The Early Growth of the European Economy: Warriors and Peasants from the Seventh to the Twelfth Century*, trans. Howard B. Clarke, pp. 102–107; Roger Grand and Raymond Delatouche, *L'Agriculture au moyen âge de la fin de l'Empire Romain au XVIe siècle*, p. 11. Actually the trade routes by way of the Alpine passes and the Rhine were not new in the early Middle Ages. They, as well as others farther east, were already in use in the early second millennium B.C.; see Stuart Piggott, *Ancient Europe from the Beginnings of Agriculture to Classical Antiquity*, p. 120. For Dorestad, see D. P. Blok and A. C. F. Koch, "De naam Wijk bij Duurstede in verband met de ligging der stad," *Mededelingen van de Vereniging voor Naamkunde te Leuven en de Commissie voor Naamkunde te Amsterdam* 40 (1964): 38–51, 189; W. A. van Es, "Excavations at Dorestad: A Pre-Preliminary Report, 1967–1968," *Berichten van de Rijksdienst voor het Oudheidkundig Bodemonderzoek* 19 (1969): 183–207; W. A. van Es, "Die neuen Dorestad-Grabungen 1967–1972," in *Vor- und Frühformen der europäischen Stadt im Mittelalter: Bericht über ein Symposium in Reinhausen bei Göttingen vom 18. bis 24. April 1972* 1:202–17; W. A. van Es et al., *Excavations at Dorestad 1—The Harbour: Hoogstraat I*.

Frisia, and Scandinavia has indicated a broad sharing of forms and decorative motifs in the pottery and jewelry that have been excavated.[48] Close cultural contacts among the North Sea lands were also indicated by the folklore tradition of the saga. In *Beowulf*, for example, a Frisian king named Finn is mentioned alongside other figures such as Beowulf and Hygelac as leaders of groups of Angles, Saxons, Danes, Jutes, Frisians, and others who were fighting their endless battles.[49]

Perhaps it was through religion, however, that the North Sea interaction sphere had its greatest cultural impact on the affairs of northwestern Europe. It remained a vigorous stronghold of paganism on the northern flank of Christian Europe and near the end of the sixth century was able to deliver a number of serious blows at the Frankish church. For example, the Bishopric of Tournai (Doornik) was forced to retreat southward to Noyen in 577, and about the same time the Bishopric of Arras (Atrecht) was combined with that of Cambrai (Kamerijk). A little earlier the bishop of Tongeren (Tongres) had had to seek safety in the *castellum* at Maastricht; around 590 this bishopric appears to have been combined temporarily with the Archbishopric of Cologne. These developments are best explained in terms of the continuing non-Christian character of the North Sea interaction sphere that was capable of stimulating a pagan resurgence deep in the Merovingian kingdom for a while. The report that Chilperic, king of the West Franks (561–84), was one of the few Merovingian rulers able to turn back a raid of Frisians and Suevi, suggests that the pagan advance in northern Francia was accompanied at least in part by force.[50]

The North Sea interaction sphere always remained vague and tenuous. Even at its height it was never more than a set of similarities, a network for communication and exchange, and perhaps some feelings of solidarity in the face of a new and aggressively expanding Christian culture. Some of these characteristics, however, soon began to disappear, ultimately leading to the demise of the entire interaction sphere. The first step

[48] Slicher van Bath, "Problemen rond de Friese middeleeuwse geschiedenis," p. 265; Jellema, "Frisian Trade in the Dark Ages," pp. 17–18; Russchen, "Jutes and Frisians," p. 32; *De "Noordzeecultuur"*, pp. 38–122, 142–43.

[49] H. Halbertsma, "De cultuur van het noordelijk kustgebied," in J. E. Bogaers et al., eds., *Honderd eeuwen Nederland*, p. 196; Blok, *De Franken in Nederland*, pp. 121–22. For an account of a Frisian who sang old songs glorifying the exploits of the Frisian kings and people, see D. A. Stracke, "Bernlef," *Historisch tijdschrift* 4 (1925): 59–70, 150–69.

[50] Blok, *De Franken in Nederland*, pp. 10, 29–30; Jellema, "Frisian Trade in the Dark Ages," p. 16; Musset, *The Germanic Invasions*, p. 13; *De "Noordzeecultuur"*, pp. 31–37.

of this process occurred with the Anglo-Saxon conversion to Christianity in the seventh century; the second, with the conquest and forced conversion of Frisia by Frankish soldiers and Anglo-Saxon missionaries in the eighth century.[51] Nevertheless, this second step, of particular importance to Rijnland, was completed only after a drawn out and at times bitter struggle that lasted about 150 years.

Frisian control of the coastal regions of the western Netherlands was challenged during the seventh century by a powerful northward expansion of the Franks.[52] The Austrasian or, East Frankish, nobility, in particular, was involved in this movement. Beginning in the late sixth century, a number of powerful clans gradually spread their influence into the middle Maas Valley and surrounding areas. One of these was the Pepin clan, the later Carolingians. During the reign of Dagobert I, the last significant Merovingian ruler of the Franks, the northward expansion down the Maas toward the Rhine-Maas delta also became royal policy. By Dagobert's death in 639, Franks had moved as far downstream as Utrecht, but the Frisians under their kings began stiffening their resistence and effectively prevented further Frankish penetration northward and westward for nearly a century. It was at this very time, in fact, that the Frisians were able to expand their territory to its maximum extent, creating a "Greater Frisia" that stretched along the North Sea from the Zwin, in Flanders, to the Weser, in Germany. They also regained Utrecht and moved up the Rhine at least as far as Dorestad, for the Frisian King Aldgisl ruled over this international commercial center in 678.[53]

Frankish royal power all but disappeared after Dagobert as groups of aristocrats took control. In Austrasia the Pepin clan emerged as the leader of the aristocratic party and, as Mayors of the Palace, became kings in

[51] Musset, *The Germanic Invasions*, p. 208, suggests that the Anglo-Saxon missionary effort to the Continent was evidence of a continued feeling of solidarity with old confederates across the North Sea.

[52] One of the best treatments of the Franks, from their first appearance until the collapse of the Carolingian empire, is Blok, *De Franken in Nederland*.

[53] F. L. Ganshof, "Het tijdperk van de Merowingen," in *Algemene geschiedenis der Nederlanden*, 1:258; Jappe Alberts and Jansen, *Welvaart in wording*, p. 23; Blok, *De Franken in Nederland*, pp. 22, 35–38; Boeles, *Friesland tot de elfde eeuw*, pp. 269–73; Gosses, *Handboek tot de staatkundige geschiedenis der Nederlanden*, pp. 29–31; J. F. Niermeyer, "Het midden-Nederlands rivierengebied in de Frankische tijd op grond van de *Ewa quae se ad Amorem habet*," *Tijdschrift voor geschiedenis* 66 (1953): 154–55.

deed if not in name. By 687, Pepin II had defeated the Mayor of the Palace in Neustria or West Francia and in the name of the Merovingian kings had begun to rule over the entire Frankish kingdom. Almost immediately he turned his attention northward toward the delta, toward what appeared to be the natural extension of the middle Maas River region, where so many of the possessions of his clan and of his Austrasian aristocratic followers were to be found.[54]

Shortly before 690, under Aldgisl's successor, Redbad, the Frisians were forced to surrender Dorestad to the Franks under Pepin II after a significant battle in the central Netherlands. By 696, Pepin was in possession of the fort at Utrecht and was minting Frankish coins at Dorestad, though the area north and west of Utrecht, including Rijnland, remained firmly in the hands of the Frisians. At approximately the same time, as mentioned earlier, the Anglo-Saxon missionary Saint Willibrord made his first attempts to convert the pagan Frisians to Christianity. This conversion effort was tied closely to Frankish dynastic interests. Those who would become Christian were promised the special protection of Pepin II. As a result, however, the notions "Christian" and "Frank" took on virtually identical meanings for most people living in the western and northern coastal regions of the Netherlands. Christianity, therefore, made little progress among the Frisians outside those areas directly under Frankish control.[55]

The death of Pepin II in 714, soon after the murder of his only legitimate son, inaugurated a struggle for control of the Frankish kingdom. Redbad took advantage of the chaotic situation by reconquering most of the former territory of "Greater Frisia." In 716 he even sailed his fleet up the Rhine and threatened the city of Cologne. By 717, however, the struggle over succession in the Frankish kingdom had ended, with Charles Martel, Pepin II's illegitimate son, at the head of a new, revitalized expansion movement. The Frisians surrendered Utrecht in 718, marking the beginning of the final and ultimately successful Frankish assault. The death of the Frisian Redbad the following year made possible the conquest of Frisia

[54] Blok, *De Franken in Nederland*, pp. 38–42.

[55] Ganshof, "Het tijdperk van de Merowingen," pp. 26–61; Blok, *De Franken in Nederland*, pp. 40–45; Gosses, *Handboek tot de staatkundige geschiedenis der Nederlanden*, pp. 31–32; Boeles, *Friesland tot de elfde eeuw*, pp. 270–78; Niermeyer, "Het midden-Nederlands rivierengebied in de Frankische tijd," p. 155; Halbertsma, "The Frisian Kingdom," p. 71; Wilhelm Levison, *England and the Continent in the Eighth Century*, especially pp. 45–69 for the early Frisian mission.

around the mouth of the Oude Rijn as well as the coastal areas on the north. By 734, Charles Martel held all of the western and part of the northern Netherlands. Once again Anglo-Saxon missionaries took up the task of converting the Frisians to Christianity, led this time by both Saint Willibrord and Saint Boniface.[56]

The task of transforming Frisians into Christians and into Frankish subjects was not easy. Both the religious and the secular branches of the new order worked through aristocracies, either by rewarding Frisian nobles who joined the Frankish cause and became Christians or by introducing Frankish aristocrats and promoting them. The lands and possessions of those who had persisted in their opposition were confiscated and reallotted to the trustworthy ruling class. As a result both the religious and the political attempts to control Frisian territory became involved in long-drawn-out social struggles. The fact that the Franks could be brutal and cause widespread destruction during their conquests, as in Charles Martel's victory along the Boorn, in the northern Netherlands, in 734, actually acted as a form of reverse propaganda, stiffening the resolve of many opposed to becoming Chistian and Frank. It was a long time before resistance was broken. For example, Saint Boniface was killed at Dokum in 754 in territory that had been under Frankish control for twenty years. Even a century later many Frisians in the western and northern Netherlands voluntarily joined the pagan Vikings in their raids on Frankish territory.[57]

Still, in time the new order prevailed. The rest of the northern Netherlands came under Frankish control through the efforts of Charlemagne, who, by defeating the rest of the Frisians as well as the Saxons toward the end of the eighth century, pushed the frontier of his empire to the Elbe and eventually the Eider River, in northern Germany. In this fashion the coastal regions of the western Netherlands, including Rijnland, ceased to be a disputed border territory and slowly became integrated into the larger sphere of western Europe. This integration came about in two ways. First, by gradually accepting conversion, the inhabitants of the western coastal regions of the Netherlands came into contact with the ideas and perspectives

[56]Blok, *De Franken in Nederland*, pp. 49–58; Ganshof, "Het tijdperk van de Merowingen," p. 261; Gosses, *Handboek tot de staatkundige geschiedenis der Nederlanden*, pp. 31–32; Boeles, *Friesland tot de elfde eeuw*, pp. 71–75.

[57]Blok, *De Franken in Nederland*, pp. 47, 67–69; Fockema Andreae, "Friesland van de vijfde tot de tiende eeuw," p. 391; M. P. van Buijtenen, *Langs de heiligenweg: perspectief van enige vroeg-middeleeuwse verbindingen met Noord-Nederland*, p. 23.

transmitted by the international culture of Christianity. Second, the long arm of imperial power came to be felt in the conquered lands through the many royal possessions acquired by conquest, including a large complex covering almost the entire northern half of the Rijnland dune region with Noordwijk as its center; through the wilderness *regalia*, or royal rights to the disposal of unexploited land and water; and through legal and political administration carried out by the counts and other appointees of the royal court. Unfortunately there is no information that refers directly to Rijnland concerning these matters. One can assume only that the situation there was similar to that in some surrounding areas.[58]

The last part of the reign of Charlemagne was a period of relative peace and quiet for Rijnland as well as for the entire coastal region of the western Netherlands. With the conquest over, the local population apparently began to grow and settlement to expand, as occurred generally in Europe at this time. It seems to have been a particularly prosperous time for Frisian merchants, who, building on a long tradition of commercial activity that went back well into prehistoric times, were able to continue trade under Frankish rule virtually without interruption.[59]

Involvement in commerce by coastal inhabitants is known from the Bronze Age, when the western Netherlands, by virtue of being part of the delta system of the Rhine and Maas rivers, suddenly found itself astride some of the most important routes of the bronze trade.[60] Since knowledge

[58] See Blok, *De Franken in Nederland*, pp. 58–137; F. L. Ganshof, "Het tijdperk van de Karolingen tot de grote Noormanneninval, 751–879," in *Algemene geschiedenis der Nederlanden*, 1:306–66; Ganshof, "Het tijdperk van de Merowingen," pp. 269–70, 272–74; Boeles, *Friesland tot de elfde eeuw*, pp. 382–86, 405–15; Gosses, *Handboek tot de staatkundige geschiedenis der Nederlanden*, pp. 33–42; H. van der Linden, *De Cope: bijdrage tot de rechtsgeschiedenis van de openlegging der Hollands-Utrechtse laagvlakte*, p. 81; de Cock, *Bijdrage tot de historische geografie van Kennemerland*, p. 260.

[59] Blok, *De Franken in Nederland*, pp. 106–12; Boeles, *Friesland tot de elfde eeuw*, pp. 359–81, 394–405; Jappe Alberts and Jansen, *Welvaart in wording*, pp. 20–38; Jellema, "Frisian Trade in the Dark Ages," pp. 17–36; and TeBrake, "Ecology and Economy in Early Medieval Frisia," pp. 23–26. For the population question, see Josiah C. Russell, *Late Ancient and Medieval Population*, p. 148; and J. C. Russell, "Population in Europe, 500-1500," in Carlo M. Cipolla, ed., *The Middle Ages*, p. 36, vol. 1 in *The Fontana Economic History of Europe*. For population in Flanders, see, for example, A. Verhulst, "Historische geografie van de Vlaamse kustvlakte tot omstreeks 1200," *Bijdragen voor de geschiedenis der Nederlanden* 14 (1959–60): 9.

[60] See, for example, J. A. Brongers et al., "Prehistory in the Netherlands: An

of boats and boating was always necessary for any movement through this wet environment, it is only natural that those who lived there and knew the lay of the land and water should take part in at least local portions or branches of these major trade routes. That such traditions should continue right into the Middle Ages is not surprising. Indeed, all evidence suggests that early-medieval Frisians were extremely competent sailors.[61] Their involvement in commerce had begun to move far beyond the local scale, however. Frisians had learned to make the most of their location and physical geography and had become the preeminent merchants of northwestern Europe during the early Middle Ages.[62]

There was more to the development of Frisian trade than simply natural conditions, however. Historical and cultural affairs played a role as well, particularly the relations between the Frisian districts and the new heartland of western Europe that was beginning to evolve under Christian and Frankish tutelage in the region between the Loire and the Rhine rivers. Trade in the Frankish world tended for the most part to follow the old Roman model, organized around networks of roads. Early-medieval roads, however, were so notoriously inadequate for year-round wagon traffic that overland commerce could never compete very well with waterborne trade.[63] When the Frisians completed their expansion into and through the western Netherlands in the early Middle Ages, they had come into control of the mouths of the Schelde, Maas, and Rhine rivers. They had easy access on these waterways to the very heart of Frankish Europe and could dominate the trade between that region and those areas accessible by way of the North Sea: the British Isles, Scandinavia, and the Baltic region.

Frisians first traded for commodities unavailable or scarce in their own districts. Chief among these were cereals and other plant foods, since the possibility of raising crops was severely limited by poor drainage in

---

Economic-Technological Approach," *Berichten van de Rijksdienst voor het Oudheidkundig Bodemonderzoek* 23 (1973): 25–33; Butler, *Bronze Age Connections across the North Sea*.

[61] Niermeyer, *De wording van onze volkshuishouding*, pp. 13, 25; Slicher van Bath, "Problemen rond de Friese middeleeuwse geschiedenis," pp. 260, 279.

[62] TeBrake, "Ecology and Economy in Early Medieval Frisia," pp. 24–25.

[63] Slicher van Bath, "Problemen rond de Friese middeleeuwsche geschiedenis," p. 269; B. H. Slicher van Bath, *The Agrarian History of Western Europe, 500-1850*, trans. Olive Ordish, p. 33; Robert Latouche, *The Birth of Western Economy: Economic Aspects of the Dark Ages*, trans. E. M. Wilkinson, pp. 132–35; Robert S. Lopez, "The Evolution of Land Transport in the Middle Ages," *Past and Present*, no. 9 (1965): 17–29; Albert C. Leighton, *Transport and Communication in Early Medieval Europe, A.D. 500–1100*.

most of the coastal regions of the Netherlands. Wood too would have been a much-sought-after item, particularly in the treeless salt marsh of the Frisian heartland, in the northern Netherlands. Even in the western coastal regions, however, tree growth was restricted essentially to the dune ridges and river levees, where it had to compete with crop raising as well. In return, Frisians could offer the surplus production of their livestock-keeping activities. This included not only the livestock themselves but also many items derived from them, such as hides, parchment, leather, wool, woolen products, and articles made from bone and horn. They also provided locally manufactured salt for shipment elsewhere. Finally, Frisian merchants acted as middlemen between most of the other groups of northwestern Europe. As such they traded in amber and pelts from Scandinavia; cloth, tin, and lead from England; cloth from Flanders; glasswork and armaments from northern France; glasswork, armaments, wine, pottery, and millstones from the German Rhineland; and slaves from wherever they could find them.[64]

Frisian trade, as well as occasional acts of piracy,[65] made the coastal districts of the Netherlands rather prosperous by early-medieval standards. This is shown by the fact that excavations of Frisian *terpen* in the northern coastal regions have uncovered substantial numbers of luxury goods from widely scattered places, as well as large hoards of coins. The discovery of the coins introduces an additional important distinction. Unlike many of their contemporaries in western Europe, Frisians never gave up the use of coins. Their economy seems to have remained monetized throughout the early Middle Ages.[66] Ironically, the Frankish conquest of Frisia during the eighth century, far from bringing an end to Frisian prosperity, apparently enhanced it further by providing expanded trading contacts with the European interior under imperial protection. In any case, Frisian commerce seems to have achieved its greatest extent between 750 and 850, when permanent Frisian trading colonies were established in Frankish lands, England, and Scandinavia.[67]

[64] TeBrake, "Ecology and Economy in Early Medieval Frisia," pp. 24–25, especially nn. 84–86.

[65] Ibid., p. 25 n. 87.

[66] Ibid., pp. 25–26.

[67] Jellema, "Frisian Trade in the Dark Ages," pp. 25–34; Koch, "Phasen in der Entstehung von Kaufmannsniederlassungen zwischen Maas und Nordsee in der Karolingerzeit," pp. 316–17.

One of the most interesting characteristics of Frisian trade in the early Middle Ages was the degree to which it was decentralized. It seems never to have been limited to certain places, such as towns, nor was it dominated by a few individuals. Instead, Frisian merchants came from innumerable little villages, where they might have been not only keepers of livestock but also part-time artisans. In fact, this practice persisted in the Frisian districts throughout the remainder of the Middle Ages and into early modern times. In some respects it seems to have resembled the custom revealed in the Norse sagas, in which young men might travel far and wide for a number of years and only later confine themselves to their farmsteads.[68]

While Frisian trade was decentralized in the sense that its practitioners came from widely scattered places, there were nevertheless a number of settlements that played particularly important roles, serving as conduits through which much of the trade passed. The best known of these was Dorestad, situated at the point where the Kromme Rijn–Oude Rijn split off from the more southerly branch of the Rhine known as the Lek. At its greatest extent in the eighth and ninth centuries it consisted of a line nearly two kilometers long of wharfs and warehouses on the waterfront of the Kromme Rijn, according to archaeological excavations that are still in progress.[69]

A place like Dorestad derived much of its importance from its apparent designation by the Carolingian rulers as an *emporium*, that is, as a sort of official port of entry into the Carolingian empire. Apparently a second *emporium* was situated near the mouth of the Maas, at a place called Witla. Outside the Frisian districts *emporia* were also established at Quentovic, somewhere on the Canche River south of Boulogne, and possibly at Rouen, on the Seine, as well. At Dorestad, as at the other *emporia*, the Carolingian rulers attempted to control and tap into the wealth derived from trade through the levying of tolls and duties and the minting of coins.[70] Besides these *emporia* there were many other important Frisian trading centers, such as at Emden, Stavoren, and Medemblik on the northeast, Deventer,

[68] Slicher van Bath, "Problemen rond de Friese middeleeuwsche geschiedenis," p. 270.

[69] There have been many preliminary reports on these excavations. See, for example, the special number of *Spieghel historiael* 13 (1978), no. 4, devoted entirely to Dorestad. The first of the final excavation reports appeared only recently: van Es et al., *Excavations at Dorestad 1—The Harbour: Hoogstraat I*, 2 vols.

[70] J. B. Akkerman, "De vroeg-middeleeuwsche emporia," *Tijdschrift voor rechtsgeschiedenis* 35 (1967): 230–83; Koch, "Phasen in der Entstehung von Kaufmannsniederlassungen zwischen Maas und Nordsee in der Karolingerzeit," pp. 313–14.

Muiden, Utrecht, and Tiel on the east, and Vlaardingen and Walcheren south of Rijnland.[71]

In Rijnland itself, Rijnsburg, on the Oude Rijn estuary, offers some indications of having had a commercial character in the early Middle Ages. Archaeological investigations have uncovered some buildings from the Carolingian phase of the settlement that did not exhibit strictly agrarian features. There is evidence of iron smelting at least and possibly of commerce as well, given the substantial quantities of imported pottery found in the settlement.[72] Further, the tenth-century compilation of the possession of Saint Martin's Church at Utrecht spoke of *sidilia* or *ofstedi* in Rijnsburg.[73] While these words may have indicated small, enclosed farmsteads, the possibility exists that in such places as Rijnsburg, closely associated with a strategic waterway, they referred instead to the living and warehousing facilities of merchants.[74]

Less than 15 kilometers upstream from Rijnsburg, at Koudekerk, further evidence has been uncovered of nonagrarian activities by Rijnlanders in the early Middle Ages. In particular it is the discovery of high percentages of imported pottery that points to commercial activity. While at first glance it may seem highly unlikely to find commercial activity at such small, largely agrarian settlements, it is well to remember that they lay along one of the major branches of the Rhine system, providing access to the more important centers at Utrecht and especially Dorestad. Even a site such as the one found at Koudekerk, sandwiched between empty bogs, could provide a welcome stopover along a lonely stretch of river.[75] The evidence shows, in short, that Rijnland too was touched by the commercial activity characteristic of the larger social system of Frisia in the early Middle Ages.

[71] W. Haarnagel, "Die frühgeschichtliche Handels-Siedlung Emden und ihre Entwicklung bis ins Mittelalter," *Friesisches Jahrbuch* (1955): 9–78; Koch, "Phasen in der Entstehung von Kaufmannsniederlassungen zwischen Maas und Nordsee in der Karolingerzeit," pp. 312–22; J. C. Besteman, "Carolingian Medemblik," *Berichten van de Rijksdienst voor het Oudheidkundig Bodemonderzoek* 24 (1974): 48; G. J. Borger, *De Veenhoop: een historisch-geografisch onderzoek naar het verdwijnen van het veendek in een deel van West-Friesland*, p. 217.

[72] Sarfatij, "Die Frühgeschichte von Rijnsburg," pp. 297, 301 n. 30; van Es, "Early Medieval Settlements," pp. 281–85; Besteman, "Carolingian Medemblik," p. 48.

[73] See above and app. A.

[74] Koch, "Phasen in der Entstehung von Kaufmannsniederlassungen zwischen Maas und Nordsee in der Karolingerzeit," pp. 319–20; Sarfatij, "Die Frühgeschichte von Rijnsburg," pp. 298, 301 n. 30.

[75] Sarfatij, "Archeologische kroniek van Zuid-Holland over 1978," pp. 331–35.

The Emergence of West Frisia–Holland

The time of peace and prosperity that the western Netherlands enjoyed after the Frankish conquest did not last very long. Before the end of Charlemagne's reign bands of people emanating from Scandinavia began to offer a new challenge to the composure and well-being of northwestern Europe. Groups known variously as Norsemen, Northmen, Danes, and Vikings, long a part of the North Sea world, apparently began behaving in an increasingly aggressive manner toward others, particularly those living along the North Sea coasts of present-day England, France, and the Low Countries. Frisian lands, including the western Netherlands, were no exception.[76]

For much of the ninth century these areas saw the presence of Scandinavian groups, particularly Danes, who ruled portions of them in varying degrees. Some of the newcomers were eventually recognized as vassals by the successors to Charlemagne, but in reality their presence brought an end to Carolingian control of Frisia. When for various reasons Viking rule began to erode in the last quarter of the ninth century, it was replaced in Frisian lands and elsewhere by new political entities that were largely autonomous and only nominally subject to royal power. The western Netherlands saw the emergence of West Frisia, ruled over by a new family of counts that began to assemble a substantial power base of its own, known since the beginning of the twelfth century as Holland.[77]

The exact reasons why various groups of Scandinavians began applying pressure to portions of England and the Frankish empire in the late eighth and early ninth centuries are unknown. Some may have left Scandinavia in search of land, because population growth or climatic deterioration had created population pressure in areas of limited agricultural potential, or because political upheaval had produced significant numbers of political exiles. Others simply used force to obtain wealth through plunder, robbery, or piracy, as a substitute for or supplement to other forms of eco-

[76] Boeles, *Friesland tot de elfde eeuw*, pp. 388–89; Fockema Andreae, "Friesland van de vijfde tot de tiende eeuw," p. 403; Blok, "Holland und Westfriesland," p. 348; Blok, *De Franken in Nederland*, p. 87; F. L. Ganshof, "Het laat-Karolingische tijdperk: het ontstaan van het graafschap Vlaanderen en van de Lotharingse gravenhuizen," in *Algemene geschiedenis der Nederlanden*, 1:367–75, 378, 383; I. H. Gosses, "Deensche heerschappijen in Friesland gedurende den Noormannentijd," in his *Verspreide geschriften*, ed. F. Gosses and J. F. Niermeyer, pp. 130–51; Gosses, *Handboek tot de staatkundige geschiedenis der Nederlanden*, pp. 42–45; Halbertsma, "The Frisian Kingdom," pp. 78–106.

[77] Blok, "De Wikingen in Friesland," pp. 29–32.

nomic activity such as trade. Still other Scandinavians became more active in the North Sea world and elsewhere as part of a search for adventure, inspired perhaps by adherence to a cult of heroism. Scholars often disagree about which of these possible explanations is most important, though all were no doubt operable.[78]

Virtually all of the Scandinavian activity in the Low Countries was associated with the mouths and lower courses of the Schelde, Maas, and Rhine rivers. As early as the reign of Charlemagne's successor, Louis the Pious, two Danish brothers, Harald and Rorik, acquired as a fief the Frisian commercial center of Dorestad, along the Rhine upstream from Rijnland. In 841, Harald also received Walcheren and vicinity in Zeeland, south of Rijnland. A short time later Harald died, and, because Rorik was often absent and neglected to carry out certain obligations, the original Dorestad fief was declared vacant. By 850, however, Rorik had returned from Denmark, and again he came into possession of Dorestad and the surrounding districts ("Dorestadum et alios comitatus") with the responsibility of collecting the taxes due the royal treasury and of keeping other Danes out of the area. His holdings, known as the Regnum Fresonum (Frisian Kingdom), included a number of areas along the lower Rhine, as well as portions of the western coastal districts; Betuwe, east of Rijnland, and Kennemerland, on the north, were specifically mentioned.[79]

Rorik appeared only intermittently in the sources after 850 because of his frequent absences while attempting to seize the Danish throne. In 862, after an absence of five years, he returned to Frisia and was converted to Christianity, though he was gone again in 867, at the time of a revolt of Frisians known as Cokingi. In 870 he was in Frisia again, more powerful than ever. In battles between Charles the Bald and Louis the German over the inheritance of Lothar II, Rorik first sided with Charles, whose vassal he became in 872, but in 873 he switched over to Louis. Rorik died sometime during the late 870s. The Viking commander Godfried, his nephew, was in charge of his possessions on the coast and along the lower Rhine in 882.

[78] Blok, in ibid., pp. 25–47, discusses most of these and many other aspects of the Scandinavian presence in Frisian lands.

[79] Halbertsma, "The Frisian Kingdom," pp. 78–85; Gosses, "Deensche heerschappijen in Friesland gedurende den Noormannentijd," pp. 134–38; and especially Blok, "De Wikingen in Friesland," pp. 29–30. Archaeologists have found a Viking grave of the early tenth century at Velsen, in Kennemerland; Jelgersma et al., "The Coastal Dunes of the Western Netherlands," p. 145.

Godfried had at least two Frisian counts under him who helped him in the administration of his kingdom, but in 885 he was murdered by a group of his Frisian subjects, presumably including one of these counts. This action prompted a general rising sufficient to expel most of Godfried's Viking accomplices, marking the end of the Regnum Fresonum. A serious harvest failure in northwestern Europe in 892 presumably persuaded most of the remaining Scandinavians to leave the Low Countries, many going to England.[80] After that, Vikings were content for the most part to carry out occasional small raids along the coast and rivers of the western Netherlands, presumably including the area along the Oude Rijn, since it was not safe for the bishops of Utrecht to resume residence in their city until approximately 920. Some sporadic Viking activity continued to plague parts of the western Netherlands into the eleventh century.[81]

While it is possible to reconstruct the bare outlines of the Scandinavians' activity in the North Sea region, problems begin to arise in trying to assess the long-term significance of their presence. Scholars disagree over the actual behavior of the people from the North in their new environments, and these differences carry over into the way they evaluate the entire period.[82] For example, traditional historical treatments have tended to mirror the contemporary narrative sources written by clerics who, alarmed by the willingness of Vikings to attack and plunder churches and monasteries, which often contained considerable treasure, described the behavior and impact of these interlopers in terms of unending violence, plunder, depopulation, and generally unmitigated disaster.

In contrast, more recent scholarship has tended to play down these aspects. Pointing to the small numbers of people involved in the Scandina-

[80] Halbertsma, "The Frisian Kingdom," pp. 82–87; Blok, "De Wikingen in Friesland," pp. 30–31; Blok, "Holland und Westfriesland," pp. 348–50, 355–56; Boeles, *Friesland tot de elfde eeuw*, pp. 388–89; W. J. de Boone, "De Vikingen in Velsen," *Westerheem* 7 (1958): 30–38; Fockema Andreae, "Friesland van de vijfde tot de tiende eeuw," pp. 403–405; Ganshof, "Het laat-Karolingische tijdperk," pp. 367–83; Ganshof, "Het tijdperk van de Karolingen," p. 308; Gosses, "Deensche heerschappijen in Friesland gedurende den Noormannentijd," pp. 138, 141–42; Gosses, *Handboek tot de staatkundige geschiedenis der Nederlanden*, pp. 42–45; Gosses, "De vorming van het graafschap Holland," pp. 239–42.

[81] Jappe Alberts and Jansen, *Welvaart in wording*, p. 48; Halbertsma, "The Frisian Kingdom," pp. 97–98; J. M. P. L. de Vries, "De laatste Wikingstochten in de gewesten van den Nederrijn, XIe eeuw," *Bijdragen voor vaderlandsche geschiedenis en oudheidkunde*, 5th ser., 10 (1923): 249–56.

[82] See, for example, A. d'Haenens, "De post-Karolingische periode," in R. C. van Caenegem and H. P. H. Jansen, eds., *De Lage Landen van prehistorie tot 1500*, p. 173.

vian forays, never more than 800 to 1,000 in an assault force, and their limited traces in the linguistic, settlement, and legal records of the areas in question, some historians have suggested that the original clerical reports enlarged the true significance of the Scandinavians out of all proportion, that often they were no more violent or unprincipled than most other groups at the time. Although not everyone agrees with all aspects of these new assessments, there is a definite tendency in Viking scholarship toward seeing the effects of Scandinavian activities in the southern North Sea region as much more limited than previously thought.[83]

The general tendency in Viking scholarship to reassess the long-term impact of the era has been echoed in the Netherlands. Recent studies have carefully weighed and sifted a wide variety of evidence and concluded that here also the older views depended too uncritically on the overstatements of narratives written by the monks. This is not to suggest that Vikings were harmless. They did in fact sack, plunder, and demand tribute. Dorestad, for example, was attacked by them and partly destroyed several times, beginning in 834, and eventually it went under for good, though hydraulic conditions may have been just as critical in Dorestad's demise.[84] In any case, sacking, plundering, and demanding tribute would cause serious disruption in economic, social, cultural, and political affairs.

Yet the long-term effect of the Vikings in the western Netherlands was negligible. There, as in eastern England and elsewhere, they established power centers along the coasts and in the estuaries. Yet, while there is evidence of an actual Scandinavian colonization in the Danelaw of England beginning in the last quarter of the ninth century, and subsequently in Normandy as well, this stage was never reached in the Low Countries. The newcomers in the western Netherlands were few and acted only as a small and intermittent ruling group that replaced Carolingian power. Although there is some evidence of attempts in the 880s at beginning a second, colonizing stage in parts of Frisia, particularly in some of the activities of the Dane Godfried, they were dashed quickly by the growing opposition of the

[83] Much of the literature is discussed in Blok, "De Wikingen in Friesland," pp. 25–28. See also A. d'Haenens, *Les invasions normandes en Belgique au IXe siècle: le phénomène et sa répercussions dans l'historiographie médiévale*; J. M. Wallace-Hadrill, "The Vikings in Francia," in his *Early Medieval History*, pp. 217–36; van Regteren Altena and Heidinga, "The North Sea Region," pp. 53–67.

[84] See, for example, W. J. van Tent, "De landschappelijke actergronden," *Spieghel historiael* 13 (1978): 214; Blok, *De Franken in Nederland*, pp. 37, 108, 123.

local population, which resulted in Godfried's death in 885 and the expulsion of his cohorts. By this time, moreover, the supposed land hunger of the Scandinavians could be satisfied by colonization in the English Danelaw and elsewhere, and the Viking presence in the western Netherlands dwindled to occasional piratical raids.[85]

With the end of Scandinavian rule in the Low Countries by 885, life for the local populations gradually returned to what it had been before the Viking interlude. Frisian commerce, for example, rebounded sharply, if in fact it ever disappeared or declined. Although Dorestad never recovered its former importance in the trade, other commercial centers, such as Stavoren, Deventer, Utrecht, and Tiel, gained larger portions of the trade of northwestern Europe.[86]

Similarly, the Christian religion, recently established in the Netherlands, survived the disruption of the largely non-Christian Vikings surprisingly well. Although Bishop Balderik (918–75) had to devote considerable attention to rebuilding churches, reorganizing diocesan administration, and reasserting control over the vast possessions of his bishopric when he reestablished his see in Utrecht sometime around 920, he could build on a Christian base that, despite some lingering paganism, had persisted throughout the Viking occupation. It was at this time, for example, that he drew up the list of the possessions of Saint Martin's Church at Utrecht.[87] In fact, instead of having to begin with conversion all over again, the church was able to bring the first glimpses of a truly literate culture to the western Netherlands. Strong religious and artistic impulses emanating from such centers as Cologne and especially Liège (or Luik) were felt throughout the Low Countries during the tenth and early eleventh centuries, and such places as Tiel, Utrecht, and the monastery at Egmond (south, east, and north of Rijnland, respectively) were no exception.[88]

[85] Van Regteren Altena and Heidinga, "The North Sea Region," pp. 53–67; and especially the discussion of Blok, "De Wikingen in Friesland," pp. 34–36. After surveying the archaeological, legal, institutional, onomastic, and linguistic evidence of the period, Blok, ibid., pp. 37–47, concludes that the total, permanent legacy of Scandinavian rule in the Frisian districts can be reduced to the *heervaart*, the organization of the countryside into districts for the purposes of providing rowers for military service, and even this remains only a possibility.

[86] Jappe Alberts and Jansen, *Welvaart in wording*, pp. 39–56.

[87] Blok, "Het goederenregister van de St.-Maartenskerk te Utrecht," pp. 89–104; and above, this chapter.

[88] See d'Haenens, "De post-Karolingische periode," pp. 170–73.

Finally, there seems to be very little evidence to support the complaints of the monk-chroniclers that the Vikings caused serious depopulation with their violence and barbaric behavior. In Rijnland, for example, the population continued to live atop the same coastal ridges and riverbanks and, in fact, was denser in the early tenth century than it had ever been before. As we saw earlier, the Saint Martin's list alone recorded forty-three inhabited places.

The real and lasting significance of the Scandinavian interlude in the Netherlands lay not in the long-term social, economic, or religious spheres but rather in the way that it contributed directly to a shift in political power. It is well known that central political authority in the middle and western portions of the Carolingian empire, specifically the kingdoms of Lothar and Charles, began splintering into smaller entities around the middle of the ninth century. The Vikings reinforced this trend by establishing temporary power bases that were independent of centralized control, despite the fiction that Scandinavians like Rorik and Godfried were vassals of the Carolingians.

When the Viking age came to an end, however, the Carolingians were never able to reassert their former authority over the coastal regions of the Low Countries. No doubt the faroff location and the inaccessibility of these wet environments contributed to the problem, but, just as important, these areas saw the appearance of new, powerful local ruling families who were able to take over where the Viking rulers they helped expel had left off. In Flanders, for example, Count Baldwin I and his successors, from the mid-ninth century onward, were able to establish a firm basis for territorial authority that took over most of the rights and powers exercised by royal government before the invasions. A new ruling clan was able to achieve much the same thing in the coastal regions north of Flanders in the western Netherlands, specifically the West Frisian lands of Kennemerland, Maasland, and Rijnland.[89]

The assassination of Godfried the Dane in 885 was an extremely important event in the history of the coastal regions of the western Netherlands. Chief among the participants in this act was one of the victim's Frisian counts, Gerulf. Partly because Gerulf and his supporters were now the

[89]Koebner, "The Settlement and Colonization of Europe," pp. 55, 57; Blok, "De Wikingen in Friesland," pp. 29, 31–32; Ganshof, "Het laat-Karolingische tijdperk," pp. 367–85; Halbertsma, "The Frisian Kingdom," p. 88; d'Haenens, "De post-Karolingische periode," pp. 161–67.

most powerful people in the region and partly as a reward for his role in helping eliminate a troublesome rival to royal authority, the Carolingians gradually invested Gerulf with comital control over the western coast of the Netherlands from the Vlie to the mouth of the Maas and some other territory inland along the Rhine delta.[90] Part of this territory was the northern half of the Rijnland dune area, described as lying between the Oude Rijn and Suuithardeshaga, and was granted in a charter by East Frankish King Arnulf in 889. These were presumably lands that had formed part of the royal complex acquired during the Frankish conquest of Frisia in the early eighth century.[91] In addition, the Saint Martin's list stated that Gerulf held the fishing rights in the lower Oude Rijn.[92] In 922, King Charles II gave additional royal possessions to Gerulf's successor, Diederik I; they too lay in Rijnland, from Suuithardeshaga to Fortrapa and Kinnem.[93]

These and other grants, as well as some outright expropriations by Gerulf and his successors, formed the basis of authority of the new ruling house of West Frisia–Holland. They included sizable blocks of more or less developed territory, that is, lands containing pastures, fields, meadows, woods, buildings, livestock, and people along the coastal dunes and the lower courses of the major rivers of the western Netherlands. Between these complexes lay the still unpopulated and impassable peat bogs, the wilderness that in theory belonged to the kings. By the late tenth or eleventh century, however, the counts of West Frisia–Holland had acquired the rights to the control and disposition of this wilderness as well.[94]

This last aspect in particular, the acquisition of the wilderness *regalia*, was extremely significant in the subsequent development of the western Netherlands because it helped create the cultural climate that favored the eventual reclamation of the wilderness. Although some of the earliest reclamation of the peat bogs may have been undertaken without

[90] Ganshof, "Het laat-Karolingische tijdperk," pp. 371, 379–80; Blok, "Holland und Westfriesland," p. 355; Blok, "De Wikingen in Friesland," pp. 31–32; Boeles, *Friesland tot de elfde eeuw*, p. 389; Fockema Andreae, "Friesland van de vijfde tot de tiende eeuw," p. 404; Gosses, "De vorming van het graafschap Holland," pp. 241–42.

[91] Koch, ed., *Oorkondenboek van Holland en Zeeland tot 1299*, vol. 1, no. 21. See also above, this chapter.

[92] "Piscatio quam Gerulfus habet in extreme parte Hreni fluminus," in Gysseling and Koch, ed., *Diplomata Belgica ante annum millesimum centesimum scripta*, pt. 1, *Teksten*, no. 195.

[93] Koch, ed., *Oorkondenboek van Holland en Zeeland tot 1299*, vol. 1, no. 28. See also above, this chapter.

[94] See, for example, van der Linden, *De Cope*, pp. 60–61.

specific permission from authorities, in the eleventh, twelfth, and thirteenth centuries the peat-bog wilderness of the western Netherlands was reclaimed by colonists with permission to do so from the counts, while the counts further aided the colonists by setting up a judicial and administrative framework that favored such development.[95] In return, Gerulf's descendants were assisted in their attempts to consolidate political power by the process of reclamation. Peat-bog reclamation in the western Netherlands created a compact, populated realm in place of scattered holdings separated by uninhabited wilderness. Thus the rise of West Frisia helped set the stage for the medieval reclamation era in the western Netherlands, while the reclamation of peat bogs ensured the survival and prosperity of West Frisia–Holland.

[95] H. van der Linden, *Recht en territoir: een rechtshistorisch-sociografische verkenning*, p. 6; van der Linden, *De Cope*, passim.

5

Early-Medieval Settlement in Rijnland: The Ecological Context

By the middle of the tenth century Rijnland contained many small settlements along the dunes and riverbanks in the vicinity of the mouth of the Oude Rijn. Although hemmed in by the still-vacant peat bogs that constituted the wilderness outside the world of human affairs, they did have connections to areas beyond this wilderness, to similar enclaves near the mouths of the other great rivers of the western Netherlands, to the coastal regions of the northern Netherlands, and upstream to Utrecht and beyond. They were small cultural units with social, political, and economic ties to Greater Frisia and, at another level, to the North Sea world as well.

To look at the cultural dimensions of these settlements alone is to see only part of the story, for they were much more than simply places of residence or commercial activity. Each settlement site was associated with a village territory containing lands and waters that constituted a unit of environmental exploitation from which residents acquired most of the things they needed for existence. By employing subsistence strategies such as hunting, gathering, fishing, or varieties of agriculture within their village territories, Rijnlanders interacted with the physical materials and biological processes that constituted the environmental context of their lives.

To measure exactly or specify in detail the manner in which early-medieval Rijnlanders interacted with the world of nature around them, and to discover the internal social mechanisms that governed such interactions, is a difficult task. Human ecology remains at best a highly tentative undertaking, even when the data are plentiful.

This problem was addressed by L. P. Louwe Kooijmans, of the Rijksmuseum van Oudheden (State Museum of Antiquities), at Leiden, at a conference on settlement archaeology in the Netherlands held in Amsterdam, 6 October 1979. Louwe Kooijmans began by pointing out that, while it is possible to identify such activities as crop production in fields, gardening, or livestock keeping, we do not know how they fit together or in what proportions each was undertaken. Nor do we know how fishing and collecting or wild or domestic root crops or leaf greens fit into early food provisioning. It is possible to be somewhat certain only about the ratios of hunting to herding, since a single source reveals both activities. Further, he continued, it is virtually impossible to be specific about a number of agrarian strategies, such as winter stalling or feeding of livestock, manuring, the use of crop rotation or fallow periods, plowing, harrowing, and weeding. Such activities, which could be very significant aspects of food provisioning, are difficult to establish with precision or consistency. Finally, he pointed to problems in classifying activities. For example, herding or pastoralism could be part of a mixed agriculture or a nomadic activity. Whether or not herding was carried out by mounted horsemen may have had far-reaching effects on the entire cultural system. The aim of herding could vary widely, from the production of meat, milk, wool, or hide, to traction or the prestige of possessing a large herd. In short, we are still a long way from completely understanding the ecological context of human existence in the past.[1]

Louwe Kooijmans was speaking of the Neolithic period, not the early Middle Ages, but what he had to say is equally true for both eras. In many respects we know no more about environmental exploitation during the ninth or tenth centuries than we do about the late Neolithic, 3,000 to 4,000 years earlier. One thing remains certain, however. The manner in which Rijnlanders used their environment was closely keyed to the physical geography of the western Netherlands, as was the location of their settlements. In this, too, there was a persistence of patterns, the rudiments of which had been established thousands of years earlier. Thus by supplementing the extremely meager evidence from written sources with information from archaeological and other kinds of research, and by broadening the range of inquiry both temporally and spatially, it is possible to place Rijnland within the broad outlines of an ecological tradition that, evolving from its origins in prehistoric times, persisted until well into the early Middle Ages.

[1]L. P. Louwe Kooijmans, "Het onderzoek van neolithische nederzettingsterreinen in Nederland anno 1979," *Westerheem* 29 (1980): 116.

The Early-Medieval Documentary Record

The evidence from written sources concerning the lands and peoples of the coastal regions of the Netherlands for the period up to the midtenth century consists of two types. First, there are the descriptions that have survived in the histories, sagas, saints' lives, and other narrative sources. Generally they help very little, since they were not usually based on direct observation. Even when they were, however, they rarely include the sort of information useful in trying to identify or reconstruct patterns of environmental exploitation. A few, nevertheless, are worth investigating further.

One of the earliest descriptions referring to the coastal areas dates from the Roman period, incorporated into Pliny the Elder's *Natural History*. Pliny made his observations while participating in a Roman military expedition into the coastal regions of the northern Netherlands and northwestern Germany lead by Corbullo in A.D. 47. He described an unending landscape of treeless coastal marsh that was alternately flooded and exposed by the tides, making it difficult to determine where the land ended and the sea began. The inhabitants built their dwellings on mounds artificially raised above the bleak plain that, surrounded by waters at high tide, seemed to resemble ships at sea. According to Pliny, these people had no apparent means of subsistence besides fishing, they used dried mud for fuel, and in general they seemed so miserable that conquering them and including them in the Roman Empire would be doing them a great service.[2] Although Pliny was speaking specifically of the land of the Chauci, who at that time inhabited the coastal marshes between the Ems and Weser rivers (the later Ostfriesland), his comments concerning physical geography could have applied to the land of the Frisians west the Ems as well, especially to the northern Netherlands but also, in certain respects, to the western coastal regions.[3]

What Pliny actually saw, most likely, were the temporary dwelling mounds and huts used by fishermen during the summer months. Their permanent settlements were farther back in the marshes, where other means of

[2]Pliny the Elder, *Natural History*, trans. H. Rackham, 16.2–5. See the helpful biographical and geographical notes in W. Byvanck, ed., *Excerpta Romana: de bronnen der Romeinsche geschiedenis van Nederland*, vol. 1, pp. 132–33, 151–52.

[3]For the location of the Chauci and Frisians at the beginning of our era, see the map in H. Jankuhn, *Vor- und Frühgeschichte vom Neolithikum bis zur Völkerwanderungszeit*, p. 130.

subsistence prevailed.[4] That a representative of Roman civilization should have so little understanding or appreciation of the patterns of environmental exploitation of these far-off people is, of course, understandable. Pliny was, after all, a native of northern Italy, accustomed to firm, dry soil under his feet. This landscape of mud, marsh, and water no doubt made him uncomfortable. Despite his cultural bias, however, Pliny provided a valuable early description of the physical geography of the coastal marshes along the North Sea. More important to my concerns here, he described the physical remoteness and cultural peculiarity of the people who lived there.

Nearly eight centuries later another observer suffering from another cultural bias expanded on Pliny's theme of physical and cultural seclusion. An anonymous biographer of Saint Boniface, writing in the early ninth century, described the Frisians, who by then inhabited the western as well as the northern coastal regions of the Netherlands, as a people who lived in water like fish and rarely traveled outside their home territory unless they could do so by ship. Because they lived in such a remote region, cut off from ready contact with other people, he thought of them as stupid and barbaric.[5] The probable author of this piece was Frederik, bishop of Utrecht (ca. 820–35),[6] a great admirer of Saint Boniface, who had been killed by recalcitrant, pagan Frisians about seventy-five years earlier. As such, he was more interested in showing the saintliness of Boniface than in carefully describing the human ecology of the area. Perhaps he was trying to indicate that the Frisians lacked all civilizing and socializing tendencies until they were converted to Christianity by missionaries like the martyred

[4] See W. Jappe Alberts and H. P. H. Jansen, *Welvaart in wording: sociaal-economische geschiedenis van Nederland van de vroegste tijden tot het einde van de middeleeuwen*, p. 10; P. C. J. A. Boeles, *Friesland tot de elfde eeuw: zijn vóór- en vroege geschiedenis*, p. 83; H. Halbertsma, *Terpen tussen Vlie en Eems: een geografisch-historische benadering*, pp. 11–12; I. H. Gosses, *Handboek tot de staatkundige geschiedenis der Nederlanden* 1:12–14.

[5] "At primum Frisonibus, quibus jam antea praedicaverat, navigio revectus est; qui fere, quemadmodum et pisces, morantur in aquis, quibus ita undique concluduntur, ut raro ad exteras regiones accessum habeant, nisi navibus subvehantur. Hos remotos a ceteris nationibus, ideoque brutos ac barbaros, coelestis semini verbius adiit," Anonymous of Utrecht, *II vita S. Bonfacii auctore presbyterro S. Martini Ultrajecti*, in *Acta Sanctorum Iunii, tomus primus*, 1.9, vol. 21 in *Acta Sanctorum quotquot toto orbe culuntur, vel a catholicis scriptoribus celebrantur et Latinis et Graecis*, new ed., ed. J. Carnandet, pt. 1, p. 471.

[6] D. P. Blok, *De Franken in Nederland*, 3d ed., p. 119; J. Romein, *Geschiedenis van de Noord-Nederlandse geschiedschrijving in de middeleeuwen: bijdrage tot de beschavingsgeschiedenis*, no. 2.

saint. This brief statement nevertheless contains a kernel of truth, pointing once again to the physical remoteness and cultural difference of the coastal areas.

Both Pliny and the Utrecht cleric were referring primarily to the salt-marsh areas of the northern Netherlands and northwestern Germany. In such a landscape, lacking a coastal dune belt to protect the land from the sea, the inhabitants lived atop the artificial dwelling mounds, or *terpen*, they had built along natural ridges of marine sand and clay covered by a vegetation of salt-tolerant grasses, protruding slightly above an extensive landscape of salt marsh and sand and mud flats intersected by tidal creeks and gullies. They were physically separated from mainland Europe by a broad, low-lying band of lagoons in which considerable peat growth had taken place. The lagoons were traversable only by boat along natural waterways. In many respects this region must have resembled the western coastal regions before the completion of the coastal-barrier system.[7]

The descriptions of both of these writers concerning the physical remoteness of the coastal inhabitants could have applied as well to the entire western coast of the Netherlands. The existence of the Older Dunes, however, eliminated the need for any *terpen* there. Roman *castella* such as the one at Valkenburg, in Rijnland, were frequently rebuilt on the same site, and eventually they grew to resemble artificial mounds as well, but because their purpose was almost entirely restricted to providing military protection, they should not be considered *terpen*. The inhabitants lived on the ridges of the dune system or the natural levees of the Oude Rijn and other large rivers.[8] They too were separated from mainland Europe, in this

[7]W. Roeleveld, *The Holocene Evolution of the Groningen Marine-Clay District*, pp. 95–118; Halbertsma, *Terpen tussen Vlie en Eems*, pp. 2, 99; William H. TeBrake, "Ecology and Economy in Early Medievl Frisia," *Viator: Medieval and Renaissance Studies* 9 (1978): 2–9; L. J. Pons, "De zeekleigronden," in *De bodem van Nederland: toelichting bij de bodemkaart van Nederland 1:200,000*, pp. 27–34, 62–63; Boeles, *Friesland tot de elfde eeuw*, pp. 81–82; *Bodemkaart van Nederland, schaal 1:200,000*, sheets 1–2; T. Edelman, "Oude ontginningen van de veengebieden in de Nederlandse kuststrook," *Tijdschrift voor economische en sociale geografie* 49 (1958): 239–40; T. Edelman, *Bijdrage tot de historische geografie van de Nederlandse kuststreek*, pp. 26, 32–42; the section on the formation of the coastal-barrier system in chap. 3.

[8]See above, chap. 3; W. A. van Es, *De Romeinen in Nederland*, pp. 65–69; Glasbergen et al., *De Romeinse castella te Valkenburg Z.H.: de opgravingen in de dorpsheuvel in 1962*; Haalebos, *Zxammerdam-Nigrum Pullum: Ein Auxiliarskastell am niedergermanischen Limes*; Boeles, *Friesland tot de elfde eeuw*, p. 71.

instance by a wide expanse of impassable peat bogs. Their only connection with the interior was along the natural waterways of the Rhine delta.[9]

A third description, from the tenth century, was based on the observations of a Spanish-Arabian merchant named Tartosi as he traveled through the western Netherlands on his way to Utrecht and Saxony. He described the land he saw as a *sebcha*, a sort of dried-up salt lake on which no seeds or plants flourished. The people residing there lived from the milk and wool of livestock. Their environment was treeless, and they were forced to burn dried mud for fuel.[10] Although Tartosi's route presumably took him through the Rhine delta, his descriptions of a dried-up salt lake do not fit well. Except at the mouths of the great rivers, the western Netherlands was dominated by fresh water. Perhaps he saw some mud flats in the Zeeland area and generalized from there, not fully understanding that the large, treeless expanses of peat bog behind the coast were part of a freshwater regime. Nevertheless, his references to livestock keeping and peat burning tell us something about how the residents of the coastal areas used their natural environment.[11]

The final description of relevance here, and the best one, comes from a Viking saga compiled in the twelfth century but based on much older material. Around the middle of the tenth century, it was reported, the warrior-poet Egil led a nighttime plundering expedition up a large river in the land of the Frisians. It was difficult for Egil's force to land because the river was very shallow over great distances. The land consisted of open plains with woods nearby. After landing their ship, most of the group proceeded farther inland between the river and woods until they encountered a series of villages. As the raiding party approached each village, the inhabitants, who were said to be numerous, fled into the woods with the invaders in pursuit. The land continued to be flat, and the open plains were large.

[9] Well expressed, for example, by S. J. Fockema Andreae, "'Aen't ende van den lande': de Hollands-Utrechtse grensstreek bij Woerden," *Zuid-Hollandse studiën* 1 (1950): 84. See also T. Edelman, "Oude ontginningen van de veengebieden in de Nederlandse kuststrook," pp. 239–40; T. Edelman, *Bijdrage tot de historische geografie van de Nederlandse kuststreek*, pp. 26–42; A. T. Clason, *Animal and Man in Holland's Past: An Investigation of the Animal World Surrounding Man in Prehistoric and Early Historical Times in the Provinces of North and South Holland*, p. 105; the sections on soil formation and prereclamation settlement in Rijnland in chaps. 3 and 4 above.

[10] Georg Jacob, ed. and trans., *Arabische Berichte von Gesandten an germanische Fürstenhofe aus dem 9. und 10. Jahrhundert*, 1:26.

[11] Blok, *De Franken in Nederland*, pp. 123–25; Jappe Alberts and Jansen, *Welvaart in wording*, p. 57.

Ditches had been dug over a large area, and water stood in them; the inhabitants had enclosed their fields and meadows with these ditches and, where necessary, had thrown beams or wooden bridges over them. The account goes on to tell about some fighting and ditch vaulting during which Egil became separated from the rest of his party. Fortunately for him, he was near the woods, which he followed back to the landing place.[12]

Egil had led his party into a landscape which, according to this description, must have resembled closely the area near the mouth of the Oude Rijn in the midtenth century. The presence of forest nearby, especially of gallery forest so close to the river that Egil could follow it back to his ship, indicates that they had entered a largely freshwater environment characteristic of the western Netherlands behind its protective coastal dunes. The Oude Rijn, therefore, may well have been the river that Egil described.[13]

The only other possibilities would have been the former IJ branch of the Rhine delta on the north or the mouth of Maas south of the Oude Rijn, if the assumption is correct that this landscape was protected from direct influence of the sea by coastal dunes. The IJ, however, had silted shut by the early Middle Ages,[14] thus eliminating it as a possibility here. The Maas can be eliminated as well, because the protective dunes between the mouth of the Maas and the mouth of Schelde had been extensively breached and eroded during the period between approximately A.D. 250 and 600, opening up a large area behind them to extensive penetration by salt and brakish water.[15] Finally, the fact that the Oude Rijn itself eventually silted shut helps make it the most likely route for Egil's excursion. The Oude Rijn steadily lost significance with respect to the Lek and Waal, the more southerly branches of the Rhine, which increasingly received more and more of the river's water. Although this process may have begun considerably earlier, the time of final closure was most likely between 800 and 1000.[16] Per-

[12]See *Egils saga Skallagrímssonar*, chap. 69. I have used the Dutch translation provided by Blok, *De Franken in Nederland*, pp. 125, 146 n. 260.

[13]D. P. Blok, for example, thinks the gallery forests make the Oude Rijn the most likely route for Egil's excursion, *De Franken in Nederland*, p. 125; personal communication, 23 April 1976.

[14]W. Zagwijn, "De ontwikkeling van het 'Oer-IJ' estuarium en zijn omgeving," *Westerheem* 20 (1971): 17.

[15]L. J. Pons et al., "Evolution of the Netherlands Coastal Area during the Holocene," in *Transactions of the Jubilee Convention*, enclosures 6–9; chap. 3 above.

[16]D. P. Blok, "Probleme der Flussnamenforschung in den alluvialen Gebieten der Niederlande," in Rudolf Schützeichel and Matthias Zender, eds., *Namenforschung: Festschrift für Adolf Bach zum 75. Geburtstag am 31. Januar 1965*, p. 222; S. J. Fockema An-

haps the problems that the Vikings had with sandbars and shallow water over great distances is an indication that the process was well under way. Although it will never be possible to make a firm identification with the Oude Rijn, it does remain the most likely possibility.

Whether or not Egil and his party actually sailed into Rijnland, the description of the villages along the river in question adds a few more useful details about how Rijnlanders may have exploited their environment. Again there is reference to both pastoral and arable agriculture, evidenced by both meadows and fields. Further, the meadows and fields were not part of temporary or shifting agricultural practices but were permanent installations, since the parcels were said to be enclosed by ditches. Finally, by digging the ditches, possibly for drainage as well as for marking off individual parcels, the inhabitants may have been attempting to deal with that greatest of all obstacles to human presence in the western Netherlands, too much water.

Besides the general descriptions, often lacking in specific detail, in the narrative sources of the prereclamation period, there is another body of written sources that provides specific detail without a general picture in which to place it. I have in mind here the items of property mentioned in the various administrative documents of the early Middle Ages, the charters, cartularies, and lists of possessions mentioned in chapter 4.[17] For Rijnland specifically there are the charters granting possessions to Counts Gerulf (889) and Diederik I (922), as well as the list of possessions of Saint Martin's Church at Utrecht (918–48). The grant to Gerulf consisted of a woods, several fields, and seven complete estates or manors and all that belonged to them, including clearings, fields, pastures, meadows, woods, waters, mills, and fisheries, as well as serfs. The grant to Diederik men-

---

dreae, "De Oude Rijn: eigendom van openbaar water in Nederland," in *Rechtskundige opstellen op 2 november 1935 door oud-leerlingen aangeboden aan prof. mr. E. M. Meijers*, pp. 699–700; L. P. Louwe Kooijmans, *The Rhine/Meuse Delta: Four Studies on Its Prehistoric Occupation and Holocene Geology*, pp. 120, 122; A. C. F. Koch, "Phasen in der Entstehung von Kaufmannsniederlassungen zwischen Maas und Nordsee in der Karolingerzeit," in Georg Droege, Peter Schöller, Rudolf Schützeichel, and Matthias Zender, eds., *Landschaft und Geschichte: Festschrift für Franz Petri zu seinem 65. Geburtstag am 22. Februar 1968*, p. 323; W. C. Braat, "Early Medieval Glazed Pottery in Holland," *Medieval Archaeology* 15 (1971): 113.

[17] For the western Netherlands in general, see the first 60 or so documents in A. C. F. Koch, ed., *Oorkondenboek van Holland en Zeeland tot 1299*, vol. 1; as well as the relevant portions of Otto Oppermann, ed., *Fontes Egmundenses*.

tioned lands with serfs (thus arable land), meadows, woods, pastures, and waters. The list of possessions of Saint Martin's Church, besides mentioning individual woods and fisheries, included 128½ manors, a villa, and 3⅓ villages.[18] The manors, villas, and villages, of course, represented all of the specific features mentioned in the two charters.

The total harvest of information from written sources concerning the lands and peoples of the coastal regions of the western Netherlands for the period up to the midtenth century is thus not very impressive. The evidence shows that the settlements were associated with such landscape features as fields, meadows, pastures, woods, plains, and ditches and that their inhabitants engaged in such activities as livestock keeping, crop raising, fishing, hunting, gathering, and drying peat for use as fuel.

The Early-Medieval Archaeological Record

The findings of recent archaeological research in large part confirm the information derived from written sources concerning the uses that early-medieval Rijnlanders made of their environment. For example, excavations at Rijnsburg, occupied at various times from the seventh to the tenth centuries, have uncovered traces of plowing, a building with stalling facilities, and the bones of cattle and also of pigs, sheep, goats, beavers, elk, sand seals, whales, red deer, and sturgeon and other fish.[19] The inhabitants of this site, therefore, not only raised crops and kept livestock but also hunted and fished. Although archaeologists know of other early-medieval settlement sites in the Oude Rijn district, I have not seen mention of additional material concerning environmental exploitation besides some weights for fishing nets found at Leiderdorp, a short distance upstream from Rijnsburg.[20]

This meager list of archaeological finds can be supplemented with the results of excavations made into other contemporary settlements of the

[18] Koch, ed. *Oorkondenboek van Holland en Zeeland*, vol. 1, nos. 21, 28; M. Gysseling and A. C. F. Koch, eds., *Diplomata Belgica ante annum millesimum centesimum scripta*, pt. 1, *Teksten*, no. 195; the discussion in chap. 3, above.

[19] W. A. Van Es, "Early Medieval Settlements," *Berichten van de Rijksdienst voor het Oudheidkundig Bodemonderzoek* 23 (1973): 281–85; H. Sarfatij, "Die Frühgeschichte von Rijnsburg (8.-12. Jahrhundert): Ein historisch-archäologischer Bericht," in B. L. van Beek, R. W. Brandt, and W. Groenman–van Waateringe, eds., *Ex Horreo: I.P.P. 1951–76*, pp. 295–97; Clason, *Animal and Man in Holland's Past*, pp. 23–24.

[20] W. C. Braat, "Leython," *Leids jaarboekje* 44 (1950): 85.

western Netherlands, especially in the areas on the north. The first of these is Velsen, situated along the dunes near the banks of the former IJ River. Pollen analyses carried out in conjunction with excavations revealed considerable arable and pastoral agricultural activity in the vicinity of Velsen during the early Middle Ages.[21] Investigations at another dunal site at Schoorl, about 25 kilometers farther north, have yielded distinct traces of a field layer, covered over with drifted dune sand and containing, among other things, fragments of pottery of the Carolingian period.[22] Two more sites with relevant finds are known from the Medemblik area, about 30 kilometers northeast of Schoorl, along creek banks in a peat-bog landscape near a large lake. The first site produced a pollen profile showing that wheat, barley, and increasingly rye were raised in fields that may have been surrounded by a network of ditches.[23] Excavations at the second site, a short distance southwest, near Droge Wijmers, uncovered a pit containing animal manure.[24] Finally, excavations at a place known as Het Torp, near Den Helder, at the tip of North Holland, have revealed a settlement on a silted-up sand ridge in a brakish-to-salt-marsh setting just east of the coastal dunes. Finds included large numbers of plant remains associated with agricultural activity, such as barley and common vetch (normally a fodder crop but occasionally grown for human consumption as well) and various field weeds. The site also yielded fruit stones of rose, suggesting the collection of rose hips, as well as signs of grazing and trampling by what may have been livestock.[25]

To date the most productive excavations of an early-medieval settlement have been carried out not in the coastal districts but at the site of Dorestad, the early-medieval Frisian trading center in the river-clay region of the central Netherlands. The findings of this research have relevance to

[21] S. Jelgersma et al., "The Coastal Dunes of the Western Netherlands: Geology, Vegetational History and Archaeology," *Mededelingen van de Rijks Geologische Dienst*, n.s., no. 21 (1970): 127, 131.

[22] P. J. Woltering, "Archeologische kroniek van Noord-Holland over 1977," *Holland: regionaal-historisch tijdschrift* 10 (1978): 270; P. J. Woltering, "Archeologische kroniek van Noord-Holland over 1978," *Holland: regionaal-historisch tijdschrift* 11 (1979): 272.

[23] J. C. Besteman, "Carolingian Medemblik," *Berichten van de Rijksdienst voor het Oudheidkundig Bodemonderzoek* 24 (1974): 43–106; P. J. Woltering, "Archeologische kroniek van Noord-Holland over 1975," *Holland: regionaal-historisch tijdschrift* 8 (1976): 255–57.

[24] Woltering, "Archeologische kroniek van Noord-Holland over 1977," pp. 267–68.

[25] W. A. van Zeist, "The Environment of 'Het Torp' in Its Early Phases," *Berichten van de Rijksdienst voor het Oudheidkundig Bodemonderzoek* 23 (1973): 349–53.

my purposes, even though the place of origin is about 50 kilometers upstream from Rijnland. Throughout the prereclamation period the central river-clay region had always been a natural extension of the coastal-lowland landscape because both lay at similar elevations and were subject to flooding at similar times and under similar conditions. Good drainage, or the lack of it, was always the ultimate determinant of a human presence in the river-clay region as it was in the coastal districts.[26] Excavations show that, in addition to having a strong industrial and commercial character, early-medieval Dorestad was, at least in part, an agrarian community. Some of the farmsteads, for example, included stalling facilities for livestock, while others were associated with granaries. Large quantities of seeds were found, including barley, oats, club wheat, rye, pea, celtic bean, lentil, and common vetch, along with both field and pasture weeds. More than 90 percent of all identifiable faunal remains came from domesticated mammals, over half of which were cattle followed by pigs, sheep, horses, goats, dogs, and cats. The remainder consisted of chickens, ducks, geese, hares, roe deer, red deer, wild swine, whales, fish, and molluscs. A number of fish traps and a fish corf, or holding pen, made of wickerwork were also uncovered.[27]

The results of archaeological research, though adding much detail to the information already derived from written sources, do not tell us all we need to know about early-medieval environmental exploitation in the coastal regions of the western Netherlands. The problem is how to proceed from a

[26]Louwe Kooijmans, for example, places both the river-clay and coastal regions within the same lowland landscape; *The Rhine/Meuse Delta*. See also L. P. Louwe Kooijmans, "The Neolithic at the Lower Rhine: Its Structure in Chronological and Geographical Respect," in Sigfried J. de Laet, ed., *Acculturation and Continuity in Atlantic Europe Mainly during the Neolithic Period and the Bronze Age*, pp. 150–73; Waterbolk, "The Lower Rhine Basin," in Robert J. Braidwood and Gordon R. Willey, eds., *Courses toward Urban Life: Archaeological Considerations of Some Cultural Alternates*, pp. 227–53.

[27]W. A. van Zeist, "Agriculture in Early Medieval Dorestad: A Preliminary Report," *Berichten van de Rijksdienst voor het Oudheidkundig Bodemonderzoek* 19 (1969): 209–12; W. A. van Es, "Vis uit Dorestad voor mijnheer Calkoen," *Westerheem* 23 (1974): 89–94; W. A. van Es, "Excavations at Dorestad: A Pre-Preliminary Report: 1967–1968," *Berichten van de Rijksdienst voor het Oudheidkundig Bodemonderzoek* 19 (1969): 183–206; W. A. van Es, "Die neuen Dorestad Grabungen, 1967–1972," in *Vor- und Frühformen der europäischen Stadt im Mittelalter: Bericht über ein Symposium in Reinhausen bei Göttingen vom 18. bis 24. April 1972*, pp. 202–17; Wietske Prummel, "Vlees, gevogelte en vis," *Spieghel historiael* 13 (1978): 282–93.

simple listing of landscape features and human activities to an accurate picture of how all the bits and pieces fit together. What is needed is a general framework into which the material can be placed, but no such framework exists ready-made.

As we saw earlier, most studies that examine patterns of environmental exploitation during the early Middle Ages do so almost exclusively in terms of fields and field systems. From the point of view of a strictly economic history, of course, this makes a certain amount of sense. Fields were very important in the early Middle Ages because control of the cereals they produced formed the basis of all power and wealth. Such a narrow conceptual framework hardly lends itself to my purposes, however. For example, it appears that fields often produced only a small portion of the total diet; during the early Middle Ages crop yields were at times appallingly low. It is well to keep in mind that crop production was never practiced by itself but was undertaken side by side with other varieties of agriculture that were part of a whole set of subsistence strategies. The early-ninth-century capitulary, *De villis*, confirms this by showing that early-medieval food provisioning could be extremely broad-based, at least in the idealized setting of royal estates.[28] In other words, there was much more to early-medieval environmental exploitation than activity in the fields.

Even if an emphasis on fields were appropriate for some of the light-soiled and better-drained portions of the Frankish world, however, it might be irrelevant to the low-lying Netherlands. Things were much different there, something that the first two eyewitnesses referred to above sensed. Both Pliny and the cleric from Utrecht spoke rather disparagingly of the cultural level of the inhabitants of the remote North Sea coasts of the Netherlands and Germany. If the condescending tone of these descriptions is stripped away, however, an element of truth remains. Coastal peoples in fact exhibited a culture that was substantially different from what was known in more inland areas. The authors were wrong, however, in assuming that this difference could be removed with the elimination of the element of remoteness, whether by including them in the Roman Empire or by converting them to Christianity and integrating them into the political sphere of the Franks. The difference was much more fundamental than that. It stemmed ultimately from a set of interrelationships between the inhabitants of the coastal areas and the natural conditions in force there that

[28] See the discussion above, chap. 2.

had evolved over a very long period of time. Because natural conditions in the western Netherlands were very different from those farther away from the coast, the results of such culture-nature interactions were very different as well.

Early Land Use in the Coastal Regions

Ironically, it is often easier to be precise about certain aspects of land use in the coastal areas for prehistoric times than for the Middle Ages. This is due primarily to the coming of age of the field of prehistory, the result of using the most sophisticated modern archaeological practices and dating techniques and of placing a much greater emphasis on interpreting artifacts in terms of their total context, environmental as well as cultural. Thus an archaeologist now treats a charred seed or an animal-bone fragment with as much care as a pottery sherd or a flint sickle. Further, biologists, soil scientists, and other experts often work alongside archaeologists in analyzing the site and the material from an excavation. As a result a much clearer picture can be presented of the ways in which prehistoric human communities fit into their environments.[29]

A number of prehistoric sites in the coastal regions of the western Netherlands have been excavated and analyzed extensively. These settlement sites belong to periods ranging from the late Neolithic Age through the Iron Age, but the Bronze Age of West Friesland is particularly well represented. Medieval archaeology, in comparison, remains in a stage of infancy. Although the pace definitely is quickening,[30] the results of such research are a long way from being incorporated into mainstream medieval historiography. Nevertheless, because the archaeological evidence available so far points to an unbroken continuity from the late Neolithic through the early Middle Ages, not only for the location of settlement but also for land-use patterns, some of the lacunae in the early-medieval archaeological and written record can be sketched in by analogy with similar situations

[29] W. A. van Es, "De Nederlandse archaeologie na 1945," *Westerheem* 25 (1976): 295–99; Colin Renfrew, *Before Civilization: The Radiocarbon Revolution and Prehistoric Europe*, p. 254.

[30] See, for example, the wide range of medieval archeological publications included in the bibliography of J. C. Besteman and H. Sarfatij, "Bibliographie zur Archäologie des Mittelalters in den Niederlanden 1945 bis 1975," *Zeitschrift für Archäologie des Mittelalters* 5 (1977): 163–231.

of an earlier age. This is not to suggest that individual settlement sites were continuously inhabited throughout. Rather, as I demonstrated earlier, coastal settlements were established in certain types of places, along the dune ridges or the clay and sand banks of rivers and tidal creeks, with areas of peat growth remaining empty. Similarly, each generation of coastal inhabitants tended to exploit the environment along lines practiced and perfected by previous generations once the basic patterns had been established.[31]

THE ELEMENTS OF SUBSISTENCE IN THE COASTAL LOWLANDS

The archaeological record indicates that the first permanent settlements of the western Netherlands were established along river and tidal creek banks and dune ridges during the late Neolithic period, around 2450 B.C.[32] The possibility exists, however, that there were earlier settlements that have thus far escaped discovery. In any case, it would be wrong to assume that the coastal regions of the western Netherlands were entirely beyond the sphere of human affairs until late Neolithic times. On the contrary, what is known about late Mesolithic and early Neolithic coastal populations from Portugal to Denmark, along with the results of archaeological research in areas inland from the Netherlands coast, suggests that the coastal regions of the western Netherlands too must have seen some collecting, hunting, and fishing activities during late Mesolithic and early Neolithic times.[33] In fact, this view is supported by the recent discovery of what appears to have been an early Neolithic fishing and hunting camp at Bergschenhoek, on the north side of Rotterdam, in a peat landscape less than 10 kilometers south of Rijnland.

The Bergschenhoek find, the oldest known settlement site from the western Netherlands, was a temporary encampment, not a permanent settlement. It was apparently occupied repeatedly for short periods of time around 3450 B.C. from five to eleven times over a period of three to six years. The finds include large quantities of bird bones, fish remains, and wild-plum pits, as well as some fish traps made of cord and willow shoots.

[31] Louwe Kooijmans, *The Rhine/Meuse Delta*, p. 42.

[32] For much of what follows, see the relevant sections of chap. 3 and 4 above.

[33] Louwe Kooijmans, "Het onderzoek van neolithische nederzettingsterreinen in Nederland," p. 110.

The site, atop a peat layer near a freshwater lake, was made passable by the frequent laying down of bundles of reeds. This was, presumably, a fowling and fishing outpost or extraction camp associated with more permanent settlement elsewhere.[34] In many respects the Berhschenhoek camp may have served a function similar to that of the temporary dwelling mounds and huts used by fishermen in the northern coastal regions that Pliny the Elder observed three and a half millennia later.[35]

It is tempting to wonder whether the Bergschenhoek camp was a unique phenomenon or in fact a typical means of exploiting peat-bog wilderness that may have continued throughout the prereclamation period. In fact, the latter is suggested by the recent discovery of a number of temporary summer fishing camps from the late Neolithic period near Spijkenisse, about 20 kilometers southwest of Bergschenhoek in a peat-bog landscape.[36] So far, however, no later sites of this type have been discovered. Still, some of the documentary evidence of the early Middle Ages contains mention of fisheries. While most of these would have been situated in the river-clay, estuarine, or dune portions of the western Netherlands, at least one site appears to have been a fishery in the peat-bog wilderness north of the Oude Rijn, namely, the fishery "in flumine Fennepa" mentioned in the list of possessions of Saint Martin's Church (918–48).[37] Fennepa is a pre-Germanic water name associated with bog conditions (the English word *fen* is related to the root) that in this case referred to a stream east of the innermost dune ridge in an area that was later engulfed by the Haarlemmermeer.[38] Whether the Fennepa fishery actually was exploited from a temporary campsite built up, as described above, with layers of reeds placed on top of the soggy peat surface will never be known. It may have been close enough to existing dune or riverbank settlements that no temporary campsites were needed at the location. Besides, the remains of a tem-

[34] The Bergschenhoek site was investigated by L. P. Louwe Kooijmans. See H. Sarfatij, "Archeologische kroniek van Zuid-Holland over 1976," *Holland: regionaal-historisch tijdschrift* 9 (1977): 245–47; H. Sarfatij, "Archeologische kroniek van Zuid-Holland over 1977," *Holland: regionaal-historisch tijdschrift* 10 (1978): 298–99; Louwe Kooijmans, "Het onderzoek van neolithische nederzettingsterreinen in Nederland," pp. 108–10, 112.

[35] Pliny, *Natural History*, 16.2–5.

[36] Excavated by L. P. Louwe Kooijmans; see H. Goudappel, "Uit de kranten," *Westerheem* 29 (1980): 311.

[37] Gysseling and Koch, eds., *Diplomata Belgica ante annum millesimum centesimum scripta*, pt. 1, *Teksten*, no. 195.

[38] Maurits Gysseling, *Toponymisch woordenboek van België, Nederland, Luxemburg, Noord-Frankrijk en West-Duitsland (vóór 1226)*, p. 1003.

porary camp, never very substantial to begin with, would have been destroyed long ago by the Haarlemmermeer; the Bergschenhoek finds survived primarily because they were buried under 8 meters of peat.[39]

The fact remains, however, that both fishing and fowling were important early subsistence strategies and remained so throughout the prereclamation period, according to the faunal remains that have survived, particularly when it is remembered that bird and fish material, which is much more susceptible to the ravages of time than are larger, more durable remains, may well be underrepresented in the faunal inventories.[40] The huge peat bogs of the western Netherlands, extremely rich in fish and waterfowl, may well have been a source of such items. An extraction camp of the Bergschenhoek type, therefore, was always a possible if not a necessary setting for fishing and hunting in the more remote portions of the peat-bog wilderness.

Except for the Bergschenhoek finds, most of the early evidence of environmental exploitation in the western Netherlands is associated with populations who resided there permanently beginning during the late Neolithic period. As we saw earlier, the first of these have been identified on the basis of pottery and house types as representatives of the Vlaardingen Culture.[41] The evidence shows, however, that they did much more in their estuarine, dunal, and marine-clay environments than simply build houses and fabricate pottery. Along with their contemporaries in higher-lying areas on the east and south, Vlaardingen Culture people were truly Neolithic in the sense that they used agricultural knowledge and skills alongside other varieties of environmental exploitation.[42]

Two of the most important Vlaardingen Culture sites were along creek banks at Vlaardingen and Hekelingen, on the north and south sides of the Maas estuary, respectively. Both were adjacent to extensive and at least partly wooded swamp, where residents hunted red deer and wild pig, the remains of which figure prominently among the faunal inventories that have been assembled. The skeletal remains of beaver as well as pike indi-

[39] Sarfatij, "Archeologische kroniek van Zuid-Holland over 1976," p. 245.

[40] Prummel, "Vlees, gevolgelte en vis," p. 285; Louwe Kooijmans, *The Rhine/Meuse Delta*, p. 333.

[41] See above, chap. 3.

[42] For a discussion of how the Vlaardingen Culture fit into the larger Neolithic context of northwestern Europe, see Louwe Kooijmans, "The Neolithic at the Lower Rhine," pp. 150–73.

cate that fresh running water was available in the vicinity. The inhabitants also consumed many varieties of waterfowl and evidently hunted sea mammals along the nearby seashore. Between one-quarter and one-third of all the bones uncovered, however, came from various species of domesticated animals, the most numerous being cattle, with evidence of sheep, goats, pigs, horses, and dogs or wolves as well. There seems to have been little opportunity for the first residents of Vlaardingen and Hekelingen to grow cereals in their marsh-and-swamp surroundings. Presumably these were imported from inland areas with which they maintained contact.[43]

Three other Vlaardingen Culture discoveries reveal a slightly modified land-use pattern in the dune landscape during the late Neolithic. At Leidschendam and Voorschoten, in what became Rijnland, and at adjacent Voorburg, in eventual Maasland, settlements were established along the low, sandy ridges of the oldest and easternmost set of dune ridges. The ridges were covered with an alder-elm vegetation complex, with peat growth on the east and halophytic, or salt-tolerant, grasses on the west. Such an environment offered a number of opportunities for exploitation.

In the wooded portions of the sand ridges, along the streams, and along the nearby beach residents presumably could hunt the aurochs, wild pigs, red deer, roe deer, pine martins, gray seals, sperm whales, and beavers represented among the animal remains. Further, fishing and fowling would have been possible throughout the area. Sturgeon, most likely from waters associated with the nearby Oude Rijn estuary, were particularly well represented in the faunal remains of Voorschoten. Despite the richness of such natural resources, however, residents of these settlements, from the initial stage of occupation onward, placed a much greater emphasis on livestock keeping than on hunting. The evidence from Leidschendam and Voorschoten, and less clearly from Voorburg, shows that over 85 percent of all animal remains came from domesticated animals, chiefly from cattle but also including pigs, sheep, goats, and dogs. The flanks of the dune ridges and the strips between the ridges would have provided ample grazing opportunities for cattle, sheep, and goats. Finally, the pollen profiles devel-

[43] Clason, *Animal and Man in Holland's Past*, pp. 10–11, 102; Louwe Kooijmans, *The Rhine/Meuse Delta*, pp. 21–23; Waterbolk, "The Lower Rhine Basin," pp. 241–42; J. A. Brongers et al., "Prehistory in the Netherlands: An Economic-Technological Approach," *Berichten van de Rijksdienst voor het Oudheidkundig Bodemonderzoek* 23 (1973): 11–13; W. Glasbergen et al., "Neolithische nederzetting te Vlaardingen (Z.H.)," in *In het voetspoor van A. E. van Giffen*, pp. 44, 58, 60–61.

oped for these sites show that cereals were grown in the immediate vicinity of the settlements. While it is impossible to establish the exact location of fields, it is safe to assume that they lay atop the dune ridges, since the surrounding peat-and-clay areas would have been too wet for crops.[44]

Although the Vlaardingen Culture residents of the coastal dune sites exploited their surroundings in a variety of ways, the important point to remember here is that they had begun to show a greater reliance on varieties of agriculture than had their contemporaries at Vlaardingen and Hekelingen along the Maas estuary, a reliance that was increasingly reinforced as time went on. In particular, as human presence intensified in the western Netherlands, the first real attempts were made at actually restructuring the environment to suit an expanded agricultural focus. While the first Vlaardingen Culture settlers at Voorschoten and Leidschendam grazed their cattle on naturally occurring pastures along and between the dune ridges, newcomers during a second occupation stage began using fire to clear large pastures from the forest vegetation of the dunes themselves for their presumably larger herds of cattle. Such burning, a coastal extension of a practice known to have been employed by culturally related cattle-rearing groups on the higher-lying sandy areas of the eastern Netherlands and elsewhere, is evident from the widespread occurrence of the pollen of lanceolate plantain weed and other grasses and herbs associated with burning and grazing. The remains of cereal pollen and charred seeds indicate that they cultivated crops as well. Most evidence suggests, however, that pastoralism was fast becoming the major subsistence activity.[45]

While livestock keeping was growing in importance as a subsistence strategy in coastal-dune settlements during the late Neolithic, the archaeological record shows that crop raising too was beginning to see some important development. The clearest evidence for this comes from a number of settlement sites in West Friesland, the northeastern section of the mod-

[44] W. Glasbergen et al., "Settlements of the Vlaardingen Culture at Voorschoten and Leidschendam," *Helinium* 7 (1967): 3–31, 97–120; W. Groenman–van Waateringe et al., "Settlements of the Vlaardingen Culture at Voorschoten and Leidschendam (Ecology)," *Helinium* 8 (1968): 105–30; and Louwe Kooijmans, *The Rhine/Meuse Delta*, pp. 23–26.

[45] Groenman-van Waateringe et al., "Settlements of the Vlaardingen Culture at Voorschoten and Leidschendam," pp. 107–10; Waterbolk, "The Occupation of Friesland in the Prehistoric Period," *Berichten van de Rijksdienst voor het Oudheidkundig Bodemonderzoek* 15–16 (1965–66): 19; Brongers et al., "Prehistory in the Netherlands," p. 10; Jelgersma et al., "The Coastal Dunes of the Western Netherlands," p. 131; Louwe Kooijmans, *The Rhine/Meuse Delta*, p. 26.

ern province of North Holland. During late Neolithic times, a broad sea arm stretched from the western coast eastward as far inland as the vicinity of modern-day Medemblik. Alongside this sea arm were broad mud flats that were silting up into salt marsh raised above normal tidal action. Vlaardingen Culture people and their successors began establishing settlements along the highest segments of the sandy ridges of tidal creeks and silted-up creeks that were part of this salt-marsh landscape, expecially near Zandwerven, Aartswoud, and eventually Oostwoud.[46]

The evidence from these sites shows a typical late Neolithic mixture of subsistence activities. Hunting is represented by the remains of beavers, ducks, and common porpoises, while fishing is represented by the remains of sturgeon, pike, and gray mullet. Large quantities of mussel shells, as well as remains of acorns, hazelnuts, blackberries, wild apples, and marshmallows, show that collecting too was practiced. The residents of these settlements were also agriculturalists, however. For example, most of the faunal evidence derived from domesticated animals, especially cattle but including sheep or goats, pigs, and dogs. Some of the most interesting information, however, derives from the crop-raising activities of these settlers. Literally thousands of barley seeds, as well as numerous seeds of emmer wheat and flax, have been found in the settlement site near Aarswoud.[47] Excavations at the Zandwerven site, meanwhile, uncovered the distinct traces of plowing marks in the soil, among the oldest such evidence in the Netherlands.[48] Finally, the somewhat younger late Neolithic settlement site excavated at Oostwoud also yielded fossilized plowing marks in the soil; in this instance there was evidence of plowing as well as cross plowing. Actually, the possibility exists that some of the soil marks at Oostwoud may have been made by an evolved type of plow because they were wider and deeper than those normally encountered in that period and

[46] J. F. van Regteren Altena and J. A. Baker, "De neolithische woonplaats te Zandwerven," in *In het voetspoor van A. E. van Giffen*, p. 39; A. E. van Giffen, "Nederzettingssporen van de vroege Klokbekercultuur bij Oostwoud (N.H.)," in *In het voetspoor van A. E. van Giffen*, p. 66; communication from F. F. van Iterson Scholten in Woltering, "Archeologische kroniek van Noord-Holland over 1975," pp. 240–41.

[47] Communications by F. F. van Iterson Scholten in Woltering, "Archeologische kroniek van Noord-Holland over 1975," pp. 239–41; Woltering, "Archeologische kroniek van Noord-Holland over 1977," pp. 254–55; Woltering, "Archeologische kroniek van Noord-Holland over 1978," pp. 249–50.

[48] Van Regteren Altena and Bakker, "De neolithische woonplaats te Zandwerven," pp. 34, 38–39, 171–72.

locale.[49] Taken together, the evidence from these three West Frisian sites clearly shows that crop raising was taking its place, alongside hunting, fishing, collecting, and especially livestock rearing, as a significant subsistence strategy during the late Neolithic period.

THE PATTERNS OF SUBSISTENCE IN THE COASTAL LOWLANDS

It is possible to carry this discussion of late Neolithic environmental exploitation one step beyond merely stating that residents of the western Netherlands used a variety of subsistence activities. Archaeological evidence points to a distinct pattern of spatial organization within some settlements that was associated with variations in the quality of drainage. Those soils with the best drainage were reserved for residence and crop raising. Less well drained soils were used for grazing livestock. Those soils not drained well enough for crops or livestock grazing were nevertheless part of the hunting, collecting, and fishing territory of a settlement. Such a pattern is seen, for example, in the settlement site excavated at Zandwerven, in West Friesland. At this site the best-drained soils were along the sandy ridge, the very ridge on which the plowing marks were found. This ridge, of course, lay in an essentially salt-marsh landscape, above normal tidal influence, that was eminently suited to grazing;[50] salt marshes have always been prized for grazing and hay cutting, in both prehistoric and historic times.[51] Finally, on the lower portions of the marsh, along the nearby sea arm, in the tidal creeks traversing the marsh, and in the freshwater swamps inland beyond the influence of salt or brackish water, the Zandwerven residents could practice the hunting, fishing, and collecting activities that continued to be essential to their livelihood.[52]

[49] Van Giffen, "Nederzettingssporen van de vroege Klokbekercultuur bij Oostwoud (N.H.)," pp. 66, 68; Woltering, "Archeologische kroniek van Noord-Holland over 1978," pp. 250–51. For early plows their evolution, and their uses, see Bernard Wailes, "Plow and Population in Temperate Europe," in Brian Spooner, ed., *Population Growth: Anthropological Implications*, pp. 154–79.

[50] Van Regteren Altena and Bakker, "De neolithische woonplaats te Zandwerven," pp. 34, 38–39.

[51] TeBrake, "Ecology and Economy in Early Medieval Frisia," pp. 6, 19; Jappe Alberts and Jansen, *Welvaart in wording*, pp. 58–59.

[52] It is important to remember that, generally speaking, agriculture did not become the predominant means of human subsistence until the Bronze Age; see Waterbolk, "The Lower Rhine Basin," p. 246.

Similar patterns of exploitation are suggested by the evidence from nearby Aartswoud and Oostwoud, both situated along well-drained ridges in a salt-marsh landscape,[53] and from the sites near Voorschoten, Leidschendam, and Voorburg, along the coastal dunes.[54] For the dunal sites, however, some allowance must be made for slightly different local conditions: perhaps more extensive sand ridges permitting not only residence and crop raising but also some pasturage, instead of a salt marsh an essentially freshwater context that nevertheless provided extensive grasslands, and the nearby raised peat bogs, Oude Rijn and Maas estuaries, and seashore that provided considerable fishing, hunting, and collecting opportunities.

A closer examination of the archaeological evidence from late Neolithic settlements in the western Netherlands thus reveals that the quality of drainage, besides determining the location of settlements, also determined the location of subsistence activities within these settlements. Further, these patterns were repeated and reinforced over and over again throughout the remainder of the prereclamation period. Thus a settlement such as the one associated with early-medieval Rijnsburg showed the same kind of internal spatial organization as did those of the late Neolithic period.[55] In fact, such a pattern had certain predictive possibilities after it was initially established. It becomes possible, for example, to predict that wherever well drained places would be found in the western Netherlands, from the late Neolithic through the early Middle Ages, there too would most likely be found settlements, each with an internal spatial organization resembling the patterns visible from late Neolithic West Frisian and dunal sites.

The focus of human activity in each location would be the ridge or outcrop that made it attractive as a settlement. On the crest of the ridge or outcrop would be found residence, gardens, and fields. Lower down, on the flanks of the ridge or outcrop and also in adjacent areas where it still remained just below the surface of the surrounding swamp or marsh, would be found pastures and meadows. Farther away from the ridge or outcrop, in

[53] Van Iterson Scholten, in Woltering, "Archeologische kroniek van Noord-Holland over 1975," p. 241; van Giffen, "Nederzettingssporen van de vroege Klokbekercultuur bij Oostwoud (N.H.)," pp. 66, 68.

[54] Glasbergen et al., "Settlements of the Vlaardingen Culture at Voorschoten and Leidschendam," pp. 7, 98, 114; Groenman-van Waateringe et al., "Settlements of the Vlaardingen Culture at Voorschoten and Leidschendam," pp. 109–10, 116.

[55] See, for example, Sarfatij, "Die Frühgeschichte von Rijnsburg," pp. 290–302.

the wetter portions of marsh and swamp, would be found the hunting, fishing, and collecting grounds of the settlement, equivalent to the *saltus* of early-medieval communities.[56]

SOME COMPOSITE PICTURES OF SUBSISTENCE

By the end of the late Nolithic period, sometime around 1700 B.C., the broad patterns of environmental exploitation that were to persist for the next two and a half millennia were beginning to take shape. What appears dimly from the settlements along the Maas estuary and more clearly from dunal sites in southern Rijnland and neighboring Maasland and from the salt-marsh landscape of West Friesland is a population that had learned not only to read and understand but also, wherever possible, to use effectively the natural environment of the western Netherlands. Late Neolithic settlers were able to recognize the various possibilities that this environment held for human existence and had become experienced and adept at applying the exploitative knowledge and skills appropriate to such possibilities.

While broad patterns of environmental exploitation persisted after the late Neolithic period, what did change was the ratio of agricultural subsistence activities, both arable and pastoral, to nonagricultural subsistence activities. For example, during the Bronze Age the Netherlands in general achieved fully effective food productiion for the first time. From then on, pastoral and arable agriculture increasingly became the predominant means of subsistence, with only a small, continuing reliance on hunting, fishing, and collecting.[57] The same became true of the western Netherlands.

Inventories of animal remains from eight Bronze Age settlements from the western Netherlands indicate that more than 90 percent, in some instances 100 percent of the remains, came from domesticated animals. Bones were found at the following sites: at Hoogkarspel (six find concentrations), Oostwoud, Stedebroec (formerly Bovenkarspel), and Wevershoef (two find concentrations), in West Friesland; at Langeveld and

[56] See the graphic representations in Louwe Kooijmans, *The Rhine/Meuse Delta*, p. 110 fig. 29, p. 277 fig. 121, for the late Neolithic–early Bronze Age settlements at Ottoland-Kromme Elleboog and Molenaarsgraaf, respectively, about 20 kilometers southeast of Rijnland in the central river-clay region of the Netherlands.

[57] Waterbolk, "The Lower Rhine Basin," pp. 246–47; Brongers et al., "Prehistory in the Netherlands," p. 12; Louwe Kooijmans, *The Rhine/Meuse Delta*, p. 30; Clason, *Animal and Man in Holland's Past*, pp. 5–6, 206. This was true for central Europe as well; see Jankuhn, *Vor- und Frühgeschichte vom Neolithikum bis zur Völkerwanderungszeit*, pp. 68–69.

Vogelenzang, along the dunes near the north end of Rijnland; and at Molenaarsgraaf, in the central river-clay region about 20 kilometers southeast of Rijnland. The remaining faunal material consisted mostly of fish, occasionally fowl, and virtually no terrestrial wildlife.[58]

The inhabitants of the western Netherlands also raised crops wherever possible along the ridges or outcrops on which they lived, and the evidence for such activities becomes increasingly more visible in the archaeological record. Fields are known from the fossilized traces of plowing and spading found in the soil of a number of Bronze Age sites: Andijk, Hoogkarspel, Medemblik, Stedebroec (Bovenkarspel), and Wevershoef, in West Friesland; and dunal sites at Monster, in Maasland, and Velsen, in Kennemerland, south and north of Rijnland, respectively. At each site the plowing or spading marks were found on the well-drained ridges that formed the focus of the settlement.[59] Fields and field systems are further indicated by the discovery of extensive networks of ditches at Hoogkarspel, Medemblik, Stedebroec (Bovenkarspel), and Velsen, dug for purposes of drainage as well as for marking off land parcels, the coastal equivalent of hedges and fences in drier environments.[60] While no fields have been found in associa-

[58] Hoogkarspel: J. A. Bakker et al., "Hoogkarspel-Watertoren: Towards a Reconstruction of Ecology and Archaeology of an Agrarian Settlement of 1000 B.C.," in van Beek, Brandt, and Groenman–van Waateringe, eds., *Ex Horreo: I.P.P. 1951–76*, pp. 204–208. Langveld, Wevershoef, and Oostwoud: Clason, *Animal and Man in Holland's Past*, pp. 12, 14, 16. Stedebroec: Woltering, "Archeologische kroniek van Noord-Holland over 1977," p. 255. Vogelenzang: W. Groenman–van Waateringe, "Nederzettingen van de Hilversumcultuur te Vogelenzang (N.H.) en Den Haag (Z.H.)," in *In het voetspoor van A. E. van Giffen*, pp. 83, 87, 176; Clason, *Animal and Man in Holland's Past*, p. 13. Molenaarsgraaf: Louwe Kooijmans, *The Rhine/Meuse Delta*, p. 278.

[59] Ankijk: F. Baars, "Afdelingnieuws: afdeling Noord-Holland Noord," *Westerheem* 27 (1978): 237. Hoogkarspel: J. A. Bakker, "Een grafheuvel en oud akkerland te Hoogkarspel (N.H.)," in *In het voetspoor van A. E. van Giffen*, p. 104; Bakker et al., "Hoogkarspel-Watertoren," pp. 189, 191, 195; Woltering, "Archeologische kroniek van Noord-Holland over 1978," p. 249. Medemblik: Besteman, "Carolingian Medemblik," p. 53. Stedebroec: Janneke Buurman, "Cereals in Circles—Crop Processing Activities in Bronze Age Bovenkarspel (the Netherlands)," in Udelgard Körber-Grohne, ed., *Festschrift Maria Hopf zum 65. Geburtstag am 14. September 1979*, p. 22. Wevershoef: P. J. Woltering, "Archeologische kroniek van Noord-Holland over 1976," *Holland: regionaal-historisch tijdschrift* 9 (1977): 193. Monster: Jelgersma et al., "The Coastal Dunes of the Western Netherlands," p. 143. Velsen: Woltering, "Archeologische kroniek van Noord-Holland over 1978," p. 253.

[60] Hoogkarspel: Bakker, "Een grafheuvel en oud akkerland te Hoogkarspel," p. 104; Bakker et al., "Hoogkarspel-Watertoren," pp. 214–22. Medemblik: Besteman, "Carolingian Medemblik," p. 53. Stedebroec: Woltering, "Archeologische kroniek van Noord-Holland over 1977," p. 257. Velsen: Woltering, "Archeologische kroniek van Noord-Holland over 1978," p. 253.

tion with the Bronze Age settlement at Molenaarsgraaf, in the central Netherlands river-clay area, high values of *Cerealia* and track-weed pollen in the pollen diagrams indicate that crops were raised in the immediate vicinity.[61]

The importance of crops for subsistence in the western Netherlands is suggested by the often sizable finds of crop remains at Hoogkarspel, Molenaarsgraaf, and Stedebroec (Bovenkarspel), finds that included the seeds of einkorn, emmer wheat, bread wheat–club wheat, hulled barley, flax, and field weeds.[62] The discovery of crop-processing and storage paraphernalia, meanwhile, show that Bronze Age residents of the western Netherlands had incorporated the fruits of arable agriculture into their subsistence economies. These included a number of millstone fragments from Molenaarsgraaf,[63] cereal storage pits from Hoogkarspel,[64] and traces in the soil of a number of storage circles for grain sheaves from Andijk, Hoogkarspel, and Stedebroec (Bovenkarspel). These storage circles, no doubt an attempt to avoid the crop rotting that would easily occur in a soggy environment, consisted of small, slightly raised areas constructed of soil dug from an encircling drainage ditch.[65]

In recent years a number of prehistorians of the Instituut voor Prae- en Protohistorie of the University of Amsterdam and the Rijksdienst voor het Oudheidkundig Bodemonderzoek (State Service for Archaeological Investigations) in Amersfoort have begun assembling a composite picture of human existence in West Friesland during the middle and late Bronze Age. Drawing on the results of extensive archaeological research since World War II, but particularly since about 1965, these scholars have been paying particular attention to subsistence activities, placing them firmly within their environmental and social contexts. The result is the most nearly complete picture to date of prereclamation life in the western Netherlands, a picture that satisfies many of the requirements for a general framework

[61] Louwe Kooijmans, *The Rhine/Meuse Delta*, p. 277.

[62] Hoogkarspel: Bakker et al., "Hoogkarspel-Watertoren," pp. 200–204. Molenaarsgraaf: Louwe Kooijmans, *The Rhine/Meuse Delta*, p. 277. Stedebroec: Woltering, "Archeologische kroniek van Noord-Holland over 1977," pp. 256–57; Woltering, "Archeologische kroniek van Noord-Holland over 1978," p. 246; Buurman, "Cereals in Circles," pp. 22–23, 30, 34–36.

[63] Louwe Kooijmans, *The Rhine/Meuse Delta*, p. 277.

[64] Bakker et al., "Hoogkarspel-Watertoren," p. 214.

[65] Andijk: Baars, "Afdelingnieuws: afdeling Noord-Holland Noord," p. 237. Hoogkarspel: Woltering, "Archeologische kroniek van Noord-Holland over 1978," p. 249. Stedebroec: ibid., p. 246; and especially Buurman, "Cereals in Circles," pp. 21–37.

into which the discrete information concerning prereclamation environmental exploitation derived from archaeological and eventually written sources can be placed. It not only provides the basic parameters of human existence in Bronze Age West Friesland, especially in the vicinity of Hoogkarspel and Stedebroec (Bovenkarspel), but also shows many generations of a rural society going through a gradual process of adaptation to its environment.[66]

After a hiatus that separated the late Neolithic and early Bronze Age occupation phases from the middle and late Bronze Age phases, migrants from the sandy areas on the southeast and perhaps from the dunes on the southwest recolonized the West Frisian marine-clay and salt-marsh area beginning around 1300 B.C. Like their predecessors, they possessed agrarian knowledge and skills which they quickly applied, together with more traditional activities, such as hunting, fishing, and collecting, to the generally wetter environment of West Friesland. They soon learned to read and understand drainage and soil patterns in their new setting, establishing their fields on the highest portions of the sandy and sandy-clay ridges emerging slightly above the surrounding clay and salt marsh. They established their farmyards at the edges of the fields on the flanks of the ridges, using the lower, wetter clay lands for grazing livestock and cutting hay. The wet conditions of the West Frisian landscape made it necessary for the residents to encircle their farmyards and fields with ditches to aid drainage.[67]

The residential portion of the typical West Frisian Bronze Age settlement consisted of a small collection of large, rectangular houses of about 30 by 7 meters. These buildings were of the three-aisled type, a housing form that lasted until well into the Middle Ages and was characterized by two parallel rows of upright posts, standing within the living space, that supported the roof beams.[68] Such a house performed a dual function. First, it provided shelter and living space for a household averaging six people,

[66] R. W. Brandt, "De kolonisatie van West-Friesland in de bronstijd," *Westerheem* 29 (1980): 139–42; Woltering, "Archeologische kroniek van Noord-Holland over 1975," pp. 187–92; H. T. Waterbolk, "Siedlungskontinuität im Küstengebiet der Nordsee zwischen Rhein und Elbe," *Probleme der Küstenforschung im südlichen Nordseegebiet* 13 (1979): 7.

[67] Waterbolk, "The Occupation of Friesland in the Prehistoric Period," p. 25; Buurman, "Cereals in Circles," p. 22; Bakker et al., "Hoogkarspel-Watertoren," p. 222; Brandt, "De kolonisatie van West-Friesland in de bronstijd," pp. 142–49; Woltering, "Archeologische kroniek over 1975," p. 191.

[68] R. W. Brandt, "Landbouw en veeteelt in de late bronstijd van West-Friesland," *Westerheem* 25 (1976): 38; Bakker et al., "Hoogkarspel-Watertoren," p. 208.

according to estimates made by R. W. Brandt, of the Instituut voor Prae- en Protohistorie, derived from material of both prehistoric and medieval provenance. Short life expectancies meant that the typical household had a high ratio of children to adults.[69] In addition, the house contained stalling facilities at one end for twenty to thirty head of cattle as well as space for smaller livestock, such as pigs and sheep or goats.[70] This combination of living and stalling space under the same roof was a characteristic feature of housing forms in northwestern Europe well into the Middle Ages.[71]

At first cattle made up about 65 percent of the livestock, but the ample grazing opportunities on the salt-marsh-covered clay soils surrounding the ridges on which the fields and houses stood eventually made it possible for cattle to comprise 85 percent or more of the average stock of domesticated animals, the remainder consisting of sheep or goats, pigs, dogs, and chickens. In fact, cattle, by supplying meat, milk, and traction, became central to the local economy. As time went on, there was also a significant expansion of the area of cultivation under relatively dry conditions. Crops consisted of hulled barley and emmer wheat, sometimes sown separately and at other times mixed, as well as some flax. Summer weeds in cereal samples suggest that residents used a spring-planting and summer-harvesting schedule. The presence of weeds associated with fallowing, along with other indicators, points to a system of alternating cropping and fallowing periods.[72] At the high point of Bronze Age settlement in West Friesland, around 1000 B.C., entire field systems were marked off by drainage ditches containing individual fields that ranged in size from less than one-fourth hectare to 6 hectares.[73]

According to R. W. Brandt, a single household of six people would have required about 12,000 calories a day,[74] and he wonders whether such a daily requirement could have been acquired within close proximity to the

[69] Brandt, "Landbouw en veeteelt in de late bronstijd," p. 59.

[70] Ibid., p. 63.

[71] Waterbolk, "Evidence of Cattle Stalling in Excavated Pre- and Protohistoric Houses," in A. T. Clason, ed., *Archaeozoological Studies*, pp. 383–94.

[72] Brandt, "De kolonisatie van West-Friesland in de late bronstijd," p. 148; Woltering, "Archeologische kroniek van Noord-Holland over 1975," p. 191; Baker et al., "Hoogkarspel-Watertoren," pp. 200–208, 222.

[73] Bakker et al., "Hoogkarspel-Watertoren," pp. 214–22; Brandt, "De kolonisatie van West-Friesland in de late bronstijd," p. 148.

[74] Brandt, "Landbouw en veeteelt in de late bronstijd van West-Friesland," pp. 59, 64; TeBrake, "Ecology and Economy in Early Medieval Frisia," pp. 16–17 n. 52.

farmstead. After considering various possibilities, he decided that the average diet during the Bronze Age would have consisted of approximately 50 percent meat, 40 percent cereals, and 10 percent milk products, vegetables, and fish. The meat component apparently could have been supplied by the annual consumption of 1,160 kilograms of beef or 780 kilograms of pork or 750 kilograms of mutton; actually some sort of combination of these would have been more likely. Considering the size of the animals, based on bone measurements, and the estimated herd sizes, based on stalling facilities, he concluded that an annual slaughter of five cattle, five pigs, and five sheep or goats would have supplied 50 percent of the caloric requirements. The cereal component, meanwhile, would have consisted of an annual consumption of 550 kilograms. If 170 kilograms of seed sown per hectare resulted in a very modest yield of 340 kilograms at harvest, 170 kilograms of which would be set aside as seed stock, then 3.5 hectares would have been needed for cultivation each year. Since long fallowing may have been the only means of maintaining soil fertility, perhaps as little as one-fifth of the potential arable would have been planted at any time. A total of 17.5 hectares of potential arable land would have been required, therefore, for each household. After carefully studying the soil map of Bronze Age Hoogkarspel, Brandt calculated that 17.5 hectares of sandy or sandy-clay ridges suitable for arable agriculture could certainly have been found in the 78 hectares that were within a radius of 500 meters of a farmstead. In fact, according to his calculations, the sort of diet postulated above fits not only the environmental possibilities of Bronze Age West Friesland but also the labor potential of a six-member household using the cattle-drawn *ard* and various stone and bronze tools and implements.[75]

Toward the end of the Bronze Age deteriorating drainage conditions began to affect much of West Friesland and eventually human subsistence patterns as well. The gradual silting shut of the broad West Frisian sea arm sometime earlier meant that the original marine-clay area covered by salt marsh came increasingly under the influence of fresh water that could no longer find easy egress to the sea. As a result, the ground-water table began to rise, accompanied by extensive peat growth. Peat growth in turn led to a worsening of local drainage conditions.

The late Bronze Age residents of West Friesland initially responded

[75] Brandt, "Landbouw en veeteelt in de late bronstijd," pp. 58–66. See also J. A. Brongers, *Air Photography and Celtic Field Research in the Netherlands*, pp. 65–68.

by placing their houses higher on the ridges that were the focus of their settlements and later by raising house sites before the houses were built and digging deeper and wider ditches around their farmyards and fields. They also reduced the area of cropland and placed a greater emphasis on livestock breeding. At the same time the livestock component came to be represented by a growing proportion of sheep and goats, which were less likely than cattle to sink into waterlogged pastures and reduce them to muck and mire. Fishing too became a more important though still minor activity, including fishing in the drainage ditches of the settlements themselves. Finally, around 700 B.C., much of the area was abandoned because of the continued rise in water levels.[76] With a simple technology and with no compelling reasons for taking a stand, the population left for more suitable areas, such as the coastal dunes on the southwest but also, perhaps, on the newly emerged salt-marsh areas on the northeast, in the modern provinces of Friesland and Groningen.[77]

With only slight modifications to suit local variations, the surprisingly detailed picture of Bronze Age land use in West Friesland could apply to land use in other parts of the western Netherlands as well. For example, many of the details provided by Louwe Kooijmans concerning early Bronze Age land use at Ottoland and Molenaarsgraaf, in the river-clay area about 20 kilometers southeast of Rijnland, would easily fit the scheme developed for West Friesland. In particular, his calculations of the number of people that could have been fed by the arable land available along the former stream ridges certainly accords well with Brandt's calculations, though Brandt presupposes a longer fallow period. Both agree on the central importance of livestock, especially cattle, in the Bronze Age rural economy.[78]

Besides what has become known in recent years about Bronze Age West Friesland, the only other fairly clear composite picture of life in the coastal Netherlands before reclamation comes from the northern coast, present-day Friesland and Groningen. Beginning in the early Iron Age, residents of

[76] Brandt, "De kolonisatie van West-Friesland in de late bronstijd," pp. 148–50; Woltering, "Archeologische kroniek van Noord-Holland over 1975," pp. 191–92; Louwe Kooijmans, *The Rhine/Meuse Delta*, p. 43.

[77] Waterbolk, "Siedlungskontinuität im Küstengebiet der Nordsee," pp. 8, 12; Waterbolk, "The Occupation of Friesland in the Prehistoric Period," p. 26.

[78] Louwe Kooijmans, *The Rhine/Meuse Delta*, pp. 108–11, 274–78; Brongers, *Air Photography and Celtic Field Research in the Netherlands*, pp. 65–68.

the extensive salt marshes of this region, who eventually came to be known as Frisians, developed a pattern of environmental exploitation that not only resembled what is known from Bronze Age West Friesland but also persisted relatively unchanged until reclamation began during the High Middle Ages. Once again it was a combination of livestock keeping and crop raising, supplemented by fishing, that provided the wherewithal for a human presence.

The salt marsh of the northern Netherlands provided ample grazing opportunities. The Frisians, living on their *terpen*,[79] were literally surrounded by vast expanses of halophytic grasses, rushes, and sedges that could be harvested successfully by livestock. From the beginning, therefore, they kept livestock, with cattle always the most important, according to surviving faunal remains. The cattle were stalled in the byre end of the large rectangular houses that are known from the early Iron Age through the early-medieval period; in most houses there was space for twenty or more head of cattle, and some provided room for as many as fifty head. The discovery of churns and cheese molds shows that livestock were kept not only for meat but also for milk products.[80]

The Frisians also raised crops, not only on the edges of their *terpen* but also on the highest portions of the salt marsh during relatively dry times. Cultivation is represented archaeologically by the remains of stalks, hulls, pods, and other types of organic material and by some fossilized plow traces in the soil. The possibility of cultivation on the marsh surface has been made more believable by some modern crop-raising experiments on undiked salt marsh in both northwestern Germany and the Netherlands. The experiments show that a certain amount of desalination can occur on the highest marsh portions.[81] Fishing, meanwhile, was possible not only in the tidal creeks but also in the broad band of peat bogs that separated the salt marsh from the higher-lying areas on the south and east.[82]

It is possible to get a rare glimpse of the dietary habits of Frisians from the discovery of some early-medieval cooking pots containing remnants of food in one of the *terpen* at Leens, Groningen. Upon analysis the contents were found to consist of wheat, oats, rye, unidentifiable vege-

[79] See above, chap. 3; TeBrake, "Ecology and Economy in Early Medieval Frisia," pp. 2–9.

[80] TeBrake, "Ecology and Economy in Early Medieval Frisia," pp. 17, 19, 22–23.

[81] Ibid., pp. 6–7, 18–19.

[82] Ibid., pp. 2–5.

tables, and a large proportion of animal fat.[83] Other evidence suggests, however, that meat and dairy products were the most important items in the Frisian diet and that cereals, vegetables, and fish merely supplemented livestock products.

Wilhelm Abel, of the University of Göttingen, has made the following calculations of the quantities of livestock products in coastal diets based on the excavation reports for the third-to-fourth-century *terp* at Feddersen Wierde, near Bremerhaven, Germany. The average herd size, estimated from the amount of stall space available in each farmstead, was between twenty and twenty-five head and consisted of six or seven cows and one ox or steer at least three and a half years old, two heifers of two and a half years, and eleven to fourteen young cattle or calves. Such a herd, he maintains, could have provided 5,000 to 6,000 kilograms of milk products per year. Further, by butchering one calf, two or three steers or oxen, and two cows each year, the average household would acquire 420 to 520 kilograms of beef. With the addition of another 80 kilograms of pork or mutton the total would come to 500 to 600 kilograms of meat per household per year. For a five- or six-member household such quantities of meat and milk products would provide about 1,600 calories per person per day, according to Abel's calculations.[84]

While estimates of the presumed daily caloric requirement of pre-industrial populations vary widely (Abel assumes 3,200 calories per adult per day), it may well be true that Brandt's figure of 2,000 calories per person per day for an average Bronze Age population in West Friesland applies here as well.[85] In any case, the yield of 1,600 calories per person per day from meat and milk products alone that Abel has tried to demonstrate would have constituted a substantial proportion of the total diet. The important point to remember when looking at diets in this fashion, however, is not to take the actual figures too seriously. After all, a quick comparison of Abel's calculations with those provided by Brandt shows that the two began with different assumptions. Brandt assumed that meat was the only livestock product consumed, while Abel included both meat and milk products. Each was trying to explore the caloric potential of a herd of do-

[83] Ibid., p. 17 n. 55.

[84] Wilhelm Abel, *Geschichte der deutschen Landwirtschaft vom frühen Mittelalter bis zum 19. Jahrhundert*, pp. 23–24.

[85] See the discussion in TeBrake, "Ecology and Economy in Early Medieval Frisia," pp. 16–17, especially n. 52.

mestic animals that could have been housed in the stalling facilities that have been uncovered. Both agree, however, on the central importance of livestock for food provisioning in the coastal areas before reclamation.

On the whole, Frisian environmental exploitation was eminently successful, at least in the sense that it could at times support a surprisingly large population. For example, estimates for Westergo alone, the northwestern corner of the province of Friesland, point to a total population of at least 20,000 and a density of 27 per square kilometer during the first century A.D.[86] By A.D. 900, the salt-marsh areas of both Friesland and Groningen apparently contained about 35,600 people, with an average density of 20 per square kilometer; in temperate Europe only the Paris basin and parts of Flanders may have known greater population densities.[87] If such estimates are at all realistic, the residents of the salt-marsh areas of Friesland and Groningen had carried the business of subsistence in wet coastal areas far beyond the modest requirements of the minuscule communities of one or a few homesteads known from the Bronze Age of the western Netherlands.

The importance of Frisian subsistence strategies for an area such as Rijnland during the early Middle Ages lies in the fact that, ever since the Iron Age, people possessing a Frisian culture had had a tremendous impact on all of western Netherlands. As we saw earlier, Frisians were there as part of a mixed population from the Iron Age through the Roman period, but in the general reoccupation beginning in the sixth century A.D., they constituted the dominant group of colonists. In fact, documentary sources as well as the evidence from linguistic and legal research show that Fri-

[86] H. Halbertsma, "Enkele aantekeningen bij een verzameling oudheden afkomstig uit een terpje bij Deinum," *Jaarverslag van de Vereniging voor Terpenonderzoek* 23–27 (1948): 243. Though B. H. Slicher van Bath, "The Economic and Social Conditions of the Frisian Districts from 900 to 1500," *A.A.G. Bijdragen* 13 (1965): 100, thinks this figure may be somewhat too high, W. A. van Es, "Friesland in Roman Times," *Berichten van de Rijksdienst voor het Oudheidkundig Bodemonderzoek*, p. 53, believes it to be quite realistic.

[87] Van Es, "Friesland in Roman Times," pp. 52–53; B. H. Slicher van Bath, "De paleodemografie," *A.A.G. Bijdragen* 15 (1970): 194; B. H. Slicher van Bath, "Le climat et les récoltes en haut moyen âge," in *Agricoltura e mondo rurale in Occidente nell'alto medioevo*, pp. 421–22; Slicher van Bath, "The Economic and Social Conditions of the Frisian Districts from 900 to 1500," pp. 98–102, 131–33, in which he assessed the total population in 900 at 42,500 for the entire area of the present provinces of Friesland and Groningen, including the higher-lying Pleistocene soils on the south. If, however, the salt-marsh areas, Westergo, Oostergo, Westerkwartier, Hunsego, and Fivelgo, are taken alone, the total estimate is 35,600. See also chap. 2, above.

sians actively expanded into and through the western Netherlands during the early Middle Ages. In doing so, they behaved as most other colonizing groups do, bringing with them much that was familiar, such as archaeologically attested pottery and housing forms.[88] It is only reasonable to assume that they brought familiar and proven subsistence patterns as well, particularly since these would have worked as effectively in the new environments as they had in the old.

Of course, the Frisian reliance on livestock keeping, crop raising, and fishing was not new to the western Netherlands. All evidence points to the fact that the essential components were present there from the late Neolithic period. Beginning in the Bronze Age, however, with the first achievement of fully effective food production, the basic components increasingly were put together into a mix of subsistence strategies that, far from being restricted to West Friesland or the salt-marsh areas of Friesland and Groningen, became the characteristic pattern throughout the coastal regions of the Netherlands for the remainder of the prereclamation period. Whenever the information is plentiful enough to suggest even the slightest hint of a pattern, it is always the same: an overwhelming emphasis on raising livestock that were grazed on the ample grasslands available everywhere in the coastal districts, cultivation of the sandy and sandy-clay ridges that were the focus of settlements, and fishing in the plentiful waters of estuaries, salt marshes, rivers, and peat bogs. There is, therefore, every reason to believe that the residents of the forty-odd little settlements known to have existed in Rijnland in the early tenth century followed the same practices.

THE ECOLOGY OF COMPLEX SOCIAL SYSTEMS

To this point I have concentrated on the ways in which residents of the coastal areas, including Rijnlanders in the early Middle Ages, presumably acquired the food that they consumed. I have presented the material in such a way as to suggest that they lived exclusively in self-sufficient communities and did little besides raising food. In doing so, however, I have neglected a very important aspect of the ways that such communities related to their environments. From the Bronze Age onward village chieftains and local warlords repeatedly altered the patterns of local self-sufficiency by living at least in part from the surplus production of those

[88] See above, chap. 4.

whom they ruled.[89] Somewhat later, during the late pre-Roman Iron Age, the Cananefates and the Batavians arrived in the western and central Netherlands. Since they apparently constituted new ruling aristocracies, they too lived from the surplus production of their subjects.[90] Such changes in the social system would of course significantly alter or skew the subsistence patterns of previously self-sufficient communities, though at times this is very difficult to document. It was only when the western Netherlands became involved in larger social systems, through the Roman military thrust into the coastal regions at the beginning of our era and the inclusion of this region in the world of Greater Frisia during the early Middle Ages,[91] that such changes began to leave a trail we can follow.

The western Netherlands experienced unusually dry conditions between roughly 200 B.C. and A.D. 250. During this time settlements were established along tidal creeks in estuarine areas, in the river-clay areas, along the dunes, and even on the edges of some peat bogs.[92] Consequently, when the Romans arrived, around the beginning of our era, building the fortifications of their *limes* system along the left bank of the Oude Rijn, the western Netherlands was already densely settled. Rijnland itself, bisected by the imperial boundary, exhibited an essentially military occupation during much of the Roman period by virtue of a broad no-man's-land on each side of the frontier. However, considerable civilian settlement is known from the Maas River area on the south and from the Velsen region and elsewhere on the north.

Undoubtedly it was the provisioning requirements of the military garrisoned along the Rhine system that most directly affected patterns of environmental exploitation in the western Netherlands during the Roman period. From about A.D. 100 onward, however, the presence of industrial and commercial centers associated with the *castella* and an urban place such as the Municipium Cananefatium, along the dunes midway between the Oude

[89] See, for example, the suggestions of Brongers et al., "Prehistory in the Netherlands," pp. 28, 35–36.

[90] Van Es, *De Romeinen in Nederland*, pp. 170–71.

[91] See the appropriate sections of chap. 3, above.

[92] Van Es, "Friesland in Roman Times," pp. 40, 44; W. Haarnagel, "De prähistorischen Siedlungsformen im Küstengebiet der Nordsee," in *Beiträge zur Genese der Siedlungs- und Agrarlandschaft in Europa*, pp. 67–71; TeBrake, "Ecology and Economy in Early Medieval Frisia," pp. 7–8. On the relative dryness of the period, see H. T. Waterbolk, *De praehistorische mens en zijn milieu: een palynologisch onderzoek naar de menselijke invloed op de plantengroei van de diluviale gronden in Nederland*, pp. 16, 131.

Rijn and the Maas, also played a significant role. This interjection of a substantial nonagricultural population meant that agriculturalists had to produce a surplus of food and other agricultural products that was extracted either by force in the form of tribute or taxation or by more voluntary means in the form of trade. While this was not a totally new phenomenon introduced during the Roman period, given the earlier appearance of surplus-consuming village chieftains, local warlords or military aristocracies, the highly bureaucratized and militarized Roman state affected life in the western Netherlands much more profoundly than its tribal or aristocratic predecessors ever could.

Contacts between the surplus-food-producing agricultural populations of the western Netherlands and the surplus-food-consuming Romans are suggested by certain aspects of the Roman *castellum* at Valkenburg, near the mouth of the Oude Rijn. Archaeological excavations of the site, along the silted-up river-bank, show that this stronghold was particularly important as both a storage facility and a transshipment point for considerable quantities of cereals. Although much of this grain undoubtedly originated in England, there is every reason to believe that considerable quantities came from the dune, estuarine, and riverbank settlements of the western Netherlands as well.[93] After all, cereal cultivation had long been a standard practice wherever and whenever conditions were dry enough in the western Netherlands. My own cursory review of the archaeological literature has turned up more than a dozen Roman-age sites in the western Netherlands with evidence of arable agriculture. While most of these were in the dunes, plowing marks beneath the earliest *castellum* layer at Valkenburg prove that the silted-up Oude Rijn bank could be cultivated as well.[94]

The two most interesting sites, however, were south of Rijnland, in a peat-bog landscape that, during a previous transgression, had received floodwaters which had deposited a layer of clay by means of creeks that communicated with the Maas estuary. The first of these was a native Roman-period farmstead, found in modern Kethel, near Rotterdam, that was occupied in three stages during the second century A.D. The main building, which combined living space and stalling facilities under the same roof, was rebuilt twice, each time on a higher location. A large quan-

[93] W. Groenman–van Waateringe, "Grain Storage and Supply in the Valkenburg Castella and Pretorium Aggripinae," in van Beek, Brandt, and Groenman–van Waateringe, eds., *Ex Horreo: I.P.P. 1951–1976*, pp. 226–40.

[94] Ibid., pp. 231, 238 n. 13.

tity of manure shows that the stalls were in fact used for livestock. As time progressed, this site came increasingly under Roman cultural influence, as shown by growing quantities of Roman pottery and, what is more important at this juncture, a *horreum*, or granary, of Roman construction with a raised wooden floor. The general impression gained from this site is that it was at best a marginal one as far as drainage was concerned and that, despite considerable effort aimed at maintaining its habitability, such as raising the house site with branches, twigs, and rubbish, it finally succumbed to the inevitable.[95]

A second, more successful find was made in modern Rijswijk, between Delft and The Hague. It was a settlement site that was occupied for most of the first three centuries A.D., starting with a single house and culminating around 150 in three houses, one of which was built of stone. The stone house was rebuilt in wood around 200 but on a stone foundation and with provisions for central heating, thus giving it the appearance of a Roman country house. Most of the houses were of the three-aisled type that included livestock stalls at one end. Each house had a granary standing alongside it that was built on legs to keep out dampness and hungry animals. The settlement also had a number of water wells constructed of large wine vats with both ends removed. Perhaps the most significant feature of this site was the large area that was investigated: 3½ hectares completely excavated, 13 hectares investigated by means of large trenches. As a result, an entire settlement, including its field complex, was subjected to careful examination and analysis. It appears that around 200 this settlement had about 13 hectares of fields that were divided with drainage ditches into 1- to 2-hectare blocks. J. H. F. Bloemers, who has made an exhaustive study of this settlement, believes that it was a native settlement, Cananefate, to be exact, that went through a clear process of expansion owing to the stimulating effects of the nearby Roman military frontier and, after 100, of the civilian city of Municipium Cananefatium, about 3 kilometers north. According to him, any surplus labor, livestock products (milk, cheese, meat, hides, bones), or cereals that this settlement might have been able to produce would have been readily absorbed by the military and urban markets.[96]

[95] P. J. R. Modderman, "A Native Farmstead from the Roman Period near Kethel," *Berichten van de Rijksdienst voor het Oudheidkundig Bodemonderzoek* 23 (1973): 149–58.

[96] J. H. F. Bloemers, "Rijswijk (Z.H.) 'De Bult,' een nederzetting van de Cananefaten," *Hermeneus: tijdschrift voor antieke cultuur* 52 (1980): 95–106. This article is based

The subsistence activities of those portions of the western Netherlands lying beyond the Rhine military frontier were affected as well by the infusion of a large nonagricultural population associated with the *castella* and the urban centers on the south. For example, in 12 B.C., the Roman commander Drusus conquered the Frisians occupying the western and northern coastal districts of the Netherlands beyond the eventual imperial boundary. As a result, they were required to pay a substantial tribute in cowhides which, though imposed upon them against their will and disputed in an unsuccessful revolt of A.D. 28, nevertheless represented the siphoning off of a significant agricultural surplus. Frisians beyond the Rhine also sold livestock and livestock products directly to the Romans, a trade that is indicated by the discovery of a wood tablet from the late first century A.D. that records the sale of cattle to Roman buyers.[97] Such scattered bits of evidence show that the residents of the North Sea coastal lowlands, whether they wanted to or not, could and did exploit their natural surroundings beyond what was essential to their own maintenance.

The generally dry conditions that were conducive to settlement in the western Netherlands during the Roman period gradually disappeared as a rising ground-water table and the resumption of extensive peat growth inhibited drainage.[98] Consequently, many sites were abandoned once again. In fact, there is no clear evidence, written or archaeological, for a continued human presence in the western Netherlands during the fourth and fifth centuries. In those isolated places where settlement may have continued, such as along the dunes near Katwijk, in Rijnland,[99] patterns of environmental

---

on his more comprehensive *Rijswijk (Z.H.), "De Bult," Eine Siedlung der Cananefaten*. A similar Roman-age settlement was found at Ockenburgh, a short distance west of Rijswijk in the dunes, and others are known from the Vlaardingen and Schiedam area near Rotterdam as well as from upstream in the central river area; see van Es, *De Romeinen in Nederland*, pp. 131–33, 135.

[97] C. W. Vollgraff, "Eene Romeinsche koopacte uit Tolsum," *De Vrije Fries: tijdschrift uitgegeven door het Friesch Genootschap van Geschied-, Oudheid- en Taalkunde* 25 (1917): 71–101; van der Poel, "De landbouw in het verste verleden," *Berichten van de Rijksdienst voor het Oudheidkundig Bodemonderzoek* 10–11 (1960–61): 181; Clason, *Animal and Man in Holland's Past*, p. 105; Boeles, *Friesland tot de elfde eeuw*, pp. 129–30; P. J. A. Mensch, "Dierresten uit de Polder Achthoven (Gem. Leiderdorp)," *Westerheem* 24 (1975): 111, 114.

[98] Bloemers, "Rijswijk (Z.H.) 'De Bult,' een nederzetting van de Cananefaten," p. 106, attributes the abandonment of the settlement at Rijswijk to deteriorating drainage.

[99] Fockema Andreae, "De Rijnlandse kastelen en landhuizen in hun maatschappelijk verband," in S. J. Fockema Andreae et al., eds., *Kastelen, ridderhofsteden en buitenplaatsen*

exploitation no doubt would have been geared once again toward self-sufficiency on the local scale. Only with the gradual repopulation of the western Netherlands in the sixth and seventh centuries did the basic subsistence patterns of such self-sufficient communities once again become altered and skewed by inclusion in a larger social system. This time they became part of the Greater Frisian world.[100]

It was the extensive commercial activity of the Frisians that most affected the patterns of environmental exploitation in the western Netherlands during the early Middle Ages. Because the Frisian districts encompassed the delta system of the Rhine and Maas rivers, its inhabitants were exposed to whatever trade impulses may have existed in northwestern Europe. It is only natural that those who lived in this soggy environment, who knew the lay of the land and water the best, should take part in at least the local portions of the major trading routes that passed through the region. Indeed, all the evidence shows that early-medieval Frisians knew boats and sailing very well. As the previously mentioned anonymous biographer of Saint Boniface put it, they lived in water like fish and traveled by boat almost exclusively.[101]

Initially, trade in cereals and other commodities that were scarce in the coastal regions of the Netherlands made possible the surprisingly large populations of some of the Frisian districts, where demographic growth far outstripped local crop-raising efforts to provide what was needed.[102] By the early Middle Ages, however, the involvement of some coastal inhabitants in commerce had begun to transcend purely local needs. Frisians in particular had learned to capitalize on their location and expertise to become the merchants par excellence of northwestern Europe.[103] The Frankish conquest of the Netherlands during the eighth century actually seems to have

---

in Rijnland, p. 1; Sarfatij, "Friezen-Romeinen-Cananefaten," *Holland: regionaal-historisch tijdschrift* 3 (1971): 175–76.

[100] See TeBrake, "Ecology and Economy in Early Medieval Frisia," pp. 10–11.

[101] Anonymous of Utrecht, *II vita S. Bonifacii*, 1.9; Blok, *De Franken in Nederland*, p. 119.

[102] Slicher van Bath, "The Economic and Social Conditions of the Frisian Districts from 900 to 1500," pp. 98–102, 131–33.

[103] B. H. Slicher van Bath, "Problemen rond de Friese middeleeuwsche geschiedenis," in his *Herschreven historie: schetsen en studiën op het gebied der middeleeuwse geschiedenis*, pp. 260, 279; Niermeyer, *De wording van onze volkshuishouding: hoofdlijnen uit de economische geschiedenis der noordelijke Nederlanden in de middeleeuwen*, pp. 13, 25; TeBrake, "Ecology and Economy in Early Medieval Frisia," pp. 24–25; Brongers et al., "Prehistory in the Netherlands," pp. 25–33.

enhanced Frisian commerce between the world accessible by way of the North Sea and the European interior. Under royal protection it achieved its greatest extent between 750 and 850, when permanent Frisian trading colonies were established in Frankish lands, England, and Scandinavia.[104] In fact, the North Sea was known as the Frisian Sea for a while during the early Middle Ages.[105]

Subsistence patterns in the coastal districts of the Netherlands contributed to the ability of Frisians to excel in the trade of early-medieval Europe. Basic provisioning in the Frankish world increasingly came to revolve around arable agriculture. Yet a predominantly arable rural economy, particularly the Frankish type known as manorialism, tended to bind cultivators to the soil and to impose upon them a very rigid regimen of work dictated by times of planting, plowing, and harvesting.[106] Pastoralism, in contrast, did not bind individuals to the soil, nor did it impose upon them a very rigorous work schedule. In fact, given the miserably low crop yields known from the early Middle Ages, the return on labor for livestock keeping may have been as much as five times that of cereal production.[107]

The most serious problem associated with a pastoral economy is the danger of overgrazing, but in the water-rich environments of the Netherlands coastal districts, this was not a serious consideration. People like the Frisians, therefore, who emphasized livestock keeping as their primary agricultural activity, could achieve high productivity without backbreaking or endless toil. Indeed, by leaving the care of their herds and their small expanses of fields to the women, children, and elderly, many Frisian males could spend substantial portions of each year away from home as merchants and sailors in the carrying trade of northwestern Europe, something the average Frankish peasant could never do.[108]

[104] TeBrake, "Ecology and Economy in Early Medieval Frisia," pp. 1, 24–25, 27; Jellema, "Frisian Trade in the Dark Ages," *Speculum* 30 (1955): 25–34.

[105] Jappe Alberts and Jansen, *Welvaart in wording*, p. 24; *De "Noordzeecultuur": een onderzoek naar de cuturele relaties van de landen rond de Nordzee in de vroege middeleeuwen*, p. 17.

[106] See, for example, Diana Shard, "The Neolithic Revolution: An Analogical Overview," *Journal of Social History* 7 (1973–74): 165–70; B. H. Slicher van Bath, "Volksvrijheid en democratie," in his *Herschreven historie schetsen en studiën op het gebied der middeleeuwsche geschiedenis*, pp. 305–15.

[107] This is the calculation of Cooter, "Preindustrial Frontiers and Interaction Spheres: Aspects of the Human Ecology of Roman Frontier Regions in Northwest Europe" (Ph.D. diss, University of Oklahoma, 1976), pp. 44, 233.

[108] Slicher van Bath, "The Economic and Social Conditions of the Frisian Districts from

The less rigorous labor demands of keeping livestock in comparison to raising crops offered an additional advantage to Frisians. Many who did not leave home each year to take part in the trade of northwestern Europe nevertheless were given sufficient time away from the work of basic subsistence to carry on various kinds of industrial activities. Of course, even self-sufficient communities would be required to devote at least some of their time to art and craft activities if they were to have tools, equipment, shelters, pottery, jewelry, and the like, and archaeology has in fact provided ample evidence for such from all prehistoric periods of the western and northern Netherlands.[109] By the early Middle Ages, however, industrial production had grown considerably in the Frisian districts. It was no longer geared strictly to local needs but had begun to produce a surplus that could be transported elsewhere by Frisian merchants.[110]

Trade, perhaps supplemented by piracy from time to time,[111] made the coastal districts of the Netherlands rather distinctive by early-medieval standards. It allowed Frisians to specialize in the kinds of environmental exploitation that were best suited to local natural conditions and to their own expertise. As a result, the western and northern Netherlands saw considerable population growth, beyond what would have been possible without trade, as well as a high degree of prosperity.

Ecological Patterns in Early-Medieval Frisia

After such a long and tortuous discussion of the various kinds of archaeological and ecological evidence of prereclamation environmental exploitation in the coastal regions of the Netherlands, it may be useful to summarize briefly what is most relevant and to draw a few conclusions. To compensate for a lack of material for the ninth and tenth centuries specific to Rijnland, I have tried to determine what the traditional patterns of land use were in the coastal regions generally and how these patterns matched the limitations and potentials posed by the wet environment of these low-lying areas. To this end I first looked at the varieties of environmental exploitation that are known to have been practiced from the late Neolithic

---

900 to 1500," pp. 104, 106; Slicher van Bath, "Problemen rond de Friese middeleeuwsche geschiedenis," pp. 270–71.

[109] See, for example, Brongers et al., "Prehistory in the Netherlands," passim.

[110] TeBrake, "Ecology and Economy in Early Medieval Frisia," pp. 24–25.

[111] Ibid., p. 25 n. 87.

period through the early Middle Ages. Next I attempted to discover the patterns of spatial location within prereclamation settlements by matching the various kinds of exploitation to specific portions of settlement territories, differentiated according to the quality of drainage. Further, I made an effort to understand the proportions of the various types of environmental exploitation that were known to have been practiced. Finally, I tried to determine how the inclusion of the coastal regions in larger social systems might have affected patterns of exploitation.

All the evidence I know of suggests that ninth- and tenth-century Rijnlanders exploited their natural surroundings along lines that had been pioneered and perfected much earlier and were known throughout the coastal regions. They engaged in activities that were well suited to the potentials offered by their watery environment. Along the sandy ridges on which they had built their residences, Rijnlanders also laid out fields for cereal cultivation, delineated by networks of ditches to aid drainage. In the depressions filled with peat and clay between dune ridges and in the transitional zones between occupied ridges and the empty peat bogs, they grazed the livestock that provided not only meat and milk products but also hides and wool. Finally, in the always-plentiful waters, they fished and no doubt trapped or hunted waterfowl. Fifty percent or more of their subsistence requirements were met by their grazing activities alone. Inclusion of Rijnland in the world of Greater Frisia, if anything, actually reinforced the central importance of livestock keeping.

What I have drawn in this chapter are the vague outlines of a way of life applicable to the coastal regions of the Netherlands during the early Middle Ages that differs significantly from the one applying to higher-lying areas derived from Frankish sources. Instead of a world revolving around manors or large estates with their all-important fields, I have described one revolving around the rearing of cattle and sheep that was not conducive to the presence of a parasitic landlord class. It is tempting to speculate on the existence of other ecological and social patterns keyed to local environments. In addition, while some of the early observers of the coastal regions rightly pointed to differences between themselves and coastal inhabitants,[112] they were wrong to think of themselves as superior in every way. In terms of material culture, Frisians had certain advantages

[112] See the comments by Pliny and the anonymous priest from Utrecht referred to at the beginning of this chapter.

in physical geography, location, and environmental exploitation that the politically dominant Franks did not have.

Frisians were able to achieve considerable wealth and well-being during the early Middle Ages by engaging in trade to a degree unknown in other places. This was made possible, first, by their location at the juncture of the most important rivers of northwestern Europe with the North Sea. By obtaining most of their subsistence from livestock and livestock products, Frisians had the collateral to import those commodities not available or scarce locally, especially cereals, permitting population densities that would have been impossible without trade. Finally, greater population density encouraged an early craft specialization that in turn produced further items for trade. They had time-tested patterns of environmental exploitation at their disposal that, despite what would seem to many a truly dismal physical setting, allowed them to prosper.

PART THREE
Change in Rural Society

6

The Reclamation and Settlement of Rijnland's Peat Bogs

As late as the early tenth century, settlement within what became Rijnland continued to be confined exclusively to the dune ridges and riverbanks. The huge, raised peat bogs north and south of the Oude Rijn, in contrast, continued to constitute the wilderness beyond the realm of normal human affairs, where the imprint of culture, if it existed at all, was fleeting and inconsequential. Such persistence stemmed from the fact that for nearly three and a half millennia, ever since the late Neolithic, the criteria for selecting specific sites for settlement had not changed. Only the dune ridges and riverbanks were drained sufficiently, by virtue of elevation, slope, and soil type, to support a full-time human presence.

By the second half of the fourteenth century, however, this situation was drastically altered. In the intervening period the peat bogs of Rijnland, as well as all others in the western Netherlands, were reclaimed and filled with dozens of settlements. Human artifice, in the form of drainage ditches, canals, dikes, dams, and sluices, was able to provide a quality of drainage that nature could not. In lowering the water table within the peat bogs by such artificial means, Rijnlanders were able to alter the pristine character of the wilderness that surrounded them and eventually to integrate it totally into the world of human affairs.

The transformation of Rijnland's landscape occurred in conjunction with a considerable growth in population. In this respect it resembled what was happening generally in temperate Europe during the High Middle Ages.[1] When it was all over, the settled area of Rijnland was nearly three times what it had been at the beginning, while the density of population increased as well. Most of the growth in population must have resulted

[1] See above, chap. 2.

from local natural increase, since there is no evidence of any significant migration into Rijnland from other areas. To all intents and purposes the reclamation and colonization of Rijnland's peat-bog wilderness was carried out by Rijnlanders originating in the nearby dune and riverbank settlements. Indeed, population growth in Rijnland, and in the western Netherlands generally, was sufficient not only to fill up local wilderness areas but also to provide colonists for numerous land-reclamation and colonization schemes elsewhere, espcially in northwestern and northern Germany.[2]

The Evidence of Change

A familiar problem presents itself once again: the lack of the kinds of directly relevant manuscript or printed sources that usually form the basis of historical inquiry. Documentary evidence becomes available in significant quantities only for the latter part of the reclamation period or thereafter. It is necessary, therefore, to rely once again on a less traditional source of historical information. In this case it is toponymics that is most helpful. An examination of toponyms, or place-names, clearly points the way to the kinds of changes that occurred in the settlement patterns of Rijnland during the era of reclamation.

The major source of prereclamation place-names is the list of possessions of Saint Martin's Church at Utrecht, drawn up, as we saw earlier, during the first half of the tenth century. This document was actually based on data from before and perhaps during the time of Viking control of the western Netherlands, thus covering approximately the second half of the ninth century and the first decades of the tenth century.[3] While it does not include all the names of that period, the selection is large enough to provide a substantial sample of place-names from times before the period of

[2]Richard Koebner, "The Settlement and Colonization of Europe," in M. M. Postan, ed., *The Agrarian Life of the Middle Ages*, 2d ed., pp. 84–88, vol. 1 in *The Cambridge Economic History of Europe*; Alan Mayhew, *Rural Settlement and Farming in Germany*, pp. 49–50.

[3]See D. P. Blok, "Het goederenregister van de St.-Maartenskerk te Utrecht," *Mededelingen van de Vereniging voor Naamkunde te Leuven en de Commissie voor Naamkunde te Amsterdam* 33 (1957): 89–102. An indispensable guide to all pre-1226 place-names from the Netherlands, Belgium, Luxemburg, northern France, and western Germany is Maurits Gysseling, *Toponymisch woordenboek van België, Nederland, Luxemburg, Noord-Frankrijk en West-Duitsland (vóór 1226)*.

reclamation. In all it lists over forty names that can be placed with reasonable certainty in what was to become Rijnland.[4]

The prereclamation place-names of Rijnland contain two types that are important here. First, there is a group that refers to various elements of the natural landscape. Some of these consist of a suffix added to a prehistoric water name, seen in *Leid*en or *Venn*apan (as in Nieuw Vennep).[5] Others contain an element designating woods, as in *Holt*lant (later, Holland), or *geest* (the Older Dune ridges), as in Osgeres*gest* (Oegstgeest today).[6] Many of these place-names have survived to the present. Second, there is a much larger group that refers to places of habitation. These are true settlement names, usually formed by adding one of the following endings to the name of a family or individual: *bur*, *buren* ("neighborhood" or "village"); *dorp* ("field," "estate," or "village"); *heem* ("house" or "place of residence"); and *ing*, *ingen* ("house" or "place of residence").[7] Many of these true settlement names did not survive the reclamation period. However, all the names from the Saint Martin's list, whether they referred to elements of the natural landscape or to settlements, had one thing in common: all referred to places in the dune or riverbank districts of Rijnland.

The modern network of place-names in Rijnland is significantly different from the ninth or tenth century one. Most of today's toponyms contain elements referring to features of the reclaimed or humanized landscape, elements such as dikes, dams, bridges, canals, and a particular type of reclamation village (for example, Woudsen*dijk*, Leidschen*dam*, Wou-

[4]See app. A.

[5]Gysseling, *Toponymisch woordenboek*, pp. 603, 1003; D. P. Blok, "Probleme der Flussnamenforschung in den alluvialen Gebieten der Niederlande," in Rudolf Schützeichel and Matthias Zender, eds., *Namenforschung: Festschrift für Adolf Bach zum 75. Geburtstag am 31 Januar 1965*, pp. 212–27. D. P. Blok, "De vestigingsgeschiedenis van Holland en Utrecht in het licht van de plaatsnamen," in M. Gysseling and D. P. Blok, *Studies over de oudste plaatsnamen van Holland en Utrecht*, p. 14, offers a list of such prehistoric hydronyms. M. Schönfeld, *Nederlandse waternamen*, pp. 113–23, 150–54, classifies them and discusses their composition.

[6]H. J. Moerman, *Nederlandse plaatsnamen: een overzicht*, pp. 70–71, 102–103; S. J. Fockema Andreae, *Poldernamen in Rijnland*, p. 12; H. van der Linden, *De Cope: bijdrage tot de rechtsgeschiedenis van de openlegging der Hollands-Utrechtse laagvlakte*, pp. 354–61; and Gysseling, *Toponymisch woordenboek*, pp. 506, 757–58.

[7]Blok, "De vestigingsgeschiedenis van Holland en Utrecht in het licht van de plaatsnamen," pp. 27–28. See the entries for these elements in Moerman, *Nederlandse plaatsnamen*, pp. 44–45, 54–55, 85–93, 113–17.

brugge, Rijp*wetering*, and Bos*koop*, respectively).[8] Names such as these, found almost exclusively in the former peat-bog districts, are unknown from the first half of the tenth century. They are attributable to the great wave of reclamation and settlement that swept across Rijnland's wilderness between the late tenth and late fourteenth centuries.[9] A comparison of toponyms mentioned in the Saint Martin's list of the early tenth century with those of today thus indicates both continuity and change: continuity in the persistence of names designating elements of the natural landscape and change in the introduction of new place-name types reflecting the reclamation and settlement of the peat bogs.

It is the changes in the place-names, in particular, that are of importance here. As a result of the reclamation and settlement of Rijnland's peat bogs between the late tenth and late fourteenth centuries, new types of place-names appeared in such quantities that they began to outnumber older forms, and many of the names in the Saint Martin's list, especially the true settlement names referred to above, disappeared during the same period. Of the forty-three names on the list applicable to Rijnland, at least thirteen were settlement names of identical construction, each ending with the suffix *heem*.[10] In every case such a name designated a dwelling place, whether an agricultural complex, a daughter settlement, or perhaps a single house.[11] Most of the other place-names mentioned in the list represented settlements that cannot have been very much larger.

[8] See the italicized elements in Moerman, *Nederlandse plaatsnamen*.

[9] The preceding analysis and comparison of place-name types is based on the examination of place-names in Holland between the Maas and the IJ by D. P. Blok, "Plaatsnamen in Westfriesland," *Philologia Frisica anno 1966*, pp. 11–14. He found that the Saint Martin's list contained 76 names of settlements attributable to that area.

[10] The early-medieval equivalent of *heem* was *hem*. The 13 names of this type in Rijnland were Upuuilcanham, Loppishem, UUatdinchem, Suthrem, Rothulfuashem or Hrothaluashem, Heslem, Osbragttashem, Hostsagnem, Osfrithem, UUestsagnem, Oslem, UUilkenhem, Lethem, and also, perhaps, Heslemaholta. For their approximate modern equivalents or probable locations, see app. A. For the meaning, etymology, and European distribution of place-names ending in *heem*, see the discussion and analysis of Karl Roelandts, "*Sele* und *Heim*," in Schützeichel and Zender, eds., *Namenforschung: Festschrift für Adolph Bach zum 75. Geburtstag am 31. Januar 1965*, pp. 273–99; A. Russchen, "Tussen Aller en Somme," *It Beaken: tydskrift fan de Fryske Akademy* 29 (1967): 95.

[11] Blok, "Plaatsnamen in Westfriesland," p. 14; D. P. Blok, "Histoire et toponymie: l'example des Pays-Bas dans le haut moyen âge," *Annales: économies, sociétés, civilisations* 24 (1969): 937; Roelandts, "*Sele* und *Heim*," p. 280; the entry for *-heem* in Moerman, *Nederlandse plaatsnamen*, pp. 85–93. The entry for Rothulfuashem in Gysseling, *Topo-*

As the people living in the old centers of habitation along the dunes and riverbanks began the reclamation and settlement of the peat-bog wilderness around them, and as Rijnland began receiving its first subdivisions in the form of parishes and administrative centers, the function of many of the old, very small settlements began to change or even to disappear entirely. This is well illustrated by the example of Rothulfuashem, mentioned twice in the list of Saint Martin's possessions. The second time it received the qualification "which is called Rinasburg." [12] The ending *burg* signified a fortification, presumably built sometime around 900 in an effort to protect the mouth of the Oude Rijn from further Viking incursions. As such it resembled similar fortifications built to protect the mouths of other rivers along the southern North Sea in Zeeland and Flanders.[13] Apparently this stronghold was constructed near the dwelling of one Rothulf. A short time later, however, when the Saint Martin's list was assembled, this *burg* near the mouth of the Oude Rijn had begun to eclipse the earlier settlement in importance. Consequently, the village that later appeared there, Rijnsburg, took its name from the fortification near the mouth of the Oude Rijn, not from the homestead of Rothulf.[14]

Such changes in the relative importance of certain old settlement nuclei no doubt go a long way toward explaining why, of the thirteen names in the Saint Martin's list ending in *heem* and referring to places in Rijn-

---

nymisch Woordenboek, p. 864, is typical and instructive: "woning van Hrōthiwulf," or Rothulf's dwelling.

[12] "Hrothaluashem quod modo dicitur Rinasburg," in M. Gysseling and A. C. F. Koch, eds., *Diplomata Belgica ante annum millesimum centesimum scripta*, pt. 1, *Teksten*, no. 195; app. A, below.

[13] D. P. Blok, "De Wikingen in Friesland," *Naamkunde* 10 (1978): 37 n. 34, p. 42; D. P. Blok, "Holland und Westfriesland," *Frühmittelalterliche Studien: Jahrbuch des Instituts für Frühmittelalterforschung der Universität Münster* 3 (1969): 353; Blok, "Plaatsnamen in Westfriesland," p. 13; H. Sarfatij, "Die Frühgeschichte von Rijnsburg (8.–12. Jahrhundert): Ein historisch-archäologischer Bericht," in B. L. van Beek, R. W. Brandt, and W. Groenman–van Waateringe, eds., *Ex Horreo: I.P.P. 1951–1976*, pp. 198–99; H. Halbertsma, "The Frisian Kingdom," *Berichten van de Rijksdienst voor het Oudheidkundig Bodemoncerzoek* 15–16 (1965–66): 932; A. E. Verhulst, "Historische geografie van de Vlaamse kustvlakte tot omstreeks 1200," *Bijdragen voor de geschiedenis der Nederlanden* 14 (1959–60): 6.

[14] Blok, "Het goederenregister van de St.-Maartenskerk te Utrecht," pp. 89–104, demonstrates that the list was drawn up sometime between 918 and 948. See also Blok, "Plaatsnamen in Westfriesland," p. 13.

land, only one name, Sassenheim, survived the age of reclamation.[15] New churches and administrative centers could also serve as the focal points for growing concentrations of population that eventually overshadowed the older, smaller settlements.[16]

The most powerful agent in the disappearance of old place-names, however, was the great wave of reclamation that began sweeping across the peat-bog wilderness around the end of the tenth century. First, reclamation profoundly changed the appearance of the landscape, the flow of waters, and the administrative structure of Rijnland. Second, it radically shifted the center of gravity of Rijnland's population as large numbers of people left the safety and security of the dune ridges and riverbanks to settle on the surrounding peat bogs. The older centers of population, if they continued to exist, could be easily absorbed or overshadowed by the new, bustling settlements that grew up in association with the waterways, dams, dikes, and bridges that were constructed during the period of reclamation.[17] The extensive toponymic changes, both the introduction of new types and the loss of many of the old forms from the early tenth century, were, therefore, symptomatic of a profound settlement revolution in Rijnland that went hand in hand with the reclamation of the peat bogs.[18]

The Chronology of Reclamation and Settlement

The reclamation and settlement of the peat bogs of the western Netherlands began in a number of widely scattered areas before the beginning of the eleventh century. For example, in West Friesland, what is today northeastern North Holland, some peat areas were occupied and therefore reclaimed during the eighth and ninth centuries, though not necessarily permanently. There is considerable evidence of reclamation activity once

[15] Hillegom, though not mentioned in the list of possessions of Saint Martin's Church at Utrecht, is also a *heem* name and thus belongs to the same period. See Roelandts, "*Sele* und *Heim*," pp. 273–99; Russchen, "Tussen Aller en Somme," p. 95. Blok, "Plaatsnamen in Westfriesland," pp. 11–13, found that only 2 of the 27 *heem* names from the Saint Martin's list applicable to Holland between the Maas and the IJ, namely, Sassenheim and Haarlem, survived the reclamation period.

[16] Blok, "Plaatsnamen in Westfriesland," p. 14.

[17] Ibid.; Blok, "Histoire et toponymie," p. 941; Blok, "Holland und Westfriesland," p. 360; M. N. Acket, "De Oude Rijn en zijn omgeving," *Leids jaarboekje* 45 (1953): 87–88.

[18] Blok, "Plaatsnamen in Westfriesland," uses the term "revolution" to describe both the toponymic and the settlement changes of this period.

again beginning in the tenth century.[19] In Kennemerland, directly north of Rijnland, the process was well under way in the raised peat bogs during the second half of the tenth century. There the inhabitants of the coastal dunes worked their way eastward up the small rivers and creeks belonging to the natural radial drainage network of the peat cushions until they arrived at the raised centers of the bogs. The eastern edge of the county eventually came to be drawn along the watershed line that divided westward- from eastward-flowing water. The pace was so rapid, especially north of the IJ, that in some areas the eastern boundary was achieved before the beginning of the eleventh century.[20] A third area of peat reclamation lay east of Rijnland on the north side of the Oude Rijn, in an area known as Miland.[21] Finally, southwest of Rijnland, in the county of Flanders, occupation of some peat lands apparently occurred at an early date, indicating reclamation there as well.[22]

Reclamation also began at an early date along the Maas River near Vlaardingen, south of Rijnland, as indicated by a description of military activities from the early eleventh century. Apparently Count Diederik III (995–1039) had built a small fortification along the Merwede (the name given to the wide channel of the Maas at that point) and, using it as a base of operations, had attempted to collect tolls from merchants on their way to and from England. Because this activity conflicted sharply with imperial policy, which reserved such tolls for the emperor as one of the *regalia*, or royal rights, and because it hindered trade along the Maas, Bishop Adelbold of Utrecht and the merchants of Tiel brought charges against the presumptuous Diederik at the imperial diet convened by Emperor Henry II at Nijmegen in 1018. The emperor gave the task of restoring the imperial prerogative along the Merwede-Maas to Godfried, duke of Lower Lotharingia, and to the bishops of Utrecht, Liège, and Cologne. In the fall of 1018 a military force landed near Vlaardingen for that purpose. The expedition

[19] J. K. de Cock, "Veenontginningen in West-Friesland," *West-Frieslands oud en nieuw* 36 (1969): 154–71; G. J. Borger, *De Veenhoop: een historisch-geografisch onderzoek naar het verdwijnen van het veendek in een deel van West-Friesland*, pp. 215–17; J. C. Besteman and A. J. Guiran, "Het middeleeuws-archeologisch onderzoek in Assendelft, een vroege veenontginning in middeleeuws Kennemerland," *Westerheem* 32 (1983): 146.

[20] J. K. de Cock, *Bijdrage tot de historische geografie van Kennemerland in de middeleeuwen op fysisch-geografische grondslag*, pp. 254–59 and passim; Besteman and Guiran, "Het middeleeuws-archeologisch onderzoek in Assendelft," pp. 144–76.

[21] C. J. van Doorn, *Het oude Miland en zijn waterstaatkundig ontwikkeling*, p. 14.

[22] Verhulst, "Historische geografie van de Vlaamse kustvlakte," pp. 6–9.

was a dismal failure: Diederik and his Frisians routed the invaders and captured the duke. A contemporary describing the engagement said that one of the factors affecting the outcome of the battle was a vast network of ditches which he said had been dug either to aid in drainage of the low-lying land or to confound the imperial forces.[23] No doubt their purpose was to lower the water table in this wet landscape so that the surface could dry out. During the chaotic retreat, however, many of the hapless invaders fell prey to these peculiar ditches.

Rijnlanders proved to be no less capable than were their coastal contemporaries at transforming wilderness into land fit for human use and occupation, though exactly when they began doing so is not easy to say. By the early eleventh century, however, they had already reclaimed and settled significant portions of Rijnland's massive peat bogs. A document dating from around 1040 referred to three villages furnished with churches that were situated in the heart of the raised peat bog north of the Oude Rijn: Leimuiden, Rijnsaterwoude, and Esselijkerwoude, listed as daughter churches of the church at Oegstgeest (or Kirichwereve).[24] They were mentioned again in a charter of 1063.[25]

All three of these settlements lay well outside the areas of traditional settlement, the coastal-dunes or river-clay areas. Even today, after more than nine centuries of human occupation in the form of arable and pastoral agriculture and the digging of peat for fuel, the soil is still predominantly peat.[26] It is important to recall at this juncture some characteristics of

[23] Alpertus Mettensis, *De diversitate temporum*, 2.20: "Una res erat illis magno usui: quod campum omnem fossis prefoderant, sive ad defendendum majorem estum maris, qui in plenilunio validior solet fieri, sive ad impendiendum iter hostium. . . .celeri cursu [Frisii] prevenerunt ad litus et Complures qui per crepidinem littoris in aqua manibus reptabant, jaculis confodiunt." For the politics behind this struggle, see Blok, "Holland und Westfriesland," pp. 347–61; I. H. Gosses, *Handboek tot de staatkundige geschiedenis der Nederlanden* 1:56–58; I. H. Gosses, "De vorming van het graafschap Holland," in his *Verspreide geschriften*, ed. F. Gosses and J. F. Niermeyer, pp. 306–307, 330–34; Otto Oppermann, *Die Grafschaft Holland und das Reich bis 1256*, pp. 12–15, vol. 2 in *Untersuchungen zur nordniederländischen Geschichte*; H. Halbertsma, *Terpen tussen Vlie en Eems: een geografisch-historische benadering*, pp. 199–204.

[24] See the list in Otto Oppermann, ed., *Fontes Egmundenses*, p. 255, dated to ca. 1040. See also D. P. Blok, "De Hollandse en Friese kerken van Echternach," *Naamkunde* 6 (1974): 167–84; Otto Oppermann, *Die Egmonder Fälschungen*, pp. 82–84, vol. 1 in *Untersuchungen zur nordniederländischen Geschichte des 10. bis 13. Jahrhunderts*; de Cock, *Bijdrage tot de historische geografie van Kennemerland*, p. 112.

[25] A. C. F. Koch, ed., *Oorkondenboek van Holland en Zeeland tot 1299*, vol. 1, no. 84.

[26] *Bodemkaart van Nederland, schaal 1:200,000*, sheet 6.

raised peat bogs, or cushions. Capillary action within the bogs would keep their surfaces perpetually wet, the prerequisite for peat growth, rendering them totally unfit for human presence. For this reason a considerable amount of artificial drainage, sufficient to lower the water table in the peat cushion so that its surface could dry out, was necessary before the villages of Leimuiden, Rijnsaterwoude, and Esselijkerwoude could be established.[27]

That Rijnlanders living in these villages mastered the art of drainage at an early date is suggested by a certain transaction that took place in the early twelfth century. In 1113, Friedrich I, archbishop of Hamburg, announced that, after careful consideration, he was going to grant some swampy, uncultivated land near Bremen to a group of colonists who promised to reclaim it and put it under cultivation. According to the archbishop, the colonists were known as Hollanders who came from "this side" (that is, the Hamburg side) of the Rhine.[28] At the beginning of the twelfth century, the name Holland was used to refer only to a portion of Rijnland; in fact, it is possible to link it directly to the newly reclaimed peat bogs on either side of the Oude Rijn. Further, an area that included Leimuiden, Rijnsaterwoude, and Esselijkerwoude was often referred to during the Middle Ages as "*veenland beoosten Rijn*", that is, the *veenland*, or peat, east of the Rhine. It is this portion of Rijnland that must be seen as the place of origin of the Bremen colonists.[29] What is important at this junc-

[27] G. J. Borger, "Ontwatering en grondgebruik in de middeleeuwse veenontginningen in Nederland," *Geografisch tijdschrift* 10 (1976): 345–46; and the section on peat growth in chap. 3, above.

[28] The text of the contract between Archbishop Friedrich and the colonists is in Koch, ed., *Oorkondenboek van Holland en Zeeland tot 1299*, vol. 1, no. 98; also, among many others, in G. Franz, ed., *Quellen zur Geschichte des deutschen Bauernstandes im Mittelalter*, no. 67. The contract began with "Pactionem quandam quam quidam cis Renum commanentes, qui dicuntur Hollandi, nobiscum pepigerunt, omnibus notam volumus haberi. Prefati igitur viri misericordiam nostram convenerunt obnixe rogantes, quatenus terram in episcopatu nostro sitam actenus incultam paludosamque, nostris indigenis superfluam, eis ad excolendam concederemus. Nos itaque tali petitione nostrorum usi consilio fidelium, perpendentes rem nobis nostrisque successoribus profuturam, non abnuende petitioni eorum assensum tribuimus." On the date 1113, instead of 1106, as is usually given for this document, see A. C. F. Koch, "Die Datierung des Vertrags Friedrichs I., Erzbischofs von Hamburg, mit den holländischen Ansiedlern bei Bremen," in *Miscellanea mediaevalia in memoriam Jan Frederik Niermeyer*, pp. 211–15. In the English version of the text in Georges Duby, *Rural Economy and Country Life in the Medieval West*, trans. Cynthia Postan, pp. 392–93, the phrase "cis Renum" is rendered "beyond the Rhine"; it should read "this side of the Rhine," that is, the Hamburg, or east, side of the Oude Rijn.

[29] Van der Linden, *De Cope*, pp. 17–69, very convincingly establishes this Rijnland connection.

ture is that these Hollanders were able to convince an archbishop that it would be in his best interests to grant them the lands for reclamation and settlement. It is doubtful they would have been so convincing if they had not already achieved a certain amount of success at home, in the area in and around Leimuiden, Rijnsaterwoude, and Esselijkerwoude.

These three villages were not the only peat-bog settlements in Rijnland by the eleventh century. A number of others were indicated in later lists of villages that paid a special tax to the counts of Holland. In the early years of comital authority in the western Netherlands, that is, during the ninth and tenth centuries, the West Frisian counts personally traveled from community to community to preside over a judicial process called the *botding*. On such occasions the inhabitants of each community were responsible for the costs of maintaining the count and his retenue.[30] By the beginning of the twelfth century, the counts had evidently discontinued the practice of traveling in person to each community. The maintenance responsibilities were apparently converted to a fixed money payment called the *botting*, since the *Liber sancti Adalberti*, a collection of charters and property lists compiled in the monastery at Egmond, in Kennemerland, shortly after 1120, spoke of the *botding* as a money payment.[31] This tax was levied in the original *botding* villages for the remainder of the Middle Ages, but it was never collected in villages established later. Thus the *botting* lists contained only the names of settlements that existed at least by the early twelfth century.[32]

Besides mentioning the names of sixteen communities from the oldest settled areas of Rijnland, the coastal dunes and riverbanks,[33] the *botting*

[30]The exact origins, nature, and purpose of this judicial proceeding remain something of a problem. See Gosses, "Vorming van het graafschap Holland," pp. 293–305; Oppermann, *Die Egmonder Fälschungen*, pp. 7–9; Oppermann, *Die Grafschaft Holland und das Reich bis 1256*, pp. 224–25; W. J. Diepeveen, *De vervening in Delfland en Schieland tot het einde der zestiende eeuw*, pp. 14–15; van der Linden, *De Cope*, p. 18; de Cock, *Bijdrage tot de historische geografie van Kennemerland*, pp. 5–7; Blok, "Holland und Westfriesland," p. 356.

[31]"Quandam pensionem . . . que buttink dicitur," in Oppermann, ed., *Fontes Egmundenses*, p. 84. Gosses, "Vorming van het graafschap Holland," p. 293, provides the date.

[32]Gosses, "Vorming van het graafschap Holland," p. 293; de Cock, *Bijdrage tot de historische geografie van Kennemerland*, pp. 5–7; Acket, "De Oude Rijn en zijn omgeving," p. 87.

[33]The oldest surviving complete list of *botting* communities in Rijnland appeared in the comital accounts of 1334; see H. G. Hamaker, ed., *De rekeningen der graafelijkheid van Holland onder het Henegouwsche huis*, 1:161–62. Located on the river clay along the Oude

lists include the names of six villages associated with the peat bogs on either side of the Oude Rijn: Aarlanderveen, Oudshoorn, and Esselijkerwoude, north of the river, and Hazarswoude, Zoetermeer, and Zoeterwoude, south of the river.[34] Surprisingly, however, neither Leimuiden nor Rijnsaterwoude was included, even though we know that both definitely were in existence from 1040 or even earlier.

Attempts have been made to explain the omission of Leimuiden and Rijnsaterwoude from the *botting* lists, though none is very convincing. For example, I. H. Gosses, in his important study of the emergence of the county of Holland, argued that these places were so deep in the peat-bog wilderness that the counts never traveled there to hold the *botding*. Thus when the *botding* was later converted to the *botting*, the tax was not levied there.[35] Actually, however, these villages were no farther into the peat bogs than was Zoetermeer, south of the Oude Rijn, and the *botting* was collected there. Nor is it easy to imagine that Leimuiden and Rijnsaterwoude would have been more difficult to reach than a place like Rietwijk, a *botting* enclave deep in the wilderness in southern Kennemerland.[36]

More likely Leimuiden and Rijnsaterwoude were not included in the *botting* lists because they came into existence only after the *botding* fell into disuse and was converted to a fixed money payment. This means, of course, that the counts must have discontinued traveling from community to community to preside over the *botding* before 1040, since we know that Leimuiden and Rijnsaterwoude existed by that time. Documentary evidence indicates only that this change occurred sometime after 1120.[37] In

---

Rijn were Leiden, Leiderdorp, Koudekerk, and Alphen aan den Rijn; along the ridges of the dunes, Phillips ambacht Wassenaar, Dirks ambacht Zuidwijk, Valkenburg, Poelgeest, Oegstgeest, Warmond, Klein Sassenheim, Lisse, Hillegom, Voorhout, Heer Jans ambacht Noordwijk, and Jans ambacht Noordwijk (i.e., Noordwijkerhout); ibid. 1:19, 22, 161–63; Gosses, "Vorming van het graafschap Holland," p. 295 n. 11.

[34] Hamaker, ed., *De rekeningen der graafelijkheid van Holland*, 1:161. In addition, the *botting* payments of Esselijkerwoude and Zoetermeer were mentioned in a slightly earlier (1290s) document; see S. Muller Hz., ed., "Oude register van graaf Florens," *Bijdragen en mededelingen van het Historisch Genootschap* 22 (1901): 190, 220. Today the center of the village of Oudshoorn is on the river clay along the Oude Rijn. Nevertheless, van der Linden, *De Cope*, p. 18 n. 3, has been able to show conclusively, on the basis of a study of a number of old maps, that it was moved there from the peat bog on the north during the seventeenth and eighteenth centuries.

[35] Gosses, "Vorming van het graafschap Holland," pp. 293–9.

[36] See the map in Gosses, "Vorming van het graafschap Holland," facing p. 294; de Cock, *Bijdrage tot de historische geografie van Kennemerland*, p. 112.

[37] Oppermann, ed., *Fontes Egmundenses*, p. 84.

the following paragraphs I shall try to show that an early-eleventh-century date for the conversion of the *botding* into the *botting* is eminently reasonable, since the only objections to such an early date have always been based on a serious misunderstanding of what the prereclamation peat bogs were like.

Until recently the primary objection to placing the establishment of the *botting* as a fixed tax at the beginning of the eleventh century was the realization that it would mean that all *botting* villages existed by that time. Many scholars thought that this was simply too early for extensive peat-bog settlement, assuming that such settlement could have taken place only after a system of dikes and dams had been constructed to protect the inhabitants and their belongings from flooding by the sea or rivers. Further, it was believed, such a system of dikes and dams would presuppose the existence of a strong central authority that could marshal the laborers and material that would be needed, but the Holland counts possessed no such authority much before the late twelfth or early thirteenth centuries. Consequently, much time and energy were expended trying to show that dikes were built by some strong central authority, such as the Carolingian Franks or even the Romans, at a time early enough to allow for some settlement of the peat bogs of the western Netherlands by the year 1000. Gradually, however, it became obvious that no such early system of dikes and dams could be found, that there was no evidence for such construction in the Netherlands much before 1200.[38]

To avoid this apparent impasse, some scholars began suggesting that central portions of some of the peat bogs of the western Netherlands might have been occupied from time to time on a temporary basis. Leimuiden and Rijnsaterwoude, in Rijnland, presumably would have been examples of this: small, temporary settlements of hunters, fishermen, and probably

[38] See, for example, S. J. Fockema Andreae, *Het Hoogheemraadschap van Rijnland: zijn recht en zijn bestuur van de vroegsten tijd tot 1857*, p. 11; van Doorn, *Het oude Miland*, p. 68; the discussion of van der Linden, *De Cope*, pp. 62–69; Halbertsma, *Terpen tussen Vlie en Eems*, p. 207; J. F. Niermeyer, "Dammen en dijken in Frankish Nederland," in *Weerklank op het werk van Jan Romein: Liber Amicorum*, pp. 109–15; J. F. Niermeyer, "De vroegste berichten omtrent bedijking in Nederland," *Tijdschrift voor economische en sociale geografie* 19 (1958): 226–31; the literature review by A. E. van Giffen, "De ouderdom onzer dijken," *Tijdschrift van het Koninklijk Nederlands Aardrijkskundig Genootschap*, 2d ser., 81 (1964): 273–86. Verhulst, "Historische geografie van de Vlaamse kustvlakte," pp. 20–32, says that the first systematic and more than purely local construction of dikes in Flanders began during the twelfth century in response to deteriorating hydraulic conditions.

cattle keepers who occasionally used the peat bogs during the summer months, when evapotranspiration exceeded precipitation. If there were any artificial drainage works associated with these places, they were assumed to have been wholly inadequate for guaranteeing a permanent, year-round presence and perhaps quickly fell into disrepair. Such temporary settlements would not have been important enough for the counts to visit for purposes of holding the *botding*. Extensive permanent settlement, these writers insisted, was not possible before the thirteenth century, when the system of dikes, dams, and sluices came into being.[39]

In fact, occasional use of the peat-bog wilderness may well have been possible before reclamation, as indicated by the archaeological evidence of a temporary extraction camp dated to the Neolithic period recently uncovered at Bergschenhoek.[40] However, such temporary emplacements, or any other kind of less-than-permanent settlement, for that matter, would not have been equipped with churches, as Leimuiden and Rijnsaterwoude were. In fact, the existence of churches in these places can be used to suggest that reclamation, settlement, and the laying out of parishes had already taken place there.[41]

New findings in geological, geographical, and pedological research since World War II have seriously undermined the notion that permanent settlement of the peat bogs of the western Netherlands was impossible before a system of dams, dikes, and sluices was constructed. This view was too closely tied to the hydraulic conditions that prevailed in the peat regions at the end of the Middle Ages or even later. It was mistakenly assumed that those areas in which the surface was the lowest after centuries of artificial drainage were also the lowest before reclamation began. Much new information is now available concerning the growth and characteris-

[39] See, for example, Gosses, "Vorming van het graafschap Holland," p. 300; S. J. Fockema Andreae, ed., *Rechtsbronnen der vier hoofdwaterschappen van het vasteland van Zuid-Holland (Rijnland, Delfland; Schieland, Woerden)*, pp. ix-x; the literature cited by van der Linden, *De Cope*, pp. 62–65. Verhulst, "Historische geografie van de Vlaamse kustvlakte tot omstreeks 1200," p. 8, expresses a similar opinion with respect to Flanders.

[40] H. Sarfatij, "Archeologische kroniek van Zuid-Holland over 1976," *Holland: regionaal-historisch tijdschrift* 9 (1977): 245–47; H. Sarfatij, "Archeologische kroniek van Zuid-Holland over 1977," *Holland: regionaal-historisch tijdschrift* 10 (1978): 298–99; L. P. Louwe Kooijmans, "Het onderzoek van neolithische nederzettingsterreinen in Nederland anno 1979," *Westerheem* 29 (1980): 108–10, 112; Borger, "Ontwatering en grondgebruik in de middeleeuwse veenontginningen in Nederland," pp. 345–46.

[41] C. Dekker, "De vorming van aartsdiakonaten in het diocees Utrecht in de tweede helft van de 11e en het eerste kwart van de 12e eeuw," *Geografisch tijdschrift* 11 (1977): 351.

tics of the unreclaimed bogs, making it clear that the threat of flooding by the sea or rivers was much less severe in the tenth century than it was at later times.

In short, the peat bogs of the western Netherlands were considerably higher with respect to sea level at the beginning of the Middle Ages than they were at the end of the era. It is well to remember the conditions under which bogs were formed and the shape that they took. They grew in step with the postglacial rise in sea level, and, because of capillary action within the mass of peat itself, their centers were higher than were their peripheries.[42] Drainage was accomplished by a network of ditches that lowered the water table in the bogs and started a process that resulted in a significant lowering of the peat surface. First, drainage stopped peat growth by removing the conditions under which it formed, namely sogginess, at the very time that sea level continued to rise. Second, drainage reduced the mass of the peat by extracting water, since peat could be as much as 90 percent water by volume. As a result the peat became more densely packed or compressed, resulting in a lowering of the peat surface. Third, drainage allowed the pores that were formerly filled with water to become filled with air, setting in motion a process of oxidizing the peat substance. The result was a gradual subsidence of the peat surface, which could amount to as much as 1 meter or more over a period of a century.

Such subsidence was not a uniform phenomenon but affected various parts of the bogs at different rates. The bogs originally developed in more or less bowl-shaped depressions, eventually assuming the shape of a biconvex lens. In addition, while the oligotrophic peat constituting the raised centers could consist of 90 percent water by volume, the more nutrient-rich varieties found toward the peripheries had smaller percentages and thus a smaller potential for compaction once the water was extracted. As a result, the originally raised centers of the bogs might well have become the lowest portions after prolonged drainage (see fig. 8).[43]

Before reclamation began, therefore, the peat bogs of the western

[42] See the section on peat growth above, chap. 3.

[43] T. Edelman, *Bijdrage tot de historische geografie van de Nederlandse kuststreek*, pp. 43–45; T. Edelman, "Oude ontginningen van de veengebieden in de Nederlandse kuststrook," *Tijdschrift voor economische en sociale geografie* 49 (1958): 240; van der Linden, *De Cope*, p. 68; van Doorn, *Het oude Miland*, pp. 217–18; Halbertsma, *Terpen tussen Vlie en Eems*, p. 58; W. A. Casparie, *Bog Development in Southeastern Drente*, pp. 5, 18; David B. Grigg, *The Agricultural Revolution in South Lincolnshire*, pp. 20–21, 138.

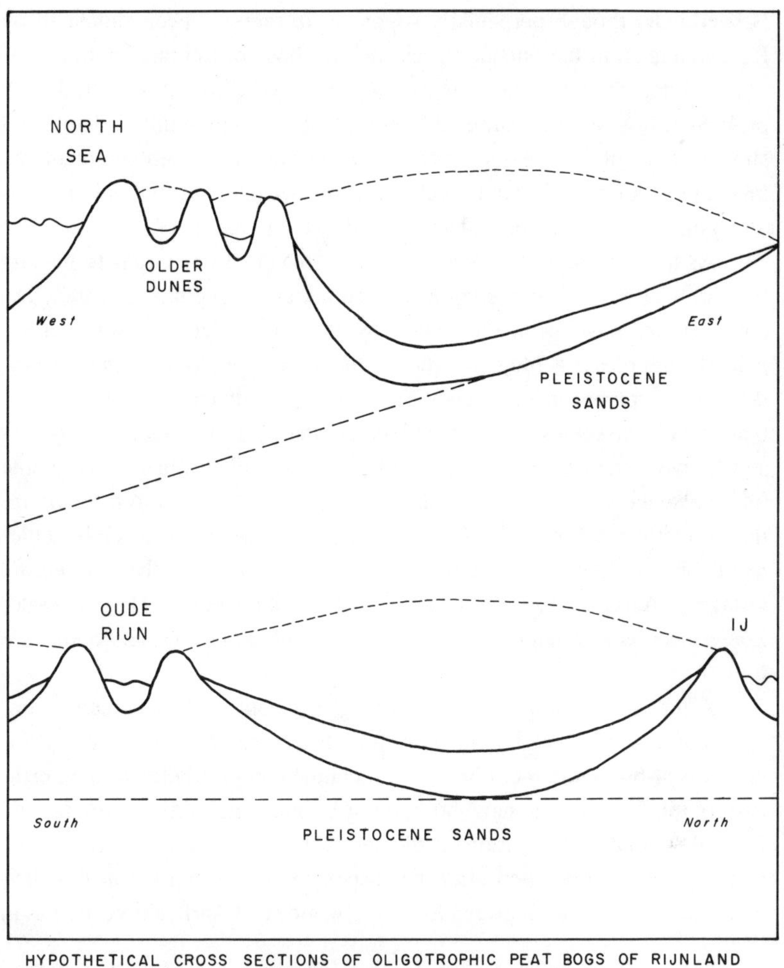

HYPOTHETICAL CROSS SECTIONS OF OLIGOTROPHIC PEAT BOGS OF RIJNLAND AFTER RECLAMATION, SHOWING INVERSION OF RELIEF CAUSED BY SUBSIDENCE AND OXIDATION. VERTICAL SCALE MUCH EXAGGERATED.

----- Original surface

——— Surface after prolonged drainage

FIGURE 8. Rijnland Bogs after Drainage

Netherlands, though perpetually soggy, were rarely if ever inundated by floodwaters from the outside. Sites such as those of Leimuiden and Rijnsaterwoude, on the higher, central portions, were certainly spared such problems. It is safe to assume, therefore, that a system of dikes, dams, and sluices was not a necessary prelude to the permanent settlement of the raised peat bogs of Rijnland. Such measures became essential only later as a response to reclamation-induced sibsidence and oxidation.[44]

As the foregoing discussion indicates, the chief objections to placing the estblishment of the *botting* as a fixed tax at the beginning of the eleventh century have been eliminated by the knowledge and perspectives gained from recent geological, geographical, and pedological research. Indeed, the apparent inconsistency between the documentation of 1040 and that of 1125 dissolves once we understand that there is no longer any reason to avoid the assumption that all the *botting* villages had in fact come into existence sometime before the *botding* was commuted to the *botting*, thus sometime before 1040. Consequently, there were at least eight settlements in the peat-bog wilderness of Rijnland by 1040: the six *botting* villages, Aarlanderveen, Oudshoorn, Esselijkerwoude, Hazarswoude, Zoetermeer, and Zoeterwoude, plus Leimuiden and Rijnsaterwoude (see fig. 9).[45]

After such an impressive beginning, the course of reclamation continued at a rapid pace. By the early fourteenth century nearly all of Rijnland's peat-bog wilderness had been reclaimed and settled.[46] In the northeastern quadrant, in the neighborhood of Leimuiden, Rijnsaterwoude, and Esselijkerwoude, three more settlements, Friezekoop, Kudelstaart, and Kalslagen, appeared soon after the suspension of the *botting* in the eleventh century. Ter Aar, between Esselijkerwoude and Aarlanderveen, was in

[44] M. C. P. Scholte, "Wat was er eerder: de dijk, de veenontginning of de polder?" *Holland: regionaal-historisch tijdschrift* 12 (1980): 1–9; T. Edelman, *Bijdrage tot de historische geografie van de Nederlandse kuststreek*, p. 57.

[45] My argument for an early-eleventh-century establishment of the *botting* as a fixed payment agrees essentially with the findings of de Cock, *Bijdrage tot de historische geografie van Kennermerland*, pp. 112, 245–59, for the area north of Rijnland. Van der Linden, *De Cope*, p. 360, says it occurred sometime before 1063, the date he accepts for the definite existence of Leimuiden and Rijnsaterwoude; on pp. 18 and 39, however, he makes reference to the list of 1040 but seemingly assumes it was contemporary to the charter of 1063.

[46] Van der Linden, *De Cope*, pp 251–72; van der Linden, *Recht en territoir*, pp. 12–14, 20–23; Gosses, "De vorming van het graafschap Holland," pp. 305–21; J. F. Niermeyer, *Wording van onze volkshuishouding: hoofdlijnen uit de economische geschiedenis der noordelijke Nederlanden in de middeleeuwen*, p. 61.

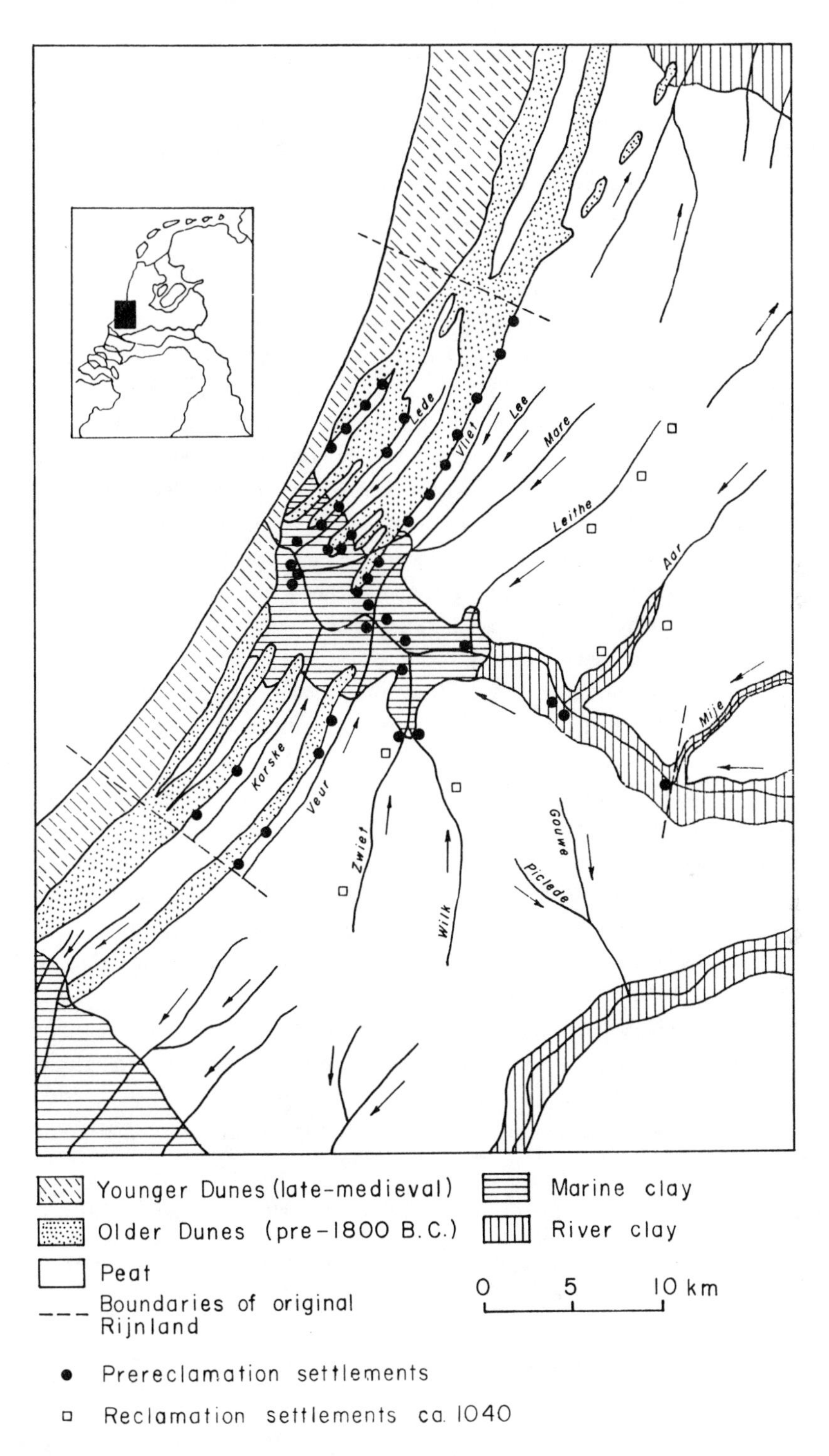

FIGURE 9. Settlement circa 1040 A.D.

existence by the late twelfth century, while Nieuwkoop, Oukoop, Achttienhoven, and Nieuwveen, farther east and north, were founded before the end of the twelfth century.[47] In the northwestern quadrant of Rijnland, west of Esselijkerwoude, Rijnsaterwoude, and Leimuiden, the communities of Hoogmade, Vrije Lage Boekhorst, Vrouweven, and Alkemade made their appearance during the thirteenth century.[48]

Across the Oude Rijn, near the peat-bog stream known as the Gouwe, Bloemendaal, north of Gouda, was in existence by the early twelfth century. Boskoop, near the source of the Gouwe, was mentioned in the early thirteenth century. By the end of the thirteenth century the entire southeastern quadrant of Rijnland, the area along both sides of the Gouwe, had been reclaimed, and Snijdelwijk, Poelien, Groensvoorde, Hubertsgerecht, Waddinxveen, Zuidwijk, Randenburg, Reyerskoop, and Middleburg had been established.[49] Also in the thirteenth century, near the *botting* communities of Zoeterwoude, Zoetermeer, and Hazarswoude, in the southwestern quadrant of Rijnland, Zegwaard, Roggeveen, Wilsveen, and Benthuizen, were settled on land reclaimed from the peat-bog wilderness.[50]

After the thirteenth century only a few widely scattered tracts of unreclaimed peat bog remained in Rijnland for allocation to colonists. De Kaag, in the northwestern quadrant, was settled around the beginning of the fourteenth century. Schoot, between Ter Aar and Nieuwkoop in the northeast, was reclaimed during the fourteenth century. In the southeastern quadrant along the west side of Waddinzveen, Coencoop was colonized in 1340, while the community of Poelien was allowed to extend farther westward in 1365. Finally, a region of peat bog later known as Hoogeveen, in the southwestern sector of Rijnland south of Hazarswoude, was reclaimed sometime before 1365 (see fig. 10).[51]

Thus virtually all of the villages established in the reclaimed peat-bog sections of Rijnland were in existence by the beginning of the fourteenth century. They were mentioned by name, for example, in the lists of communities required to pay the *bede*, a tax recorded in the surviving account

[47] Van der Linden, *De Cope*, pp. 51–56, 252–58.

[48] Ibid., pp. 267–68; Gosses, "De vorming van het graafschap Holland," p. 299.

[49] H. van der Linden *Recht en territoir: een rechtshistorisch-sociografische verkenning*, p. 22; van der Linden, *De Cope*, p. 24, 28–36, 204–205, 271–72; Gosses, "De vorming van het graafschap Holland," pp. 309–12.

[50] Van der Linden, *De Cope*, p. 270; Gosses, "De vorming van het graafschap Holland," pp. 314–15.

[51] Van der Linden, *De Cope*, pp. 29, 235–40, 258, 268.

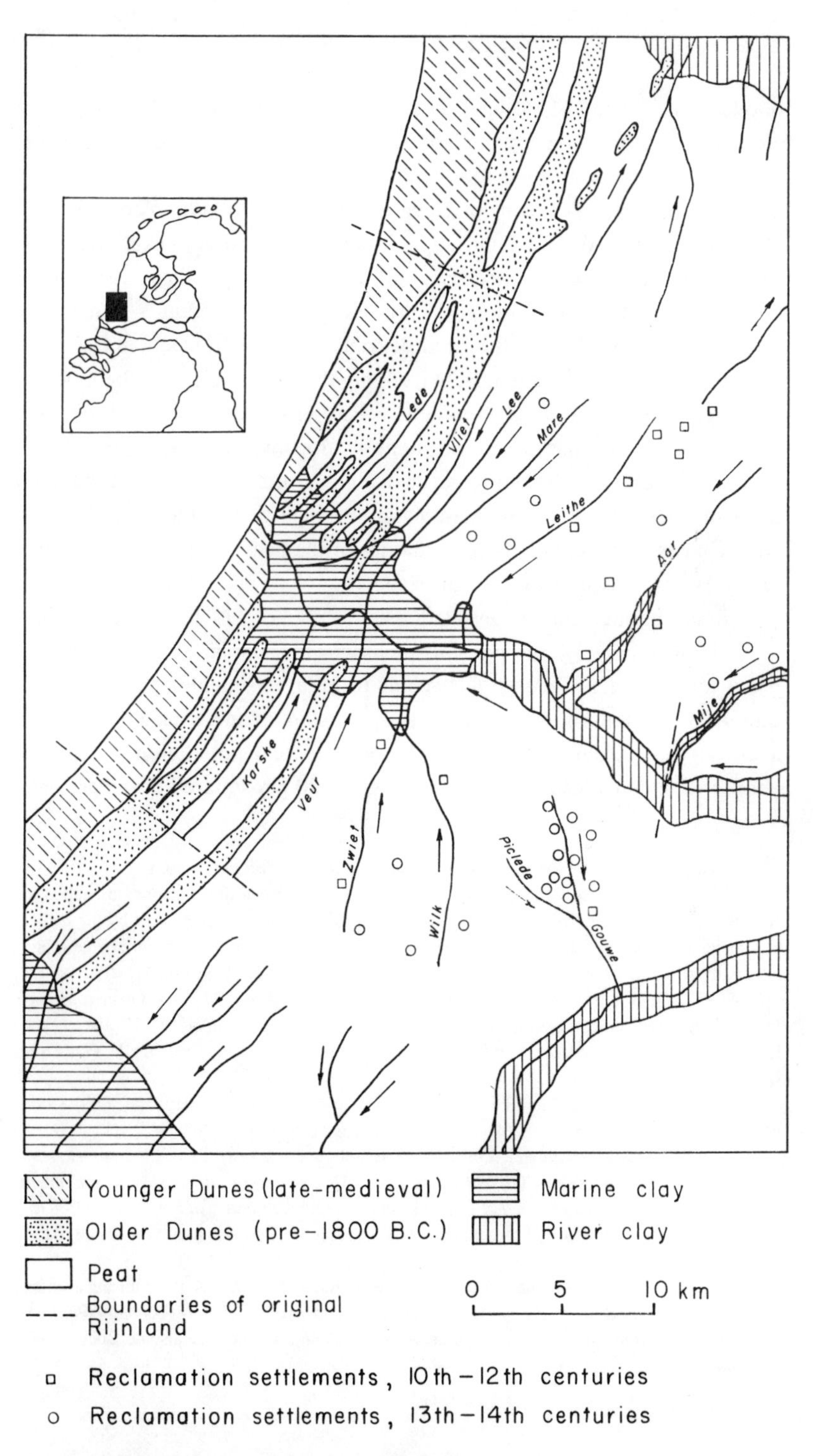

FIGURE 10. Reclamation Settlements of Rijnland

books of the county of Holland for 1316, 1334, 1343, and 1344.[52] The first clear information concerning their size and dimensions, however, dates from the second half of the fourteenth century.

In late 1368, Albrecht of Bavaria, count of Holland, Zeeland, and Hainault and lord of Friesland,[53] appointed a commission to examine complaints by a large segment of the rural population of Holland concerning the behavior of county officials. Three men from the count's immediate circle were dispatched to the rural districts of Rijnland, Delfland, Schieland, Woerden, and Haagambacht (roughly the northern half of the modern province of South Holland) between 25 February and 14 March 1369. Although its actual charge and instructions are unknown, the commission produced a register containing the answers that individual peasants gave to a set of questions about the administration of justice in the rural districts and villages. The register lists the names of the heads of household in virtually every village of the five districts.[54]

In all, about 5,500 names were listed, 2,773 of which were from Rijnland. Unfortuanately, the portions dealing with Rijnland are incomplete, but a total of 3,250 households seems to be a reasonable estimate.[55] Although we do not know how many people made up an average household, a conservative four would have produced a total population of 13,000 people with densities at least in the range of 20 per square kilometer. This

[52] Hamaker, ed., *De rekeningen der graafelijkheid van Holland*, vols. 1, 2.

[53] For details on how the dynasty founded by Gerulf in the ninth century managed to consolidate political control over most of the coastal districts of the Netherlands by the end of the thirteenth century, only to die out and be replaced by a collateral line from Hainault and eventually by Bavarian relatives, see, among others, Blok, "Holland und Westfriesland," pp. 347–61; Gosses, "Vorming van het graafschap Holland," pp. 239–44; Gosses, *Handboek tot de staatkundige geschiedenis der Nederlanden*, pp. 50–186; H. P. H. Jansen and L. Millis, "De middeleeuwen," in R. C. van Caenegem and H. P. H. Jansen, eds., *De Lage Landen van prehistorie tot 1500*, pp. 216–30.

[54] Algemeen Rijksarchief, The Hague: Archief van de graven van Holland, 889–1581, no. 676. On the date, background, and circumstances of the inquest, see D. E. H. de Boer, "De verhouding Leiden-Rijnland, 1365–1414: veranderingen in een relatie," *Economisch- en sociaal-historisch jaarboek* 38 (1975): 59–60; D. E. H. de Boer, *Graaf en grafiek: sociale en economische ontwikkelingen in het middeleeuwse "Noordholland" tussen 1345 en 1415*, pp. 47–61. In app. B, below, I present the totals for each Rijnland community as listed in this register.

[55] S. J. Fockema Andreae, "De Rijnlandse kastelen en landhuizen in hun maatschappelijk verband," in S. J. Fockema Andreae et al., eds., *Kastelen, ridderhofsteden en buitenplaatsen in Rijnland*, pp. 4, 6, 18 n. 23, argues for such an estimate. See also de Boer, *Graaf en grafiek*, p. 56.

document shows that a full two-thirds of Rijnland's households in 1369 lay on what had been uninhabited peat bog before reclamation and colonization took place.[56]

Another document, a fragment of an account from 1375, reveals that most peat-bog communities had achieved their greatest areal extent by that time. The amounts of land in each village that were subject to certain diking and draining taxes were essentially the same as those recorded elsewhere for the fifteenth and sixteenth centuries.[57] Together these two sources show the extent to which the former wilderness areas of Rijnland had been restructured and incorporated into the world of human affairs by the late fourteenth century.

The Process of Reclamation

If floodwaters from the Oude Rijn or the sea were not the constant threat to the raised peat bogs of Rijnland that some earlier researchers assumed they were, this does not mean that water was no obstacle to permanent human occupation. The bogs were perpetually soggy because the water table was always at or near the surface, the necessary precondition for peat growth in the first place. Because of this, they were much too wet and spongy to admit or support occupation and agricultural use.[58] Before they could be integrated into the realm of human affairs, Rijnland's bogs had to be subjected to a process of reclamation that would reduce the sogginess of their sur-

[56] De Boer, "De verhouding Leiden-Rijnland," pp. 60–62, calculates 67.6 percent for the former peat-bog portions.

[57] See the table in Fockema Andreae, *Het Hoogheemraadschap van Rijnland*, pp. 50–52. For the village of Grootburgerveen, col. I, the amount should read 81.600. For his first column Fockema Andreae used the original in the Rijnland archives: De oude archieven van Rijnland, no. 9508: "Register van inkomsten en uitgaven van den klerk van hoogheemraden," dated 1375. His purpose was to compute the tax base for each of the seven sluices in the Spaarndam (near Haarlem) at various times from the fourteenth to sixteenth centuries. Unfortunately, the portions for the Haarlemmersluis in col. 1 are missing, but I have found that essentially the same tax base was applied elsewhere in this document for the maintenance of the Doeswetering (a major drainage canal) and for a bridge over the Does and that the quantities listed there are virtually identical to the amounts later taxed for the Haarlemmersluis. In app. C, below, I list the area of land in each village that was subject to diking and draining taxes in 1375 for six of the seven sluices in the Spaarndam and for the Doeswetering and bridge over the Does as given in register no. 9508.

[58] See above, chap. 2; S. J. Fockema Andreae, *Willem I graaf van Holland, 1203–1222, en de Hollandse hoogheemraadschappen*, p. 1; Borger, "Ontwatering en grondgbruik in de middeleeuwse veenontginningen," p. 345.

faces and render them firm enough to support the weight of human beings and their paraphernalia.

Reclamation consisted of activities designed to lower the water table in the peat bogs. It was achieved by the application of a very simple technology: the reclaimers merely dug drainage ditches. Water from the top layers of the peat bog seeped into the ditches, which conveyed it elsewhere, until the water table in the adjacent sections of the bog had been reduced to a level corresponding to the bottoms of the ditches. The idea was to dry out not only the reed, sedge, or peat-moss plant associations growing on top of the bogs but also the top layer of accumulated dead plants constituting the peat. This process produced something resembling a sod that was capable of supporting people and livestock. To cause a 10-centimeter-thick sod to form in peat that was 80 percent water by volume, it was necessary to lower the water table by about half a meter; in peat that was 90 percent water by volume, however, such as the peat-moss peat found in the raised central portions of the bogs, a sod of those dimensions was produced only when the water table had been brought down nearly a meter.[59]

The ditches dug to lower the water level in peat areas were not scattered at random over the bog surfaces but were placed carefully according to what was to become an almost universal pattern. The reclaimers would begin by establishing a suitable reclamation base associated with a waterway that could carry away the water extracted from the bog. They might begin at the edge of an existing riverbank settlement where the occupied ridge disappeared under the bog. They might also begin by deepening a small peat-bog stream to hasten its flow and begin lowering the water table along it. In any case, at regular intervals along the reclamation base parallel ditches would be dug into the peat bog in a direction more or less perpendicular to the base. These ditches, usually spaced 30 Rijnland rods of 3.767 meters each, or just over 110 meters, served as drainage channels and also formed the side boundaries of individual parcels of land. Since houses were built at the front of each parcel, a long, narrow village was created with an axis parallel to and lying adjacent to the reclamation base.

In some of the very earliest reclamation projects in the western Netherlands the length of the ditches and hence the parcels may have been unspeci-

[59] Borger, "Ontwatering en grondgebruik in de middeleeuwse veenontginningen in Nederland," pp. 345–46.

fied. The colonists occupying the parcels would have had a right of more or less unlimited extension into the wilderness. As the pace of reclamation picked up, however, extending its tentacles into the bogs from many different directions and from many separate reclamation bases, the lengths of ditches and parcels were specified as well. By far the most common length in the western Netherlands was 1,250 meters, creating standardized parcels of just over 14 hectares (about 35 acres).[60]

The parcels thus created were designed to be reclaimed, occupied, and exploited by a single household consisting of a nuclear family. This was a rather large allotment by medieval standards. It makes sense, therefore, that only a portion of the eventual homestead was reclaimed at a time. Initially the drainage ditches marking the sides of the parcels may have extended only a few hundred meters or even less from the reclamation base into the peat bog that was to be reclaimed. To achieve a more uniform lowering of the water table within each parcel, reclaimers next dug a network of smaller ditches as needed to carry water into the side ditches. Finally, at the back of the ditched section, a low dike, known as a *kade*, or *kadijk*, would be thrown up laterally across the width of the parcel with a ditch parallel to it on the unreclaimed side. This dike plus ditch would divert runoff from the unreclaimed portions of the bog into the side ditches, preventing drained portions from flooding.[61] Once these measures had been taken, the front sections of the eventual homesteads became habitable and were ready for the final preparations preceding agriculture.

Before the newly drained land could be put to agricultural use, however, some further restructuring was required: the existing vegetation must

[60]Van der Linden, *Recht en territoir*, p. 6; van der Linden, *De Cope*, pp. 20–36; M. K. E. Gottschalk, "De ontginning der Stichtse venen ten oosten van de Vecht," *Tijdschrift van het Koninklijk Nederlandsch Aardrijkskundig Genootschap*, 2d ser., 73 (1956): 208; S. J. Fockema Andreae, *De oude archieven van het Hoogheemraadschap van Rijnland*, p. 22; de Cock, *Bijdrage tot de historische geografie van Kennemerland*, pp. 70, 154–59; Diepeveen, *De vervening in Delfland en Schieland tot het einde der zestiende eeuw*, pp. 14–15; van Doorn, *Het oude Miland*, p. 68; Pons et al., "Evolution of the Netherlands Coastal Area during the Holocene," in *Transactions of the Jubilee Convention*, p. 206; T. Edelman, *Bijdrage tot de historische geografie van de Nederlandse kuststreek*, pp. 43–50.

[61]T. Edelman, *Bijdrage tot de historische geografie van de Nederlandse kuststreek*, pp. 47–51; Gottschalk, "De ontginning der Stichtse venen ten oosten van de Vecht," p. 211; Borger, "Ontwatering en grondgebruik in de middeleeuwse veenontginningen in Nederland," pp. 348–49; de Cock, "Veenontginningen in West-Friesland," p. 159; P. J. Woltering, "Archeologische kroniek van Noord-Holland over 1978," *Holland: regionaal-historisch tijdschrift* 11 (1979): 270.

be removed. The method by which this was done varied widely from place to place because the vegetation on the bogs was not uniform. Reed, sedge, alder, and birch were common close to the Oude Rijn and some of its tributaries that were flooded from time to time with nutrient-carrying water, while peat moss, heath, and cotton grass dominated on the raised central portions of the bogs. Where woody vegetation was prominent, a pioneering mode of swidden, or slash-and-burn, agriculture would produce high yields for several years after drainage with relatively little labor input. After girdling trees to kill them and burning the underbrush and litter, colonists would rake seeds into the ash layer or work them in with a hoe or digging stick. Plowing presumably began later, made possible by the removal of stumps and made necessary by the flourish of weed growth.[62]

Away from the edges of a bog, where woody vegetation was absent, other procedures would be followed. In some places the topmost layer of the peat was peeled or pared off after drainage had induced sufficient sod to develop. Once again the final preparations for agriculture included some use of fire. Burning not only eliminated unwanted plant material but was an effective fertilizing agent as well. While peat soils initially could be very rich in nitrogen, they tended to have insufficient quantities of other essential nutrients. Fire helped counteract this problem by synthesizing or concentrating the needed nutrients.[63]

For some parts of Rijnland it is possible to construct the tentative outlines of the process of reclamation. One such area was that surrounding the

[62] Stanton Green, "The Agricultural Colonization of Temperate Forest Habitats: An Ecological Model," in William W. Savage, Jr., and Stephen I. Thompson, eds., *The Frontier: Comparative Studies*, 2:83–87; David R. Harris, "Swidden Systems and Settlement," in Peter J. Ucko, Ruth Tringham, and G. W. Dimbleby, eds., *Man, Settlement and Urbanism*, pp. 92–93; Ester Boserup, *The Conditions of Agricultural Growth: The Economics of Agrarian Change under Population Pressure*; William S. Cooter, "Preindustrial Frontiers and Interaction Spheres: Aspects of the Human Ecology of Roman Frontier Regions in Northwest Europe" (Ph.D. diss., University of Oklahoma, 1976), pp. 41–46, 233; William S. Cooter, "Ecological Dimensions of Medieval Agrarian Systems," *Agricultural History* 52 (1978): 466, 471–75; the appropriate sections of chap. 2, above.

[63] T. Edelman, *Bijdrage tot de historische geografie van de Nederlandse kuststreek*, pp. 43–47; G. J. Borger, "De ontwatering van het veen: een hoofdlijn in de historische nederzettingsgeografie van Nederland," *Geografisch tijdschrift* 11 (1977): 381–85; Borger, "Ontwatering en grondgebruik in de middeleeuwse veenontginningen in Nederland," pp. 345–49; de Cock, *Bijdrage tot de historische geografie van Kennemerland*, p. 255; C. J. Schothorst, "Subsidence of Low Moor Peat Soils in the Western Netherlands," *Geoderma* 17 (1977): 285; van der Linden, *De Cope*, p. 98; Grigg, *The Agricultural Revolution in South Lincolnshire*, pp. 21, 73.

villages of Esselijkerwoude, Rijnsaterwoude, and Leimuiden, north of the Oude Rijn. All three were apparently associated with a single peat-bog stream known as the Leithe, which began somwhere near Leimuiden and flowed in a generally southwesterly direction from there along Rijnsaterwoude and Esselijkerwoude until it joined the Oude Rijn in the village of Leiderdorp (called Leithon in the tenth century) or between Leiderdorp and the village of Koudekerk (presumably the village called Holtlant in the tenth century). The water name Leithe may well be the root of both Leimuiden and Leiderdorp.[64]

The earliest reclamation in this area would have used the clay banks of the Oude Rijn as the reclamation base and extended northward from Leiderdorp and Koudekerk into the peat-bog wilderness.[65] That these were among the earliest efforts to reclaim peat soils in Rijnland is suggested by the fact that here, as in other reclamation thrusts using the Oude Rijn banks as a starting point, the length of the parcels was not uniform, nor were the dwellings placed next to each other in a row. Apparently these were piecemeal reclamations carried out before the counts of West Frisia–Holland had begun the systematic granting of wilderness of specific dimensions to groups of colonists.[66]

Not long afterward, however, reclamation began farther into the wilderness north of Leiderdorp and Koudekerk as well. These efforts were not extensions of existing settlements but were new ventures by colonists, who used a peat-bog stream, presumably the Leithe, as the reclamation base from which they extended ditches of a uniform length, according to the conditions of a grant from the count of West Frisia–Holland. The first of these was Esselijkerwoude, which was old enough to pay the *botting*, a tax that, as mentioned earlier, was not levied in settlements established after the very early eleventh century. Before the middle of the eleventh century, however, the next two reclamation projects upstream along the Leithe were begun as well, since we know that they had churches, according to documents from 1040 and 1063.[67] Finally, because the churches at Esselijker-

[64] Blok, "Probleme der Flussnamenforschung in den alluvialen Gebieten der Niederlande," p. 226; van der Linden, *De Cope*, pp. 354–62.

[65] The pattern here was similar to what de Cock, "Veenontginningen in West-Friesland," pp. 159–60, called type "a": parcels extending into the peat bog wilderness from an existing settlement nucleus along a dune or sandy-clay ridge.

[66] Van der Linden, *De Cope*, pp. 268–70.

[67] Ibid., pp. 38–51.

woude, Rijnsaterwoude, and Leimuiden were daughter establishments of the original Saint Willibrord's church at Oegstgeest, it is possibile that the reclaimers had been recruited from the densely populated dune area around Oegstgeest.[68]

Just east of the complex of reclamation settlements associated with the Leithe lay another group that was linked by the Aar, a peat-bog stream that joined the Oude Rijn between Oudshoorn and Aarlanderveen. Both Oudshoorn and Aarlanderveen, old enough to pay the *botting*, were reclaimed, using the clay banks of the Oude Rijn as the starting point. A short time later new reclamation projects that used the Aar as their starting point were begun farther north, at Ter Aar, Nieuwveen, and possibly Kalslagen, by colonists who had received grants of specific dimensions from the counts.[69]

A third cluster of early reclamation projects can be identified south of the Oude Rijn in association with the Zwiet, a peat-bog stream that began near or in a lake known as Zoetermeer and flowed northward to the Oude Rijn just across from Leiderdorp. Both Zoeterwoude and Zoetermeer were colonized by way of the Zwiet (shown on modern maps as Weipoortse Vliet).[70] Hazarswoude, meanwhile, seems to have been reclaimed by way of the Wilk, a tributary of the Zwiet from the southeast (known as Wilck and Oude Wilck on seventeenth- and eighteenth-century maps).[71]

Protecting the New Lands

The reclamation of the peat-bog wilderness of the western Netherlands required at first only simple, straightforward procedures. Where possible, it was carried out along the rivers and streams of the natural radial drain-

[68] See Oppermann, ed., *Fontes Egmundenses*, p. 255; de Cock, *Bijdrage tot de historische geografie van Kennemerland*, pp. 68, 70, 256.

[69] Van der Linden, *De Çope*, pp. 54–56, 252–53, 258–59.

[70] S. J. Fockema Andreae, "Een verdwenen dorp? Zwieten bij Leiden," in *Varia historica aangeboden aan professor doctor A. W. Byvanck ter gelegenheid van zijn zeventigste verjaardag door de Historische Kring te Leiden*, p. 124.

[71] See the maps in Floris Balthasars, *Kaarten van Rijnland, 1615*, sheets 15–16; Jan Jansz. Dou and Steven van Broekhuysen, *Kaartboek van Rijnland*, sheet 11. See also Blok, "Probleme der Flussnamenforschung in den alluvialen Gebieten der Niederlande," pp. 221, 225. The Wilk was canalized and made into a *wetering*, a large interlocal drainage canal, at the end of the fourteenth century; De oude archieven van Rijnland, no. 12: "Het groote register," keur 88, fols. 11r-11v (dated 1399).

age system of the raised peat bogs. The streams served as the reclamation bases along which the parallel drainage ditches were marked out and dug perpendicularly into the bog. For initial reclamation it was not necessary to develop any new technology. Digging ditches to lower the water table had been practiced at least since the Bronze Age.[72] Until the twelfth century such ditches and perhaps some canals where natural streams were not available as reclamation bases were sufficient to achieve drainage.[73] As time passed, however, their effectiveness began to decline as a series of developments made necessary more complex means of water control.

The most important change in Rijnland, and also the most difficult one to assess, was the loss of the Oude Rijn as a branch of the Rhine delta. Actually the Oude Rijn lost significance gradually over a long period of time as the Lek and the Waal, the more southerly branches, received more of the Rhine's water. The actual time of final closure of what was by then a relatively insignificant stream is not known for certain though the process was no doubt largely completed by 1000.[74] Whenever it occurred, it meant a loss of tidal action in the lower Oude Rijn and a serious reduction in the quality of drainage in Rijnland. Little water from the Rhine actually entered the Kromme Rijn–Oude Rijn any longer; more significant was the fact that local water no longer had easy egress to the sea. In fact, once the final closure had occurred, water may well have begun to collect in the lowest portion of the Oude Rijn, just inside the dunes.

The drainage problems caused by the silting shut of the Oude Rijn were aggravated by the reclamation of the peat bogs on either side, not only in Rijnland but also upstream on the east. Digging drainage ditches and deepening peat-bog streams increased the rate of water flow from the bogs, or, stated differently, it decreased the ability of the raised bogs to act as reservoirs of surplus water, particularly after heavy rains or snow melts. As a result large quantities of water would accumulate along the lower

[72] See, for example, J. A. Bakker et al., "Hoogkarspel-Watertoren: Towards a Reconstruction of Ecology and Archaeology of an Agrarian Settlement of 1000 B.C.," in van Beek, Brandt, and Groenman–van Waateringe, eds., *Ex Horreo: I.P.P. 1951–1976*, pp. 214–22.

[73] Van der Linden, *Recht en territoir*, pp. 6, 9.

[74] Blok, "Probleme der Flussnamenforschung in den alluvialen Gebieten der Niederlande," pp. 222, 226; S. J. Fockema Andreae, "De Oude Rijn: eigendom van openbaar water in Nederland," in *Rechtskundige opstellen op 2 november 1935 door oud-leerlingen aangeboden aan prof. mr. E. M. Meijers*, pp. 699–700; L. P. Louwe Kooijmans, *The Rhine/Meuse Delta: Four Studies on Its Prehistoric Occupation and Holocene Geology*, p. 120.

edges of the raised bogs, near the oldest settled areas of Rijnland, unable to drain to the sea. In particular, a series of pools and ponds, normal features of unreclaimed peat bogs, gradually began to consolidate and expand along the inside of the Older Dunes north of the Oude Rijn, eventually forming the Haarlemmermeer of late-medieval and early-modern times.[75]

The reclamation of peat wilderness had the potential not only of increasing the danger of flooding at the edges of the bogs but also of causing problems in peat areas. Reclamation, as we saw, began at the edges of the bogs and worked inward in stages. Lowering the water level caused the top peat layer to begin oxidizing and subsiding.[76] Of course, the first areas to be reclaimed were the first to see these results, while those farther into the bog, which were higher to begin with, would see them only later. Those settling the highest portions of the bogs would dig their ditches to carry away the water they did not want without necessarily worrying about how this action might affect others. Thus water drained from the higher portions of the bogs could well become floodwaters in communities lower down on the bogs, where oxidation and subsidence had already begun to take their toll. Indeed, the tendency to send unwanted water downstream to become someone else's problem became a general concern in the western Netherlands.

Finally, drainage problems in the western Netherlands may have been affected by climatic change as well. Some physical scientists, for example, have suggested that a Dunkirk III-B transgressive phase of the sea was under way by the middle of the twelfth century. As evidence they point to the transformation of the Almere, a large lake-and-swamp complex, into an inland sea known as the Zuider Zee, or Southern Sea, beginning around 1170. A southwestward extension of the Zuider Zee, known as the IJ, extended from Amsterdam westward toward Haarlem at the dunes by way of the low-lying swampy area that was left over from the former IJ estuary. Thus salt water and tides penetrated deeply into Kennemerland, Rijnland's neighbor on the north.[77] Around the same time storms caused serious flood-

[75] H. van der Linden, *De Zwammerdam*, p. 5; van der Linden, *De Cope*, p. 67; Acket, "De Oude Rijn en zijn omgeving," p. 78; Blok, "Probleme der Flussnamenforschung in den alluvialen Gebieten der Niederlande," p. 224.

[76] See above, this chapter.

[77] See above, chap. 3; A. R. Güray, "De bodemgesteldheid van de IJpolders," *Boor en spade* 5 (1952): 5–7; J. Vader, "Een en ander omtrent de eerste bemaling met windwatermolens in het Hollandsche laagveengebied," *Tijdschrift voor geschiedenis* 28 (1913): 87;

ing in the Maasland area south of Rijnland.[78] Still the old question remains whether a series of serious storms qualifies as a transgression of the sea, which, presumably, entailed consistently higher mean high-water levels. Many scholars today tend to be rather skeptical of this.[79] In any case, since the Oude Rijn mouth was sealed at the coast, no storm surges could directly affect Rijnland. Nevertheless, increased precipitation may well have caused a greater incidence of river flooding. Indeed, in the period from 1135 to 1160 there was an unusually high rate of occurrence of river floods in the lands around the Netherlands, which was doubtless felt downstream in the Netherlands as well. Rijnland, in particular, would have felt the effects because of the stopped-up mouth of the Oude Rijn.[80]

In response to the drainage problems described above, the inhabitants of Rijnland began developing a more complex and comprehensive system of water management that included dams, dikes, sluices, and a series of large canals intended to carry water to points outside Rijnland. One of the first water-management measures of more than simply local significance was the damming of the Oude Rijn at the eastern edge of Rijnland, at Zwammerdam. In the Bishopric of Utrecht, east of Rijnland, the reclamation of raised peat bogs had begun in the late tenth or early eleventh century as well.[81] There too it significantly increased the flow of water out of the bogs, or, rather, decreased the bogs' capacity to act as reservoirs of surplus water. Therefore, the water from Utrecht reclamations, along with possible increased river flooding in the Rhine delta in general, caused greater amounts of water to make their way into Rijnland, the lowest segment of the Oude Rijn course, compounding an already serious drainage problem because there was no longer a direct exit to the sea. The Zwam-

---

H. van der Linden, "Iets over wording, ontwikkeling en landschappelijk spoor van de Hollandse waterschappen," *Holland: regionaal-historisch tijdschrift* 10 (1978): 101, 104–105.

[78] C. Hoek, "Schiedam: een historisch-archaeologisch onderzoek," *Holland: regionaal-historisch tijdschrift* 7 (1975): 94.

[79] A. E. Verhulst and M. K. E. Gottschalk, eds., *Transgressies en occupatiegeschiedenis in de kustgebieden van Nederland en België*; M. K. E. Gottschalk, *Stormvloeden en rivieroverstromingen in Nederland*, 2:818–23; as well as the reservations of T. Edelman, *Bijdrage tot de historische geografie van de Nederlandse kuststreek*, pp. 6–16.

[80] See above, chap. 3; Louwe Kooijmans, *The Rhine/Meuse Delta*, p. 121; Gottschalk, *Stormvloeden en rivieroverstromingen in Nederland*, 1:71–75; J. G. G. Jelles, *Geschiedenis van beheer en gebruik van het Noordhollands duinreservaat*, p. 15.

[81] See, for example, the discussions of van Doorn, *Het oude Miland*, p. 14 and passim; van der Linden, *De Cope*, p. 70.

merdam was built to limit the amount of water that entered Rijnland from the east. The immediate result of this action, however, was to transfer the problem of flooding upstream to the Utrecht side of the dam. Consequently, the emperor Friedrich Barbarossa intervened in 1165. On 25 November of that year he ordered Count Floris III of Holland (formerly West Frisia) to remove the dam and allow the Oude Rijn to flow without further obstruction so that the inhabitants of Utrecht would suffer less from flooding.[82]

Whether the Zwammerdam was actually removed as ordered by the emperor is not known. A dam in the Oude Rijn at Zwammerdam was mentioned again in 1202 in an agreement between Holland and Utrecht brought about by the efforts of the duke of Brabant. This time the count of Holland agreed to destroy the dam if the Utrechters in return would help dig three canals (*wetering*, plural *weteringen*) to carry away surplus water in Rijnland and promise not to release water downstream at inappropriate times.[83]

Two of the required *weteringen* can be identified. One was the Oudewetering, which connected the Braassemermeer, a small lake near Rijnsaterwoude and Leimuiden, with a series of lakes that had developed along the inside of the dunes in northern Rijnland and southern Kennemerland. These lakes in turn flowed by way of a peat-bog stream known as the Spaarne into the IJ, the southwestern extension of the Zuider Zee. The second known canal, the Heimanswetering, was dug from a point just southwest of the Braassemermeer, on what must have been the Leithe, to the Oude Rijn, slightly west of Alphen aan den Rijn. Somewhat earlier Rijnlanders had dug the Doeswetering, perhaps simply a canalization of the original Leithe from the Oude Rijn near Leiderdorp toward the source of

[82] The charter of Friedrich, in Koch, ed., *Oorkondenboek van Holland en Zeeland tot 1299*, vol. 1, no. 158, reads in part as follows: "Aliam quoque obstructionem Rheni, quam comes Hollandiae in loco qui dicitur Steckede sive Suadenburg iniuste et violenter erexerat et fecerat, per quam eciam innumerabilium hominum et locorum submersiones frequenter evenerunt, nos ex iudicio curiae nostrae cassavimus et penitus destrui precepimus, statuentes et imperiali iussone confirmantes, quatinus, predicta clausure destructa, aqua Rheni libera et regia strata sine omni obstaculo ibidem omni tempore fluat et decurrat, sicut antiquitus solebat."

[83] See Koch, ed., *Oorkondenboek van Holland en Zeeland tot 1299*, vol. 1, no. 246. Koch gives a date of 1200 to this charter. Van der Linden, *De Zwammerdam*, p. 28 n. 10, gives good reasons for a date of 1202. The document reads in part: "Comes ad petitionem ducis promisit destruere dammum in Suadenburch salvo iure suo, videlicet ut tres aqueductus qui vocantur weteringe preparentur, et ut aqua intempestiva in terram comitis ire non permitatur, quam si prohibere non poterunt homines episcopi tempestive, comes licite obstruet dammum tamdiu quousque aqua cohibeatur, et tunc statim destruetur dammus."

the Leithe in a series of streams and pools near the Braassemermeer that the Heimanswetering eventually tapped into. Through this network of interlocal canals water from the Oude Rijn could flow northward by way of the Doeswetering or the Heimanswetering into the Braassemermeer and from there by way of the Oudewetering into the lakes and streams that flowed into the IJ, the Zuider Zee, or Almere, and the North Sea.

The backbone of the medieval network of *weteringen* was completed by digging extensions for the Drecht north of the Oude Rijn and the Gouwe on the south around or shortly after 1200. The Drecht, a tributary of the Amstel that began near Leimuiden, was connected to the Oudewetering, thus allowing water to flow northward toward the IJ by way of the Amstel as well. The Gouwe flowed south toward the IJsel, and, by digging it through to the Oude Rijn near Alphen aan den Rijn, the Rijnlanders could release water into the IJsel as well.[84] One of the most fascinating results of the digging of the *weteringen* was that it changed the direction of water flow within the peat bogs. With the construction of the Doeswetering, Heimanswetering, and Oudewetering, for example, water that previously would have flowed in a southerly direction was made to flow northward in the direction of the IJ (see fig. 11).

The damming of the Oude Rijn at Zwammerdam and the digging of *weteringen* were not the only measures taken by Rijnlanders to keep their heads above water. After the mouth of the Oude Rijn had silted shut, and particularly after reclamation of the raised peat bogs had begun, the area on both sides of the Oude Rijn just inside the Older Dunes suffered most from flooding because it was relatively lower-lying than places upstream along the Oude Rijn. Thus at an early date, perhaps around 1200 or before, a system of interlocal dikes was thrown up to offer protection to this area. Traces of these encircling dikes, or *ommedijken*, as they were called, were still to be found in various place- and water-names on seventeenth-century maps. One dike apparently started at Leidschendam, on the innermost dune ridge, at the southern boundary of Rijnland, and ran in a generally

[84] See van der Linden, *De Zwammerdam*, p. 10; van der Linden, *Recht en territoir*, pp. 9, 36 n. 11; Blok, "Probleme der Flussnamenforschung in den alluvialen Gebieten der Niederlande," p. 224; S. J. Fockema Andreae, *Het Nedersticht*, vol. 4 in *Studiën over waterschapsgeschiedenis*, p. 8; S. J. Fockema Andreae, "De Visbrug te Leiden," *Leids jaarboekje* 43 (1951): 56–57; Fockema Andreae, ed. *Rechtsbronnen der vier hoofdwaterschappen*, pp. xiii–xiv; Fockema Andreae, *Willem I graaf van Holland, 1203–1222, en de Hollandse hoogheemraadschappen*, pp. 21, 26; Fockema Andreae, *Poldernamen in Rijnland*, pp. 8, 17, 18, 23, 34, 40.

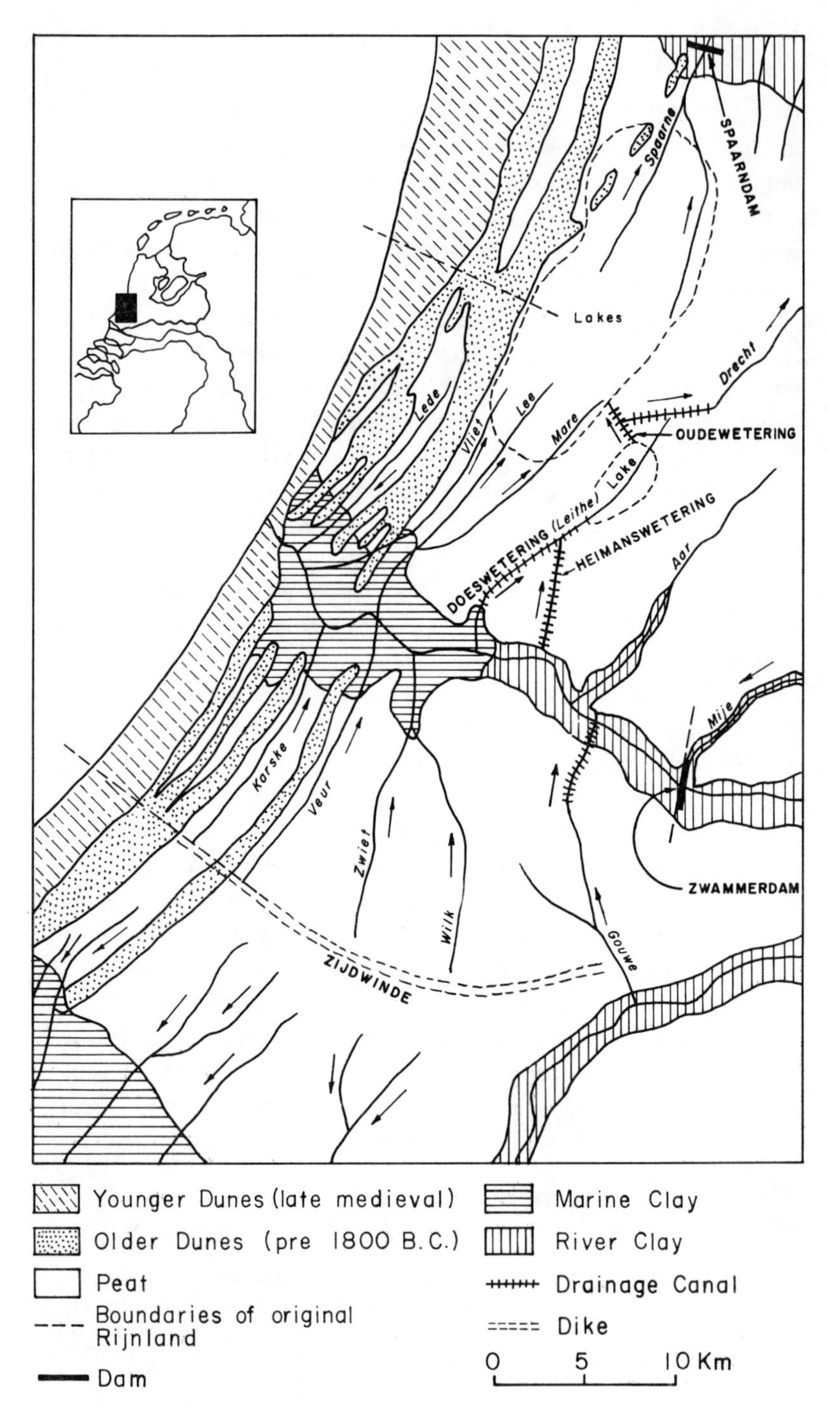

FIGURE 11. Major Hydraulic Works circa 1225 A.D.

northeasterly direction along the east side of Zoeterwoude to the Oude Rijn. Another apparently began at the Oude Rijn near Leiderdorp, from which it ran generally northward along the east side of Warmond, presumably connecting to the dunes somewhere north of Sassenheim. Associated with these dikes, it appears, were two dams with primitive sluices across the two forks of the Oude Rijn at Leiden as part of the structures of two bridges, the Visbrug and the Sint Jansbrug.[85] With the construction of the *weteringen*, however, the *ommedijken* may well have become superfluous.

In 1219 a particularly serious storm that affected the Zuider Zee and the IJ flooded with salt water much of the peat area between the IJ and Rijnland. Once again the count of Holland and the bishop of Utrecht came to an agreement over new water-control measures because both realized that cooperation was necessary to channel surplus water from the Oude Rijn northward to the IJ. Therefore, a large dam was laid across the peat-bog stream known as the Spaarne where it entered the IJ. The Spaarndam was equipped with seven sluices designed to allow drainage water from the south an exit while preventing salt water from entering the peat area. Precisely when it was constructed is not known for certain, but it was in existence by 1226. On 26 November of that year a dispute between the bishop of Utrecht and the count of Holland concerning the dam and its sluices was settled through the intervention of a papal legate.[86] The original agreement to build a dam in the Spaarne specified that Utrecht would build and maintain one-half of the sluices. Since there were seven sluices in all, Holland had suggested that Utrecht's share should be four; Utrecht protested and said it would maintain only three. The legate decided in Utrecht's favor.[87] From the early thirteenth century until the 1870s, when the IJ was drained, the dam in the Spaarne with its seven sluices, though often damaged and partly destroyed by storms, served as one of the chief defenses and drainage routes for the area between the Oude Rijn and the IJ.[88]

[85] Fockema Andreae, "De Visbrug te Leiden," pp. 56–57; Fockema Andreae, *Het Hoogheemraadschap van Rijnland*, pp. 29, 31–34; Fockema Andreae, *Poldernamen in Rijnland*, pp. 31, 39, 52; Balthazars, *Kaarten van Rijnland, 1615*, sheet 11; Dou and van Broekhuysen, *Kaartboek van Rijnland*, sheets 7, 10.

[86] See L. Ph. C. van den Bergh, ed., *Tot het einde van het Hollandsche huis*, sec. 1 in *Oorkondenboek van Holland en Zeeland*, no. 294.

[87] See van der Linden, *De Zwammerdam*, pp. 12–13; Fockema Andreae, *Willem I graaf van Holland, 1203–1222, en de Hollandse hoogheemraadschappen*, pp. 42–45; Fockema Andreae, *Het Hoogheemraadschap van Rijnland*, pp. 35–37, 39–41; Blok, "Probleme der Flussnamenforschung in den alluvialen Gebieten der Niederlande," p. 224.

[88] Güray, "De bodemgesteldheid van de IJpolders," p. 8; Fockema Andreae, *Het Hoogheemraadschap van Rijnland*.

Finally, also around 1220, the southern boundary of Rijnland was drawn by the construction of a low dike, or *kade*, called the Zijdwinde. It began in the Haagse Bos (Woods of The Hague), in the Older Dunes at Leidschendam, and ran eastward from there along the south side of Zoetermeer, Zegwaard, Benthuizen, and Hazarswoude to Waddinxveen on the Gouwe. In this manner Rijnland was marked off from Delfland and Schieland, successors to the orginal county of Maasland on the south.[89]

With the construction of the Zwammerdam, the *weteringen*, the Spaarndam with its seven sluices, and the Zijdwinde, the essential technology for peat-bog reclamation had been more or less mastered. The continued subsidence and oxidation of the peat surface, however, meant that a number of refinements were needed as time went on to ensure the continued protection of reclaimed lands. The gradual introduction of first the polder, or tract reclaimed by dikes, and eventually the windmill were just such measures.

The first polders were created on the longest-drained and lowest-lying portions of the peat bogs of the western Netherlands. They were created by encircling a parcel of land with a dike designed to divert or deflect water flowing from higher-lying portions of the bog. The encircling dike was really an extension of the *kade*, or *kadijk*, that would have been drawn at the back edge of a reclamation parcel to divert water from the unreclaimed portions to the side ditches of the parcel. Water from within the polder was collected in a network of ditches that conveyed it to one or more *weteringen* that extended through the encircling dike and were equipped with sluices or floodgates to prevent water from reentering.[90] Exactly when such

[89] "Beginnen vander dwyn sijde an biden Hagehout dair die wateringe gaet ende van't Hagehout tot Oevenzijdwijn toe ende van Oevenzijdwijn tot Soetermeer toe, ende van Soetermeer doer tot Zegwaert toe ende voirt Zegwaerde doer tot Benthusen toe. Ende Benthuussen doir tot Haessaertswoude toe ende Haessertswoude doir tot Waddinxveen toe ende Waddinxveen doir also verre als ons scouwe gaet," in De oude archieven van Rijnland, no. 12: "Het groote register," keur 83, fols. 10v–11r (dated 1394), gives its location. On its origins, see Fockema Andreae, *Willem I graaf van Holland, 1203–1222, en de Hollandse hoogheemraadschappen*, pp. 45–46, 61; Fockema Andreae, *Het Hoogheemraadschap van Rijnland*, pp. 33, 80–82.

[90] M. C. P. Scholte, "Polders en polders en polders," *Holland: regionaal-historisch tijdschrift* 13 (1981): 114–15; Scholte, "Wat was er eerder: de dijk, de veenontginning of de polder?" pp. 1–9; Diepeveen, *De vervening in Delfland en Schieland tot het einde der zestiende eeuw*, p. 21; Verhulst, "Historische geografie van de Vlaamse kustvlakte tot omstreeks 1200," p. 8; van der Linden, "Iets over wording, ontwikkeling en landschappelijk spoor van Hollandse waterschappen," pp. 108–109.

enclaves first began to appear is unknown. The term "polder" itself was not always applied until much later. Before the middle of the fifteenth century the word was commonly used only in a broad arc stretching from Flanders through northern Brabant into Holland as far as the Oude Rijn.[91] The first known use of the term in Rijnland dates from 1399,[92] but polders may well have existed before that.[93] Indeed, by 1225, Rijnland and southern Kennemerland together, defined by the Spaarndam in the north, the Zwammerdam in the east, and the Zijdwinde in the south, the area that eventually came to be known as the Hoogheemraadschap van Rijnland, already displayed many of the characteristics of a mammoth polder.

The first polders probably were not pumped, but pumping eventually became necessary because continued subsidence and oxidation of the peat surface meant that some polder levels had begun sinking to the point at which gravity-flow drainage was no longer adequate. Pumps operated by men or horses were used in the Netherlands from the fourteenth century,[94] while windmills for grinding cereals were reported at Haarlem, in Kennemerland, as early as 1274.[95] The first known combination of the windmill and the pump to remove water from a polder in the Netherlands occurred in 1408, and it had become a common feature by the 1460s.[96]

Until very recent times the methods and procedures developed during the medieval period of peat-bog reclamation remained relatively unchanged. What did change was the cumulative knowledge, wisdom, and self-assurance that was born of those early centuries and was fueled by considerable success. In this light, pumping polders with windmills was not the sort of dramatic breakthrough that we often like to ascribe to some individual inventive genius but was rather the quiet application of available technologies to changing hydraulic conditions by a group of people who were well versed in diking and draining activities.

[91] Scholte, "Polders en polders en polders," p. 115.

[92] De oude archieven van Rijnland, no. 12: "Het groote register," keuren 88–89, fols. 11r–11v.

[93] Fockema Andreae, *Poldernamen in Rijnland*, pp. 5, 10, talks of the existence of a Polder Bloemendaal at least since 1331.

[94] Vader, "Een en ander omtrent de eerst bemaling met windwatermolens," pp. 94–95.

[95] K. Boonenburg, "De oudste tot op heden bekende vermelding van windmolens in het graafschap Holland (1274)," *Bulletin van de Koninklijke Nederlandse Oudheidkundige Bond*, 6th ser., 5 (1950): 62.

[96] C. Hoek, "Heer Floris van Alkemade als oprichter van windwatermolens," *Holland: regionaal-historisch tijdschrift* 5 (1973): 95–96.

The medieval expertise of dike building and water management, gained from the long-term personal experience of many, was brilliantly summed up around 1579 in a long treatise by Andreas Vierlingh entitled *Tractaet van dyckagie* (Treatise on Diking). Among other things Vierlingh was dikemaster to William the Silent, prince of Orange, and his book was more or less a summation of virtually all that was known about diking and draining. The text sparkles with wisdom and lore and at the same time contains the kinds of practical examples and suggestions that made it a useful textbook. Vierlingh suggests throughout that only experienced dikers and drainers should be employed, that local materials should be used, and that care should be taken to work with natural forces or currents rather than trying to fight against them.[97]

[97] Andreas Vierlingh, *Tractaet van dyckagie*, ed. J. de Hullu and A. G. Verhoven, pp. 8, 41; ". . . nochtans dairinne versien dat de dammen niet te hooge en zijn, want het water en wil met geen fortse gedwongen wesen off het sal u wederomme fortse doen," and "Alle wateren en willen met geen macht wederstaen ofte verwonnen wesen, dwelcke een generael regel is."

7

Ecological and Cultural Change in Rijnland

THE landscape of Rijnland in the second half of the fourteenth century bore very little resemblance to its prereclamation appearance. By draining the raised oligotrophic peat bogs, Rijnlanders had transformed a soggy wilderness unfit for human habitation into dry land suitable for settlement and agricultural exploitation. Before reclamation the bogs were used, if indeed they were used at all, in a very incidental manner. People from the oldest settled areas, the ridges of the Older Dunes and the natural, silted-up banks of the Oude Rijn, may occasionally have fished or hunted there and perhaps grazed some livestock along the edges of the wilderness during the summer months. To all intents and purposes, however, the peat bogs lay well beyond the realm of normal human affairs. Anyone presumably had access to the wilderness, but virtually no one saw it as particularly desirable.[1] With reclamation not only did the former bogs fill with settlements, containing about two-thirds of Rijnland's population by the late fourteenth century, but they also became an integral part, perhaps even the most essential part, of a new political, social, and economic order.

In the end, one of the most significant and interesting results of reclamation and colonization was the new relationship that Rijnlanders and their neighbors came to enjoy with respect to their environment. They had become in the process the ecological dominators. In the early Middle Ages their settlement patterns and methods of obtaining subsistence were dictated almost exclusively by natural features. As we saw, only the best-drained sandy and sandy-clay ridges were occupied or subjected to arable agricultural exploitation. By the middle of the fourteenth century they

[1] J. F. Niermeyer, *Wording van onze volkshuishouding: hoofdlijnen uit de economische geschiedenis der noordelijke Nederlanden in de middeleeuwen*, p. 17; H. van der Linden, *De Cope: bijdrage tot de rechtsgeschiedenis van de openlegging der Hollands-Utrechtse laagvlakte*, pp. 60–61.

had imposed their own order on a much-expanded Rijnland with drainage ditches and canals, dikes, dams, and sluices, and they had established settlements and were raising cereal crops wherever they pleased atop the formerly empty bogs. Whereas a deterioration in the quality of drainage would eventually have led to abandonment of the land, by the fourteenth century the new settlers were digging deeper ditches, constructing polders, and ultimately installing windmills.

From West Frisia to Holland

The reclamation and colonization of peat-bog wilderness in the western Netherlands changed the relative location of the coastal districts. Until the tenth century coastal inhabitants were physically separated from the European continent by the broad band of wilderness, with only tenuous connections inland by way of the major branches of the Rhine delta. Observers in Roman and Carolingian times noted such physical separation and thought that it, as well as the perpetually wet environment of the remote coastal districts, contributed to a cultural separation as well.[2] In fact, the inhabitants of the western Netherlands had long been exposed to diverse strains of cultural influence, not only from inland areas by way of the major river routes but also from lands accessible by way of the North Sea. By the early Middle Ages, however, the maritime connections had come to predominate. The Frisians were largely oriented toward the sea. They belonged to the world of coastal Germanic-speaking peoples arrayed around the North Sea who lived by livestock keeping, trade, crafts, and occasional piracy.[3] Rijnlanders were part of this Frisian maritime world.

The gradual silting shut of the Oude Rijn between 800 and 1000 began to erode the maritime orientation of Rijnlanders, and the reclamation and colonization of the surrounding peat bogs began to reorient them to inland areas instead. In the process they created dry land between themselves and areas on the east and south. Meanwhile, the Oude Rijn, no longer a direct route of access to the sea, became more and more an artery toward the

[2]Pliny the Elder, *Natural History*, trans. H. Rackham, 16.2–5; Anonymous of Utrecht, *II vita S. Bonifacii auctore presbyterro S. Martini Ultrajecti*, in *Acta Sanctorum Iunii, tomus Primus*, 1.9, vol. 21 in *Acta Sanctorum quotquot toto orbe culuntur, vel a catholicis scriptoribus celebrantur et Latinis et Graecis*, new ed., ed. J. Carnandet, pt. 1, p. 471; and chap. 5, above.

[3]See above, chap. 4.

continental interior. Very gradually the inhabitants of Rijnland, as well as their neighbors in Kennemerland, on the north, began to pull away and dissociate themselves from the Frisian maritime world and to join the mainstream of European culture east and south of them. By the beginning of the twelfth century this process had gone far enough that Rijnlanders had begun calling themselves, and were being called by others, Hollanders instead of Frisians. Thus Count Floris II, witnessing a grant by the bishop of Utrecht in 1101, signed his name Florentius comes de Holland,[4] and the people from the Leimuiden area, who colonized some marshlands near Bremen in 1113, were referred to as Hollandi.[5] Their new name came from a wooded area along the Oude Rijn near Koudekerk, referred to in the Saint Martin's list as Holtlant (*holt*, meaning "woods," is related to the German *Holtz* and the Dutch *hout*).[6] Not surprisingly, by the beginning of the twelfth century some of the very earliest peat-bog reclamations in the western Netherlands were already well under way immediately north and south of the Koudekerk-Holtlant area.[7]

This substitution of Holland for West Frisia was merely a subtle indicator of a much more profound cultural evolution that was taking place, shown much more clearly in some toponymic changes. Most of the place-names in the Saint Martin's list, except for those that can be ascribed to an earlier name-giving pattern, were of Frisian or coastal Germanic origin.[8] The vast majority of the names of reclamation settlements, in contrast, were of Frankish or proto-Dutch origin.[9] Only a few place-names in the very oldest reclaimed areas of the western Netherlands derived from coastal Germanic: in the Leimuiden corner of Rijnland and near Vlaardingen in Maasland on the south.[10]

[4] A. C. F. Koch, ed., *Oorkondenboek van Holland en Zeeland tot 1299*, vol. 1, no. 92.

[5] Ibid., no. 98.

[6] M. Gysseling and A. C. F. Koch, eds., *Diplomata Belgica ante annum millessimum centesimum scripta*, pt. 1, *Teksten*, no. 195; app. A, below.

[7] H. van der Linden, *De Cope*, pp. 354–61.

[8] D. P. Blok, "Holland und Westfriesland," *Frühmittelalterliche Studien: Jahrbuch des Instituts für Frühmittelalterforschung der Universität Münster* 3 (1969): 359; D. P. Blok, "De vestigingsgeschiedenis van Holland en Utrecht in het licht van de plaatsnamen," in M. Gysseling and D. P. Blok, *Studies over de oudste plaatsnamen van Holland en Utrecht*, pp. 16–25.

[9] D. P. Blok, "Plaatsnamen in Westfriesland," *Philologia Frisica anno 1966*, no. 319 (1968): 15; Blok, "Holland und Westfriesland," p. 360; Blok, "De vestigingsgeschiedenis van Holland en Utrecht in het licht van de plaatsnamen," pp. 25–26.

[10] Blok, "Holland und Westfriesland," p. 359; Blok, "De vestigingsgeschiedenis van

The Zoeterwoude area, just south of Leiderdorp, offers one of the most interesting examples of the kinds of changes that occurred in the name-giving patterns of the western coastal regions during the very earliest stages of reclamation. The Saint Martin's list referred to an early-tenth-century settlement called Suetan, derived from a Frisian water name, Zwiet, from the word meaning "sweet." Although this settlement evidently disappeared, the water name lived on. A century or so later, however, when the peat wilderness nearby was reclaimed, Frisian *zwiet* was translated into Frankish or Dutch *zoet* and applied to the new settlements of Zoeterwoude and Zoetermeer.[11]

Evidence such as this indicates that Frankish had begun to replace Frisian as the language of Rijnland and surrounding areas, especially in those sections where peat-bog wilderness was being reclaimed and colonized. Yet there was no migration into the coastal regions by Frankish-speaking colonists that could have accounted for such a linguistic substitution. All evidence suggests that Rijnland's bogs were reclaimed by Rijnlanders from the old, crowded settlements along the dunes and the Oude Rijn. Therefore, we must attribute the change in language to a process of acculturation. During the eleventh century Rijnlanders and their neighbors had apparently begun to adopt a colloquial Frankish or proto-Dutch dialect as their own, no doubt greatly influenced by Flemish and Brabantine examples from the east and south. By the thirteenth century it had become the commercial and written language of the cities and chancellery of Holland.[12] It appears, therefore, that the reclamation and settlement of peat bogs in Rijnland went hand in hand with a place-name revolution that was itself an indicator of a strong de-Frisianization of the western coastal regions of the Netherlands.[13] One can only assume that changed ecological relationships, undermining the old Frisian maritime connections and ce-

---

Holland en Utrecht in het licht van de plaatsnamen," p. 25; Blok, "Plaatsnamen in West-friesland," p. 15.

[11]Fockema Andreae, "Een verdwenen dorp? Zwieten bij Leiden," in *Varia historica aangeboden aan professor doctor A. W. Byvanck ter gelegenheid van zijn zeventigste verjaardag door de Historische Kring te Leiden*, pp. 121–28; D. P. Blok, "Probleme der Flussnamenforschung in den alluvialen Gebieten der Niederlande," in Rudolf Schützeichel and Matthias Zender, eds., *Namenforschung: Festschrift für Adolf Bach zum 75. Geburtstag am 31. Januar 1965*, p. 217; Blok, "De vestigingsgeschiedenis van Holland en Utrecht in het licht van de plaatsnamen," pp. 22–23, 25.

[12]Blok, "Holland und Westfriesland," p. 360; van der Linden, *De Cope*, pp. 115–16.

[13]Blok, "Plaatsnamen in Westfriesland," p. 15.

menting new ones to the Frankish interior, not only reinforced but perhaps even inspired the linguistic change.

A New Social Order

The reclamation and colonization of wilderness in western and central Europe during the High Middle Ages characteristically inaugurated a new social order that was based on the local *Landgemeinde*, or rural commune.[14] In most cases it came into being under the direction and with the consent of a territorial prince who possessed the rights to the disposal of large parcels of uninhabited lands. As we saw earlier, such rights had traditionally been considered specifically royal rights, or *regalia*, but the disintegration of royal power during the Viking era saw these rights falling into the hands of new, local power brokers who set themselves up as territorial lords.[15]

In an attempt to exercise his new rights, a territorial prince or his representative usually entered into an agreement with a group of settlers or their representatives concerning the granting of wilderness for reclamation and colonization. Usually according to the terms of a written contract, the wilderness would be staked out in parcels of identical size and shape, provisions would be made for local government, a church with lands for its clergy would be provided for, and an agent or agents would be charged with seeing to the implementation of the plans. To procure the needed colonists, it was necessary to draw up terms that were very beneficial to potential settlers. In exchange for a small, token tax as recognition of the lord's political authority as well as a percentage of what would be produced, groups of colonists received title to the land, rights to self-government, rights to the use of woods and streams, and freedom from labor services.

Such rights and obligations belonged to the *Landgemeinde*, not to the individual. Taxes and other obligations were levied against the entire community. The land was granted to the group, and individual families, by virtue of their membership in the group, were given standard-size homesteads as free and alienable possessions in exchange for the payment of the proper

[14]"Het in cultuur brengen van woeste gronden betekent de vestiging van een nieuwe gemeenschap [The bringing of waste land into culture signifies the establishment of a new community]," according to P. W. A. Immink, "Recht en historie," in his *Verspreide geschriften*, ed. N. E. Algra, p. 36. The German term *Landgemeinde* is used widely in the literature.

[15]See above, chap. 4.

proportion of the taxes levied against the entire community. The rural commune was given legal status with the rights to exercise local self-government and to administer lower justice. In this fashion the new rural communities received a charter or constitution that was similar in many respects to the rights and privileges that formed the basis of urban communes established during the same period.[16] The *Landgemeinde*, in short, shaped a new society of free and equal agriculturalists who decided many of their own affairs themselves in the context of their communes and exploited their own property without the intervention of landlords.

Just as elsewhere in western and central Europe, the reclamation of wilderness in Rijnland produced a new social order. Indeed, there is some evidence to suggest that the county of West Frisia–Holland, and especially Rijnland, may have seen the first development of the reclamation-era rural commune from which it diffused widely throughout temperate Europe.[17] In any case, during the late ninth and early tenth centuries the counts of West Frisia–Holland had gained control of the wilderness areas adjacent to the small clusters of settlements that they ruled along the coastal dunes and riverbanks of the western Netherlands.[18] Within a century they had begun granting to groups of colonists large portions of uninhabited peat-bog wilderness in Rijnland and neighboring districts.[19]

In exchange for a tax of originally one denarius, or penny, per home-

[16] Van der Linden, *De Cope*, pp. 160–84, 247–48, 293, 349; H. van der Linden, *Recht en territoir: een rechtshistorisch-sociografische verkenning*, p. 7; H. C. Darby, "The Clearing of the Woodland in Europe," in William L. Thomas et al., eds., *Man's Role in Changing the Face of the Earth*, pp. 190–99; Georges Duby, *Rural Economy and Country Life in the Medieval West*, pp. 392–403, documents 36–46; Marc Bloch, *French Rural History: An Essay on Its Basic Characteristics*, trans., Janet Sondheimer, pp. 13–16; Richard Koebner, "The Settlement and Colonization of Europe," in M. M. Postan, ed., *The Agrarian Life of the Middle Ages*, 2d ed., pp. 84–88, vol. 1 in *The Cambridge Economic History of Europe*; Roger Grand and Raymond Delatouche, *L'Agriculture au moyen âge de la fin de l'Empire Romain au XVIe siècle*, pp. 248–49; Bryce Lyon, "Medieval Real Estate Developments and Freedom," *American Historical Review* 63 (1957): 47–61; B. H. Slicher van Bath, *Agrarian History of Western Europe, A.D. 500–1850*, pp. 155–57; *Die Anfänge der Landgemeinde und Ihr Wesen*.

[17] Van der Linden, *De Cope*, pp. 17–62, 352–53; H. Draye, *Landelijke cultuurvormen en kolonisatiegeschiedenis*, p. 65; J. M. van Winter, "Die Entstehung der Landgemeinde in der Holländisch-Utrechtschen Tiefbene," in *Die Anfänge der Landgemeinde und Ihr Wesen*, 1:439–45; J. M. van Winter, "Vlaams en Hollands recht bij de kolonisatie van Duitsland in de 12e en 13e eeuw," *Tijdschrift voor rechtsgeschiedenis* 21 (1954): 205–24.

[18] Van der Linden, *De Cope*, pp. 81–85.

[19] See above, chap. 6.

stead, as a recognition of the overlordship of the count of West Frisia–Holland, as well as about one-tenth of the income or produce, colonists received full ownership and free rights to the disposal of their land as well as complete personal freedom.[20] In Rijnland such a transaction was often referred to as a *cope* (related to the verb *kopen*, "to buy") and has survived to the present in such place-names as Boskoop, Vriezekoop, and Nieuwkoop. At the same time the size of each homestead was established. As we saw earlier, the parallel drainage ditches, dug in a direction perpendicular to the reclamation base, also served as the side boundaries of the holdings. In Rijnland they were usually spaced at 30 *roeden* (about 110 meters), while the normal length of the ditches was about 1,250 meters. This created a standardized homestead of just over 14 hectares, suitable for one family.[21]

Finally, as elsewhere, the new communities received the legal status of the *Landgemeinde*, which eventually came to be known as the *ambacht* (plural, *ambachten*) in Rijnland and surrounding districts. These *ambachten* continued to function as the basic units of rural society until the French Revolution. They looked after all collective matters of the community, from lower justice and local government to the apportionment and collection of the taxes and any other obligations that might be levied against the commune as a whole. Each *ambacht*, and occasionally two or more small ones together, was presided over by a *schout* ("sheriff"), usually a local resident who represented the count. The *schout* in association with a council or college of *schepenen* (plural of *schepen*, "alderman" or "magistrate"), chosen by and from the community of landholders, constituted the administrative and decision-making apparatus of the rural commune.[22] Thus the Rijnland *ambacht* too was a community of free and equal peasants who decided most of their affairs themselves at the community level while exploiting their own property on an individual family level without interference from landlords.

[20] Van der Linden, *Recht en territoir*, pp. 6–7; and especially van der Linden, *De Cope*, pp. 38–39, 81–119, 160–93.

[21] Van der Linden, *De Cope*, pp. 20–36, 72–79, 349; van der Linden, *Recht en territoir*, passim; S. J. Fockema Andreae, *De oude archieven van het Hoogheemraadschap van Rijnland*, p. 22.

[22] H. van der Linden, "Iets over wording, ontwikkeling en landschappelijk spoor van de Hollandse waterschappen," *Holland: regionaal-historisch tijdschrift* 10 (1978): 103; van der Linden, *De Cope*, pp. 120–24.

One of the most fascinating consequences of the establishment of this new society of free and equal landholders was the way in which it too helped undermine the old Frisian way of life. The reclaimed and colonized lands were organized into *ambachten* that functioned as the basic units of the new society, while the individual peasant exercised a voice in local affairs through sole possession and use of a farmstead. In fact, originally there were no nonlandowners in the reclaimed areas. This meant, however, that the descendants of the pastoral and commercial Frisians, lacking ties to the soil and generally having great mobility, became in the end bound to the lands they owned. If they were to leave their lands, they would lose their voices in this new society.[23]

The Regulation of Hydraulic Matters

The peculiar physical character of the western Netherlands and the special requirements associated with water control added a dimension and importance to collective action not normally found in the higher-lying regions of temperate Europe. From the very beginning of the reclamation and colonization process there was a need to coordinate diking and drainage operations in an effort to combat the natural tendency of individuals simply to send unwanted water downstream without worrying about how it might affect someone else. At the local level the *ambacht* intervened in hydraulic matters, reinforcing the community-forming tendencies already present. As the network of dikes and canals grew more complex, however, it became necessary to form new institutions that planned and coordinated hydraulic measures transcending those of individual communities.

In addition to being the basic unit in political and legal matters, the *ambacht* carried the responsibility of maintaining all community drains and dikes and coordinating whatever community efforts were needed for such maintenance. In the earliest stages it was necessary simply to ensure

[23] Van der Linden, *Recht en territoir*, pp. 7–8; van der Linden, "Iets over wording, ontwikkeling en landschappelijk spoor van de Hollandse waterschappen," p. 103; H. P. H. Jansen, "Een economisch contrast in de Nederlanden; Noord en Zuid in the twaalfde eeuw," *Bijdragen en mededelingen betreffende de geschiedenis der Nederlanden* 98 (1983): 4, 16; Diana Shard, "The Neolithic Revolution," *Journal of Social History* 7 (1973–74): 165–70; Robert B. Edgerton, "'Cultural' vs. 'Ecological' Factors in the Expression of Values, Attitudes, and Personality Characteristics," *American Anthropologist* 67 (1965): 442–47; John W. Cole and Eric R. Wolf, *The Hidden Frontier: Ecology and Ethnicity in an Alpine Valley*, p. 141.

that the river or peat-bog stream that served as the reclamation base for the village continued to function by mandating activities such as dredging and canalization. As the process of reclamation and colonization extended into those areas between recognizable rivers and streams, however, the rural commune saw to the digging as well as the maintenance of man-made drainage canals, or *weteringen*, that functioned as reclamation bases. Indeed, some of the surviving contracts establishing reclamation and colonization projects in Rijnland and adjacent districts refer to such *weteringen* along with rights of way extending through neighboring territories to a river or stream capable of handling the extra drainage water. For example, in 1244, in a document outlining to a group of colonists the rights and obligations associated with reclaiming a piece of peat-bog wilderness in southeastern Rijnland eventually named Poelien, the count of Holland gave the new community the right to drain into the upper Gouwe and, by way of a *wetering*, through the territories of Boskoop and Alphen aan den Rijn into the Oude Rijn.[24] While hydraulic matters too were sometimes entrusted to the *schout* and *schepenen* of the rural commune, especially in some areas north of the IJ, in Rijnland and surrounding districts separate officials known as *heemraden* (plural of *heemraad*, "dike reeve") were appointed by each community to work with the *schout* in supervising the construction and maintenance of all communal dikes and drains.[25]

As long as reclamation efforts were modest and affected only individual communities, the local communal apparatus of *schout* and *heemraden* saw to the planning, construction, and upkeep of particular hydraulic works. However, as the process of reclamation and colonization progressed and communities began to border on each other so that a given set of dikes and drains affected more than one settlement, matters of planning, constructing, and maintaining the hydraulic system began to require an interlocal, collective response. The general deterioration of drainage associated with

[24] L. Ph. C. van den Bergh, ed., *Tot het einde van het Hollandse Huis*, sec. 1 in *Oorkondenboek van Holland en Zeeland*, vol. 1, no. 406; as well as van der Linden, *De Cope*, p. 4: "Insuper dicta terra habebit ductum aque, qui waterganc dicitur, per aquam que Gouda dicitur usque in Renum libere absque omni pecunie extorsione." For a discussion of this and other examples, see ibid., pp. 145–50 and map on p. 29.

[25] S. J. Fockema Andreae, "Waterschapsorganisatie in Nederland en in de vreemde," *Mededelingen der Koninklijke Nederlandse Akademie van Wetenschappen*, afdeeling letterkunde, n.s., 14, no. 9 (1951): 6; van der Linden, "Iets over wording, ontwikkeling en landschappelijk spoor van de Hollandse waterschappen," pp. 102–104; van der Linden, *De Cope*, pp. 124–25.

the subsidence and oxidation of the peat surface after initial drainage, as well as the possibility of increased precipitation from time to time, made the need for joint action more urgent.[26] For example, not only individuals but also whole communities might neglect their responsibilities in such matters, thereby endangering the entire system. Cooperation between individual communities gradually became institutionalized as organizations were formed to coordinate the planning, construction, and maintenance of interlocal or regional hydraulic works.

Exactly where the initiative for such cooperation originated is not entirely clear. The tendency in the past was to ascribe it to the counts of Holland alone.[27] Recent studies, however, have stressed the possibility that, in Rijnland at least, the first moves toward broader cooperation among individual communities in hydraulic matters grew out of the local communes and the crucial role played there by the *heemraden*. The counts then encouraged such developments, defended and promoted them in negotiations and agreements with neighboring territories, and eventually gave the resulting organizations a firm legal basis.[28] Thus, beginning perhaps as voluntary associations between communities affected by particular sets of dikes and drains, these cooperative efforts eventually became permanent, autonomous institutions with legal standing. Organizations of this sort were usually known as *heemraadschappen* (plural of *heemraadschap*, "college of *heemraden*, or dike reeves") in most of Holland, while in Zeeland and Flanders on the south they were usually called *wateringen*.[29]

The members of the college of *heemraden* constituting a *heemraadschap* were notables of the districts affected, usually aristocratics but also wealthy freeholders and even some urbanites. A comital official known as the *dijkgraaf* ("dike bailiff") customarily presided over the *heemraadschap*. The college of *heemraden* along with the *dijkgraaf* carried out leg-

[26] Discussed in chap. 6, above.

[27] S. J. Fockema Andreae, *Het Hoogheemraadschap van Rijnland: zijn recht en zijn bestuur*, p. 37; S. J. Fockema Andreae, *Willem I graaf van Holland, 1203–1222, en de Hollandse hoogheemraadschappen*.

[28] C. Dekker, "The Representation of the Freeholders in the Drainage Districts of Zeeland West of the Scheldt during the Middle Ages," *Acta Historiae Neerlandicae* 8 (1975): 6–8; van der Linden, "Iets over wording, ontwikkeling en landschappelijk spoor van de Hollandse waterschappen," pp. 105–106.

[29] Dekker, "The Representation of the Freeholders in the Drainage Districts of Zeeland West of the Scheldt during the Middle Ages," pp. 5–13; van der Linden, "Iets over wording, ontwikkeling en landschappelijk spoor van de Hollandse waterschappen," pp. 106–108.

islative as well as judicial and administrative tasks. The legislative tasks grew out of a pattern of examining specific needs, procedures, or problems and issuing either orders or prohibitions with respect to them. Out of many discrete cases of this sort eventually emerged a set of rules and regulations that came to be known as *keuren* (plural of *keur*, "choice" or "selection"). The legal function was usually carried out on location and consisted of the inspection of a dike, sluice, or ditch for which either approval or disapproval was given. In the latter case after fines had been imposed, the *dijkgraaf* ensured that all deficient work was made right and the cost of it charged to those who had been negligent in the first place. In cases of deliberate dike destruction, the *heemraadschap* even had the right to mete out capital punishment. Its administrative task consisted of collecting, storing, and processing the information necessary to carry out the other tasks.[30]

Some of the earliest forms of intercommunal cooperation in Rijnland, no doubt coordinated by *heemraadschappen* of some sort, lay behind the interlocal or regional hydraulic efforts begun in the late twelfth and early thirteenth centuries. These included the following projects: the damming of the Oude Rijn at Zwammerdam before 1165 and again around 1200; the construction of a series of *ommedijken*, or encircling dikes protecting a number of communities, near the dunes and the former Oude Rijn mouth, perhaps before 1200; and the digging of the Doeswetering, Oudewetering, and Heimanswetering, as well as the extension of the Gouwe to the Oude Rijn and the Drecht to the Amstel around 1200 or shortly thereafter. Further, by 1225 all of Rijnland and the southern half of Kennemerland had joined together in a single effort to dam the Spaarne near Haarlem, to provision the dam with seven sluices, and to pursue a common hydraulic policy throughout the entire region.[31] In fact, in 1255, Willem II, count of Holland and Holy Roman emperor-elect, recognized the autonomous existence of a super-*heemraadschap*, or *hoogheemraadschap*, encompassing

[30] Van der Linden, "Iets over wording, ontwikkeling, en landschappelijk spoor van de Hollandse waterschappen," pp. 106–108; Fockema Andreae, *Het Hoogheemraadschap van Rijnland*, pp. 58–76; Dekker, "The Representation of the Freeholders in the Drainage Districts of Zeeland West of the Scheldt during the Middle Ages," pp. 4–13; P. A. Henderikx, "De oprichting van het hoogheemraadschap van de Alblasserwaard in 1277," *Holland: regionaal-historisch tijdschrift* 9 (1977): 212–22; P. A. Henderikx, "De zorg voor de dijken in het Baljuwschap Zuid-Holland en de grensgebieden ten oosten daarvan tot het einde van de 13e eeuw," *Geografisch tijdschrift* 11 (1977): 407–27.

[31] See above, chap. 6; Fockema Andreae, *Het Hoogheemraadschap van Rijnland*, pp. 26–38; H. van der Linden, *De Zwammerdam*, pp. 9–14.

all communities benefiting from the dam in the Spaarne. Originally known as the Hoogheemraadschap van Spaarndam, it has carried the name Hoogheemraadschap van Rijnland from the fourteenth century to the present.[32]

The Hoogheemraadschap van Rijnland was the first and most influential of all organizations founded during the Middle Ages for the purpose of establishing and maintaining hydraulic works at the regional level. The *dijkgraaf*, at least from 1285 onward, was none other than the highest-ranking county official in Rijnland, the *baljuw* ("bailiff"). The seven *hoogheemraden* were no less prominent. They were usually, though not always, aristocrats, resided within the territory of the Hoogheemraadschap, and had considerable landed property and were therefore personally interested in maintaining a high standard of drainage.[33] As a result, the Hoogheemraadschap van Rijnland was able to counter particularist tendencies and promote a regional water-control policy that proved both effective and rather evenhanded over the years. Through the principle of spreading the burden of diking and draining evenly over the entire area affected, it was able to marshal the manpower and material necessary to carry out its hydraulic policy.[34] In so doing, it created a man-made landscape that has survived virtually intact to the present.

New Subsistence Patterns

At the most basic level, the reclamation and colonization of peat-bog wilderness totally revolutionized subsistence patterns. Most evidence indicates that the peat bogs of Rijnland, and of the western Netherlands in general, were reclaimed primarily to provide arable land on which cereals could be grown. This is shown, for example, in the records of the counts of Holland from the first half of the fourteenth century, in which by far the largest proportion of tithe payments recorded for villages in the former peat-bog wilderness came from cereals.[35] This reflected a total reversal in the roles of arable and pastoral agriculture in the rural economy.

[32] Van den Bergh, ed., *Tot het einde van het Hollandsche Huis*, vol. 2, no. 621. See also S. J. Fockema Andreae, "De Rijnlanden," *Leids jaarboekje* 48 (1956): 45–47; Fockema Andreae, *Het Hoogheemraadschap van Rijnland*, passim.

[33] Fockema Andreae, *Het Hoogheemraadschap van Rijnland*, pp. 43–44, 46.

[34] The standard unit of land in Rijnland was the *morgen*, equal to 0.85 hectares, or about 2 acres. Each *morgen* of land was taxed by the Hoogheemraadschap at a uniform rate, "*morgen morgensgelijk*," as the sources describe it: see Fockema Andreae, *Het Hoogheemraadschap van Rijnland*, p. 104.

[35] See, for example, the accounts of 1334, in H. G. Hamaker, ed., *De rekeningen der*

As we have seen, before the wilderness was reclaimed, arable land was a scarce commodity, found only along the highest and best-drained sandy or sandy-clay ridges. Ever since Neolithic times the primary agricultural focus had been on grazing livestock in the plentiful grasslands. The livestock provided not only meat and dairy products for local consumption but also a wide variety of products that became trade items in a broad complex of commercial relations that among other things supplied imports to augment the limited supplies of locally raised cereals and other crops in the coastal regions.[36] After reclamation, in contrast, residents of the western Netherlands could establish fields virtually anywhere they wished. As a result the production of crops increased greatly. Indeed, it became the primary agricultural focus.[37]

Crops that were not consumed locally made their way to a number of nearby urban settlements, such as Leiden, in Rijnland; Haarlem and Amsterdam, a short distance on the north; and The Hague, Delft, Rotterdam, and Gouda, on the south. All these cities developed during the very period in which the wilderness was being reclaimed.[38] In fact, the appearance of so many urban places in and around the former wilderness can be seen as a measure of the success of the entire reclamation process, since cities could appear, survive, and thrive only if the surrounding countryside produced a surplus of agricultural products.

It would be wrong, of course, to see the traditional rural economy of Rijnland and surrounding areas as suddenly transformed by the reclamation and colonization process into a totally market-oriented one. Such de-

---

graafelijkheid van Holland, 1:164–70; van der Linden, *De Cope*, p. 68; D. E. H. de Boer, *Graaf en grafiek: sociale en economische ontwikkelingen in het middeleeuwse "Noordholland" tussen 1345 en 1415*, pp. 211–49; T. Edelman, *Bijdrage tot de historische geografie van de Nederlandse kuststreek*, pp. 51–53.

[36] See above, chap. 5.

[37] J. C. Besteman and A. J. Guiran, "Het middeleeuws-archeologisch onderzoek in Assendelft, een vreoege veenontginning in middeleeuws Kennemerland," *Westerheem* 32 (1983): 167–69, provide an interesting perspective on the importance of cereal production in the reclaimed bogs. At Assendelft, north of Rijnland in central Kennemerland, a number of excavated houses apparently were smaller than the houses usually built in the coastal areas at that time. There is evidence to suggest that the stalling facilities may have been absent at first, because of a predominantly arable economy, and were added only later if at all.

[38] D. E. H. de Boer, "De verhouding Leiden-Rijnland, 1315–1414: veranderingen in een relatie," *Economisch- en sociaal-historisch jaarboek* 38 (1975): 48–72; de Boer, *Graaf en grafiek*, pp. 15–17, 211–338. Indeed, many urbanlike settlements achieved urban status in law during the thirteenth century; C. van de Kieft, "Les comtes de Holland et les villes de la principauté au XIIIe siècle," *Revue de l'Université Libre de Bruxelles* 21 (1970): 479–90.

velopments did not really begin to emerge until the fifteenth century.[39] In the meantime, most agriculturalists in the western Netherlands continued to be self-sufficient by producing essentially all that they themselves required. What they did begin to do, however, was produce enough extra to sell some on the new urban markets. It is safe to assume that dairy products, especially butter and cheese, and meat made their way to cities, even though the volume of traffic is not easy to determine. The much-better-documented production of cereals in Rijnland and surrounding districts was sufficient to keep bread and porridge on urban tables in the western Netherlands at least until the last quarter of the fourteenth century, when imports from the Baltic region and France began to assume some importance.[40]

The process of peat-bog reclamation and colonization created the agricultural potential not only to feed the growing cities of the western Netherlands but also to provide raw materials and products for urban industry and commerce. For example, while the fifteenth century is generally seen as the period in which beer production in Holland began to flourish, the capability of beer brewing had become substantial by the fourteenth century. Particularly Haarlem and Gouda, but also Leiden, Delft, and a number of other towns, produced large quantities of beer brewed from local ingredients. Further, textiles made from local wool and largely a product of rural industry increasingly came to be bought, sold, and inspected for qualty in the urban *wanthuis*, or *lakenhal* ("cloth hall"). Gouda, Delft, Rotterdam, and several other cities in the western Netherlands each contained a *wanthuis* by the middle of the fourteenth century, suggesting a fully developed system of textile production and trade.[41]

The reclaimed peat bogs of the western Netherlands produced more than agricultural products. They also became the source of considerable quantities of fuel. As long as settlement had been limited to the dune ridges and the clay banks of the Oude Rijn, most fuel needs were presumably met by the woods that grew in those locations. The centers of the large, raised oligotrophic peat bogs of the western Netherlands, however, were essentially treeless. Therefore, as colonists occupied and reclaimed them, beginning during the tenth century, they had to find new local sources of fuel.

[39] H. P. H. Jansen, "Holland's Advance," *Acta Historiae Neerlandicae* 10 (1978): 1–19.

[40] De Boer, *Graaf en grafiek*, pp. 211–49, 261–71, 335.

[41] Ibid., pp. 273–308; Jansen, "Een economisch contrast in de Nederlanden," pp. 7–10.

As we know, of course, peat, especially peat moss or sphagnum peat, consisting of partly decayed plant matter, makes fuel when it is dried. Indeed, as late as the eighteenth century it was still considered one of the best fuels available, especially for cooking and baking.[42]

The digging and drying of peat, transforming it into an excellent fuel, was known from prehistoric times. In his description of the lonely people who lived on the *terpen* of northeastern Germany around the middle of the first century A.D., Pliny the Elder said that they kept themselves warm by burning some sort of "dried mud."[43] In 839, Saint Peter's Abbey, at Gent, Flanders, was given a piece of land near Aardenburg from which four cartloads of peat could be dug.[44] Also in the tenth century a Spanish-Arabian merchant traveling through the western Netherlands reported that dried mud was used as fuel near Utrecht.[45] In a charter bearing the date 1085 but actually written sometime between 1180 and 1200, the bishop of Utrecht gave to the chapter of Saint John the Baptist at Utrecht some property in western Utrecht (near Rijnland) and mentioned, among other things, rights to peat digging.[46]

One of the earliest surviving references to digging peat for fuel in Holland dates from 1320 and concerns a piece of land given to the convent at Loosduinen, near The Hague. The land in question was a piece of bog lying just south of the Zijdwinde, the southern boundary of Rijnland, from which the convent was given permission to obtain its fuel.[47] No doubt most peat-digging and burning activities were of a local nature, with the individual household or community digging and drying the fuel for its own needs,

[42] See *Tegenwoordige staat der Vereenigde Nederlanden*, vol. 4, *De beschryving van Holland*, p. 24: "Niet ten onregt wordt de Turf onder de beste brandstoffen gerekend. Men krygt 'er harde en heete Koolen van, door welker gematigde en duurzaame hitte, men veelerlei Spyzen veel beter dan door houtskoolen of eenige andere brandstoffe, kooken en stoven kan."

[43] Pliny the Elder, *Natural History*, 16.2–5.

[44] P. van Schaik, "De economische betekenis van de turfwinning in Nederland," pt. 1, *Economisch-historisch jaarboek* 32 (1967–68): 144.

[45] Georg Jacob, ed. and trans., *Arabische Berichte van Gesandten an germanische Fürstenhofe aus dem 9. und 10. Jahrhundert*, p. 26.

[46] The charter is in *Oorkondenboek van het Sticht Utrecht tot 1301*, vol. 1, no. 245. For the date of this document, see Koch, ed., *Oorkondenboek van Holland en Zeeland tot 1299*, vol. 1, no. 89, where the text is reproduced in part; W. J. Diepeveen, *De vervening in Delfland en Schieland tot het einde der zestiende eeuw*, p. 12.

[47] Frans van Mieris, ed., *Groot Charterboek der Graven van Holland, van Zeeland en Heeren van Friesland*, 2:238.

and consequently was never the subject of documentation. In many instances, indeed, the peat removed to form drainage ditches during the reclamation process may have been sufficient for fuel.[48]

As we saw earlier, the appearance of a number of cities in the western Netherlands during the twelfth century created substantial markets for many goods and materials produced in the nearby countryside. Dried peat for fuel became one of the most important of all items entering these markets.[49] In the northern latitudes of Europe, at least, an adequate source of fuel is of the same order of necessity as food. As long as settlement remained dispersed in small communities, both could be provided for locally. Once substantial numbers of nonproducers of food and fuel came together in one place, however, such commodities had to be supplied from elsewhere and were often paid for in cash. That is what had begun to occur in Haarlem, Amsterdam, Leiden, Gouda, The Hague, Delft, Rotterdam, and other emerging urban centers of the western Netherlands.[50]

By the fourteenth century supplying peat fuel for urban markets had become a very lucrative occupation for many inhabitants of the former peat-bog districts of western Netherlands. On 20 August 1341 the count of Holland, while granting certain concessions to the inhabitants of Coencoop, west of the Gouwe in southeastern Rijnland, concerning the digging of peat for fuel, attempted at the same time to introduce some controls on the production of fuel in this manner by stipulating the amount of peat that could be dug, with penalties for anyone who exceeded the limits. He also imposed a tax on peat digging.[51] On the same day he granted a piece of bogland to some citizens of the city of Gouda from which they could dig as much peat as they wished as long as they paid a somewhat higher annual

[48] Van Scahik, "De economische betekenis van de turfwinning in Nederland," pt. 1, p. 144.

[49] Ibid., pp. 144–45; Diepeveen, *De vervening in Delfland en Schieland tot het einde der zestiende eeuw*, p. 14; T. Edelman, *Bijdrage tot de historische geografie van de Nederlandse kuststreek*, pp. 65–66; van de Kieft, "Les comtes de Holland et les villes de la principauté au XIIIe siècle," pp. 479–90.

[50] De Boer, *Graaf en grafiek*, pp. 251–59, treats many aspects of the trade in peat fuel. For a discussion of the problems associated with the acquisition of fuel in medieval London, including fuel crises and substitutions and air pollution, see William H. TeBrake, "Air Pollution and Fuel Crises in Pre-Industrial London, 1250–1650," *Technology and Culture* 16 (1975): 337–59.

[51] Van Mieris, ed., *Groot Charterboek der Graaven van Holland, van Zeeland en Heeren van Friesland*, 2:654; de Boer, *Graaf en grafiek*, pp. 251–54.

tax.[52] By 1351, however, residents of Delft, Leiden, Schiedam, Rotterdam, and Gouda apparently had begun to take advantage of their peculiar legal status as citizens of urban communes and thus free from the monetary exactions imposed on the inhabitants of the countryside, by buying up reclaimed bogland and mining peat on a large scale. As a result, on 18 October of that year the count of Holland specified that in the future any citizen of a city who bought reclaimed peat lands must pay the same dues and observe the same rules and regulations pertaining to it as did a non-urbanite.[53]

The pace of production of peat fuel as well as its sale in urban centers grew considerably during the course of the fourteenth century. By the end of the century, however, it had become apparent to authorities in charge of supervising and overseeing the dike-and-drainage system in the Hoogheemraadschap van Rijnland that unlimited peat digging was beginning to have an adverse effect on the quality of drainage. Beginning in 1392, therefore, the *dijkgraaf* and *hoogheemraden* of Rijnland began issuing a long series of restrictions on peat digging, stating in particular that no one could dig within a certain distance of dikes for fear of undermining them. Further, because water collected in those places from which peat had been dug and in times of storms waves on these pools could erode the surrounding peat, severe limitations were imposed on the amount of peat that could be extracted from a given unit of land. These and many other such regulations were intended to ensure that the desire for financial gain of a few would not jeopardize the best interests of the community as a whole.[54]

Whatever else might be said about digging peat for fuel, that activity, even more than arable agriculture, represented a radical departure from the

[52] Van Mieris, ed., *Groot Charterboek der Graaven van Holland, van Zeeland en Heeren van Friesland*, 2:654.

[53] J. H. W. Unger, ed., *Regestenlijst voor Rotterdam en Schieland tot in 1425*, no. 761, p. 227: "Hertog Willem van Beyern enz., vernomen hebende dat de poorters van Delf, Leiden, Sciedam, Rotterdam en der Goude dikwijls in strijd met de daarvoor geldende bepalingen in zijn ambachten turf delven, beveelt dat deze zich bij het koopen van venen moeten gedragen volgens de keuren van heemraden, baljuwen of ambachtsheren, even alsof zij geen poorters waren en dat degenen, die ze overtreden hebben, de darop gestelde boeten zullen betalen en bezegelt het stuk. 'Ghegheven in de Haghe up sente Lucasdach int jaere ons Heren dusent driehondert seven ende vijftich [18 October 1357].'"

[54] In app. D, I assemble a number of representative peat-digging regulations issued by the Hoogheemraadschap van Rijnland during the late fourteenth and early fifteenth centuries.

prereclamation relationship between the inhabitants of Rijnland and the peat bogs. Whereas before the late tenth century the bogs were seen as a nuisance, an obstacle to human settlement and exploitation, after reclamation the peat itself was recognized as a resource and exploited as such. Rijnlanders and their neighbors had thoroughly integrated the former wilderness into the world of human affairs. The frontier, the uninhabited and unexploited zone beyond the settled area, was no more.

Reclamation and the Rise of Holland

Without a doubt the counts of Holland played a very important role in the colonization and reclamation of the peat-bog wilderness of Rijnland during the High Middle Ages. This does not mean, of course, that they were full of altruism or had only the best interests of the common people in mind when they offered vast expanses of uninhabited and unexploited lands to groups of colonists. That would be too much to expect from any ruler of that period. The counts of Holland gained a great deal from the colonization and reclamation process. Not only did they receive a steady income from the annual payments assessed against each parcel of land they gave away, but also they received the annual tithe payments, equivalent to the tenth or eleventh part of everything that the reclaimed land produced. In fact, in the middle of the fourteenth century it was reported that the reclaimed peat-bog districts of Holland were the chief cource of income for the counts of Holland.[55]

In addition, the colonization and reclamation of peat bogs represented a substantial increase in the areal extent of the county of Holland. Although it is difficult to conceive of reclamation in Rijnland without the counts of Holland, it is equally difficult to imagine the rise of comital authority without the colonization and reclamation of the wilderness areas of Rijnland and elsewhere.[56] Using both the old and new lands of the coastal areas as their base, the counts were gradually able to extend their dynastic

[55] "Et bene dispositus comes tria habere censetur jocalia, tres thesauros absconditos, paludes et moeram Hollandiae, nemora de Moermaye Hannoniae, precis mensurarum Zelandiae"; Philippus de Leyden, *De Cura reipublicae et sorte principantis*, p. 176. Since the end of the thirteenth century, Holland and Zeeland had been ruled over by the Counts of Hainault or Henegouwen (Hannonia) in modern southern Belgium and northern France.

[56] See, for example, van de Kieft, "Les comtes de Hollande et les villes de la principauté au XIIIe siècle," pp. 480–81.

control over many surrounding areas through a series of costly wars. It is very doubtful that they would have been as successful as they were had not Rijnlanders and their neighbors created the nucleus of the county of Holland by draining the wilderness and paid their hard-earned cash into the comital treasury. The labor and persistence of the colonists had transformed the counts' worthless wilderness into valuable and productive property that formed the basis of a compact and densely populated state.

8

Legacy and Significance

THE process of reclamation and colonization in Rijnland's peat bogs during the High Middle Ages created the outlines of a landscape recognizable today. By introducing dikes, dams, ditches, drainage canals, and sluices and by establishing and populating villages, with all that they contained, the people replaced what had been until then a natural landscape with the elements of a cultural landscape. From then on, human artifice took its place alongside natural process as one of the most important forces in the development of the region. What had started out as a frontier zone, the peat-bog wilderness, or *silva*, adjacent to the old settled areas, became the heartland of a new ecological and cultural order.

From a human point of view, the medieval reclamation and colonization process in Rijnland seems to have been a successful venture in most respects. Initially, at least, it offered additional space to a growing population that, confined to the scarce ridges of well-drained soils along the dunes and riverbanks, might indeed have begun to feel the adverse effects of the silting shut of the Oude Rijn. In other words, it offered a solution of sorts to what could otherwise have become an increasingly serious ecological dilemma. Once under way, however, the process of reclamation and colonization developed a momentum that carried it far beyond the initial goals. It became in the end one of the most potent factors shaping life in the coastal districts. By the late fourteenth century not only was Rijnland's population many times greater than would have been possible even under the very best of conditions four centuries earlier, but this population had, on the average, larger quantities of food, fiber, and manufactured products at its disposal.

One of the most interesting aspects of this apparent success story was that reclamation and colonization in Rijnland, and generally in the western Netherlands, did not seem to displace or dispossess any other groups of people. In this respect it differed markedly from the patterns that prevailed at the same time in other parts of temperate Europe. In the woods of central and eastern Europe, for example, reclamation and colonization were accompanied by the military conquest and subsequent domination of indigenous populations by invading groups.[1] Indeed, there is no evidence of any bog people in the western Netherlands who were forced to suffer so that reclaiming and colonizing Rijnlanders might prosper. The social costs were apparently felt by the participants themselves, first in the form of uprooted traditions and the loneliness and monotony of frontier existence and gradually in loss of mobility as they become absorbed into a much larger and more impersonal social system or territorial state.

When we shift attention away from an exclusively human perspective to one that attempts to account for the condition of the ecological system as well, things begin to look quite different, however. Human activity in the peat-bog wilderness of Rijnland during the High Middle Ages quickly destroyed a natural order that had developed gradually over a very long period of time. Reclamation halted bog development, while subsequent activities such as agriculture attempted to divert natural nutrient and energy cycles into human cultural systems. Ultimately reclamation-induced subsidence and oxidation of the peat surface as well as the large-scale digging of peat for fuel led to the destruction of the once massive peat bogs of the western Netherlands, uncovering in the process the underlying clays and sands, most of which lie several meters below sea level.

The alteration of the conditions under which peat bogs formed and the destruction of the peat material represented a significant ecological change that in turn had a tremendous impact upon human culture in the western Netherlands. The gradual sinking of the surface, from several meters above sea level to several meters below sea level, required ever more complex and expensive drainage schemes. The introduction of polders and windmills during the fourteenth century are indicators that this process was already well under way. By the end of the fifteenth century, in fact, it had become so serious that cereal production virtually disappeared from most of the western Netherlands, and much of the population of the region was obtaining inexpensive imports from France and the Baltic Sea area. Local agriculturalists, unable to compete in cereal production given the increas-

[1]Richard Koebner, "The Settlement and Colonization of Europe," in M. M. Postan, ed., *The Agrarian Life of the Middle Ages*, 2d ed., pp. 83–91, vol. 1 in *The Cambridge Economic History of Europe*.

ingly expensive drainage requirements, were forced to find other modes of production. Some resorted to digging peat and selling it for fuel, thereby aggravating the drainage problem, and ultimately abandoning the land and migrating to nearby cities, especially to Leiden. Others responded by changing over to intensive livestock production, commencing the cheese- and butter-making industries characteristic of the western Netherlands today. Still others continued to raise crops, though the crops tended increasingly to be those needed in urban industries, such as hemp, hops, and dye plants, which brought in enough income to finance the deeper drainage required for arable agriculture. In short, the reclamation of the peat-bog wilderness of the western Netherlands, originally designed to produce arable land for cereal production, unintentionally introduced a sequence of events that two to three centuries later brought an end to the raising of cereals.[2]

Such changes, although virtually unexplored and certainly deserving of their own investigation, lie beyond the scope of this book. It is sufficient for our purposes simply to make clear that the original uses of the reclaimed peat bogs did not persist to the present even though the fact of reclamation did. Ironically, the ecological dilemma of the rural sections of the western Netherlands during the late Middle Ages contributed directly to the increasingly urban and commercial character of the region, making Holland one of the most highly urbanized and commercialized portions of all of Europe by the end of the Middle Ages.[3]

An examination of the reclamation and colonization of the wilderness, or *silva*, of Rijnland and surrounding areas during the High Middle Ages has a significance that transcends the Low Countries. First, thousands of Rijnlanders and others from the western Netherlands took part in reclamation schemes throughout western, central, and eastern Europe. By doing so, they not only helped fill in the map of settlement in places far removed from their homeland but also took with them the skills, knowledge, and institutions that had proved effective in the western Netherlands. To this day their efforts are reflected in the legal, toponymic, and linguistic records of the areas they inhabited.[4]

[2]See, for example, H. van der Linden, *De Cope: bijdrage tot de rechtsgeschiedenis van de openlegging der Hollands-Utrechtse laagvlakte*, pp. 62–69; D. E. H. de Boer, *Graaf en grafiek: sociale en economische ontwikkelingen in het middeleeuwse "Noordholland" tussen 1345 en 1415*, passim; T. Edelman, *Bijdrage tot de historische geografie van de Nederlandse kuststreek*, pp. 35–75.

[3]H. P. H. Jansen, "Holland's Advance," *Acta Historiae Neerlandicae* 10 (1978): 1–19.

[4]See, for example, J. M. van Winter, "Vlaams en Hollands recht bij de kolonisatie van Duitsland in de 12e en 13e eeuw," *Tijdschrift voor rechtsgeschiedenis* 21 (1953): 205–24.

In addition, what happened in Rijnland was part of a movement that ultimately engulfed much of temperate Europe. A better understanding of what occurred in this instance can help reveal much about the general phenomenon. For example, the medieval movement of reclamation and colonization vastly extended the realm of human affairs. In Europe generally, as in Rijnland, it created the geographical outlines visible to the present. In Europe generally, as in Rijnland, it helped raise standards of living and produce economic diversification. In Europe generally, as in Rijnland, it was related to changes in population size and density, to various forms of social relations, and to political affairs. In Europe generally, as in Rijnland, it was intimately associated with the emergence of a new, urban-based civilization. In short, it is impossible to appreciate the vitality and creativity of the High Middle Ages without coming to grips with the reclamation and colonization movement.

Finally, to move beyond the reclamation and colonization endeavors of the High Middle Ages, the exploration of ecological relationships offered here can provide historians of preindustrial Europe with important perspectives that are unavailable when the physical and biological world is seen simply as a stage upon which human events are played out. As I have been able to show, for example, an examination of the connections between human population density and subsistence strategies (including technology), whether from Ester Boserup's or some other point of view, has tremendous implications for medieval social and economic history. It may well be necessary eventually to rewrite much of the agricultural history of the early Middle Ages in such a way as to incorporate ecological considerations. For the moment I have sketched a preliminary framework within which historians can begin to replace some of the persistent notions about the unchanging character of rural existence, or of peasant stupidity or lack of technical ability, with an understanding of the dynamic nature of premodern rural life.

Appendices

Appendix A

The following constitutes a translation from the Latin of those portions of the list of possessions of Saint Martin's Church at Utrecht that referred to early Rijnland settlements. In the translation I have included the locations of the mentioned places in brackets; those with question marks are probable.

> In Alfna [Alphen aan den Rijn] part of two villas [belongs to Saint Martin's]. . . . In Upuuilcanhem [near Hazerswoude] three manors. In Suetan [near Zoeterwoude] the same. In Hanatce [near Leiden] the same. In Holtlant [Koudekerk] 4 manors. In the first part of Leithon [Leiderdorp] 2, in the second 1, in the third 1. In Rodanburg [Zoeterwoude] five manors. In Legihan [Leiden?] 5. In Loppishem [near or in Leiden] two. In Lippinge [Ter Lips, Voorschoten] the same. In UUatdinchem [Voorschoten] three. In Fore [Veur] the same. In Foreburg [Voorburg] 2. In Forschate [Voorschoten] three. . . . Similarly the Church which is called Ualcanaburg [Valkenburg] with everything pertaining to it is totally and integrally Saint Martin's. Hoverathorp [Katwijk] totally. In Suthrem [Zuidwijk-Wassenaar] four manors. In Helnere [Wassenaar] 5. . . .In Elfnum [near Alphen aan den Rijn?] 2. . . .In Langongest [Langeveld near Katwijk?] a manor and a half. In Houarathorpa [Katwijk] one-half of the entire villa, which was handed down from the inheritance of Erulfus and Radulfus. In Rothulfuashem [Rijnsburg] half of the entire villa belongs to Saint Martin's, over and above the inheritance that Aldburge has there. Of the fishery that Gerulfus has in the lower end of the Hreni [Old Rhine], the fifth part pertains to Saint Martin's. In Heslem [Elsgeest?] 5 manors. In Osbragttashem [near Noordwijk] 2. In Heslemaholta [woods near Elsgeest?] 3. In Lux [Lisse?] 3. In Hostsagnem [Sassenheim] 5. In Scata [Schoot, Noordwijk] 2. In Osfrithem taglingthos [near Osbragttashem, Noordwijk?] the third part and 6 manors. In UUestsagnem [Sassenheim] all but two manors. In Luisna [Lisse?] 5. In UUarmelde [Warmond] 3. In Osgeresgest [Oegstgeest] 2. In Polgest [Poelgeest, Oegstgeest] 2. In Husingesgest [north of the Oude Rijn mouth?] 3. In Oslem [near Voorhout?] 2. In UUilkenhem [near Voorhout] all that pertains to it belongs to Saint Martin's. The same in Burem [north of the Oude Rijn mouth?] with all appurtenances, and in Taglingi [Teilingen, Sassenheim] similarly everything belongs to Saint Martin's. . . . In Norhtgo [Noordwijk] 7. Lethem [Ter Lee, Warmond?] totally and integrally belongs to Saint Martin's along with woods and all appurtenances. . . . In Suattingabvrim [Zwammerdam?] 10. . . . In Vennapan [Vennep] all is Saint Martin's. . . . In Getzeuuald in the River Fennepa the entire fishery belongs to Saint Martin's. In Hrothaluashem, which now is called Rinasburg [Rijnsburg], 13 manors from the inheritance of Radulfus and Aldburga belong to Saint Martin's, which they themselves handed down to Saint Martin's; these manors lie north of the Vliet, also the residences that are

known as ofstedi [narrow, enclosed buildings], south of the Vliet [residences] in which the above-mentioned also lived.[1]

[1]The Latin text is from M. Gysseling and A. C. F. Koch, eds., *Diplomata Belgica ante annum millesimum centesimum scripta*, pt. 1, *Teksten*, no. 195. The suggested locations are based on the following: D. P. Blok, "Het goederenregister van de St.-Maartenskerk te Utrecht," *Mededelingen van de Vereniging voor Naamkunde te Leuven en de Commissie voor Naamkunde te Amsterdam* 33 (1957): 89–104; D. P. Blok, "Holland und Westfriesland," *Frühmittelalterliche Studien: Jahrbuch des Instituts für Frühmittelalterforschung der Universität Münster* 3 (1969): 13; D. P. Blok, "Probleme der Flussnamenforschung in den alluvialen Gebieten der Niederlande," in Rudolf Schützeichel and Matthias Zender, eds., *Namenforschung: Festschrift für Adolf Bach zum 75. Geburtstag am 32 Januar 1965*, pp. 212–29; D. P. Blok, "De vestigingsgeschiedenis van Holland en Utrecht in het licht van de plaatsnamen," in M. Gysseling and D. P. Blok, *Studies over de oudste plaatsnamen van Holland en Utrecht*, pp. 13–33; J. K. de Cock, *Bijdrage tot de historische geografie van Kennemerland in de middeleeuwen op fysisch-geografische grondslag*, pp. 55–56; S. J. Fockema Andreae, *Poldernamen in Rijnland*; S. J. Fockema Andreae, "Een verdwenen dorp? Zwieten bij Leiden," in *Varia historica aangeboden aan professor doctor A. W. Byvanck ter gelegenheid van zijn zeventigste verjaardag door de Historische Kring te Leiden*, p. 125; Maurits Gysseling, *Toponymisch woordenboek van België, Nederland, Luxemburg, Noord-Frankrijk en West-Duitsland (vóór 1226)*; H. van der Linden, *De Cope: bijdrage tot de rechtsgeschiedenis van de openlegging der Hollands-Utrechtse laagvlakte*, pp. 354–55; H. J. Moerman, *Nederlandse plaatsnamen: een overzicht*; Rob Rentenaar, "De Nederlandse duinen in de middeleeuwse bronnen tot omstreeks 1300," *Geografisch tijdschrift* 11 (1977): 368.

Appendix B

The list below indicates the number of households in each Rijnland community in 1369.[1]

Community	No. Households
Leyderdorpe	36
Coudekerke	44
Nuwencoep ende die van den Woirde, Actenhoeven ende Oudencoep	191
Reynsterwoude, Leydemuden ende Burchgraven veen ende Vennep	45
Vriesencoep	15
Alcmade	102
Scoenendorp	19
Calfsla	13
Her Jacobs wout of Esslickerwoude	48
Arlenderveen	177
Outshorne	35
Are	75
Nuwenveen	69
Zevenhoeven	73
Zoeterwoude	88
Zoetermeer	82
Zegwaert	154
Gheldrijxwoude	24
Hazertswoude	199
Zwadeburdamme	67
Middelburch	17
Alphen	117
Reewijc ende Sluupic ende Nuwenbroec	107
Gravencoep	19
Waddinxveen, Groensvorde, Snidelwijc ende Poelien	168
Voerscoten	141
Valkenburch ende Catwijc (and other Katwijkers)	161
Warmonde	29
Oestgheest	67
Lisse	87

[1] Algemeen Rijksarchief, The Hague: Archief van de graven van Holland, 889–1581, no. 676, fols. 13r–44v, 46r–46v.

Nortigherhoute (plus others)	69
Hilleghem	46
Sassenem	33
Voorhoute	51
Leyden	11
Wassenaer ende Zudwijc (plus others)	94

Appendix C

The following gives the amount of land that was assessed in each of Rijnland's communities in 1375 for the maintenance of six of the seven sluices in the Spaarndam and of the Does Canal and the Does Bridge.[1] The amounts are in *morgens* and *honts*; one *morgen* equals about 0.85 hectares, and one *hont* equals one-sixth *morgen*.

Sluice	*Morgen*	*Hont*
Hazarswoude Sluice		
Randenburch	168	
Hasertsoude	3,271	2.5
Aelssemaer	1,787.5	2
Sassenem	956	2
Alkemade Sluice		
Voerhoute	1,036	2.5
Lisse	932	2.5
Vennep	81	6
Oestgheest	1,410.5	
Alcmade	2,555.5	
Suutwijc Splinters ambocht	76	
High Sluice		
Vriesencoep	901.5	
Heren Gherts ambochte van Heemskerc	2,700	
Are	1,966.5	
Scote	481	1
Aelbrechts vierrendel vanden bosch	270	
Outsoern	1,529	4
Great Sluice		
Voerscoten	1,748	1.5
Wassenaer	1,301	1.5
Hoeghemade	200	
Zoeterwoude	5,280	
Boscoep	550	

[1]Oude archieven van Rijnland, Leiden, no. 9508: "Register van inkomsten en uitgaven van den klerk van hoogheemraden (1375)," fols. 1r–2v, 3v–4r, 7v, 14r. See above, chap. 6; S. J. Fockema Andreae, *Het Hoogheemraadschap van Rijnland: zijn recht en zijn bestuur*, pp. 50–52.

Rensterwoude	605	2
Leydemuden	871	5
Snijdelwijc	306.5	2.5
Poelgen	407.5	1.5
Heslikerwoude	1,900	2
Land Sluice		
Arleveen	2,025	1.5
Zoetermeer	1,180	3
Noertich	1,838	4
Zegwarde	853	4.5
Alphen Sluice		
Leyderdorp	1,460	4
Warmonde	1,288	3
Alfen	2,400	
Groensvoerde	257	
Hillegim	461.5	
Does Canal and Does Bridge		
Soeterwoude	5,284	
Benthusen	478	2.5
Alfen	2,400	
Coudekerc	926	1.5
Hasertsoude	3,271	2.5
Poelgen	407.5	1.5
Boscoep	550	
Middelburch	312	
Gronsvoerde	257	
Randenburch	168	
Snijdelwijc	307	
Danels ambocht uten waerde	76	
Noertic	1,838	4
Leyderdorp	1,460	4
Zoetermeer	1,180	3
Zegwaerde	853	
Valkenburch	568	
Suutwijc heren Dircs ambocht	880	4.5
Voerscoten	1,748	1.5
Wassenaer	1,301	0.5
Are	1,966.5	
Scote	481	1
Arleveen	2,025	0.5
Outsoern	1,529	4

Appendix D

The following are representative peat-digging limitations issued by the Hoogheemraadschap van Rijnland during the late fourteenth and early fifteenth centuries.

Int jaer ons heren dusent ccc xcii [1392] des donredages na sinte martijns dach translatio, cuerde die dycgrave ende de hyemreders van Rijnlant Heer Bartholomeeus van Raporst, Heer Reyner Dever, Jan van Woude, Jan van Freest, Symon vander Scivyr, datmen geen turf delven en moet in die Nukerc op die westzide vanden weghe tuuysken Riedwycker Nuwen Wech ende Kerclaen. Voirt, so en sel nyenment geen turf graven op die oest zide van den wege tuuysken der groter Zijdwijn ende daes Jaghers Werf ende hier en boven en sel nyement delven meer dan twee dage turfs ende der meer niet naerre te delven dan veertich roeden. Voirt, so en moet nyement genen turf vercopen buten ambocht noch wyt voeren vander Nuwer Kerc ende wie hier boven dede dat waer boven der kuer. Voirt, so en sel nyement geen soden slaen op viertich roeden nader meer opter hyemrader cuer dat is bijder hoechster boet.

Int jair ons heren mcccxciiii [1394] des vridages na Sinte Pieters ende Sinte Pouwels dach, cuerden die dijcgrave ende die hiemraden voirs. Als dair die landsceydinge leit twijsken Scielant ende Rijnlant datmen dair niet barnen en sel noch delven ellyc eygen op x lb ten x roeden na der sceydinge, ende dat sel beginnen vander dwyn sijde an biden Hagehout dair die wateringe gaet ende vant Hagehout tot Oevenzijwijn toe ende van Oevenzijdwijn tot Soetermeer toe, ende vand Soetermeer doer tot Zegwaert toe ende voirt Zegwaerde doer tot Benthusen toe. Ende Benthuussen doir tot Haessaertswoude toe ende Haessertswoude doir tot Waddinxveen toe ende Waddinxveen doir also verre als ons scouwe gaet. Ende dese kuer hebben wij gecoirt iegens Delflant ende Scielant ende dit selmen scouwen tot allen tijden als die dijcgrave ende die hiemraden van Rijnland goet duncket.

Item. So sellen alle ambocht bewarers condigen een sonnendag naestcommende elke in sinen kerke inden venen dat nymant turf en delve after vrydage naestcomende. Hij en come een saterdag bijden heemraden tot Leyden ende bewijse wair hij gepotet heeft ende wye dat niet doet ende after die tijt dolve, die soude verbueren x lb. Anno xxi [1421].

Item. Soe en sel nyemant turff graven binnen vijftich roeden nader meeren gelegen binnen merken op een boete van x lb, sonder op Bennebroeck dat sel wesen xxv roeden. Anno xxxvi [1436].

In the year of Our Lord one thousand ccc xcii [1392], the Thursday after Saint Marti y *translatio* [i.e., the festival celebrating the translation of the body of Saint Martin of Tours, 4 July], the *dijkgraaf* and the *heemraden* of Rijnland, Bartholomeeus van Raporst, Reyner Dever, Jan van Woude, Jan van Freest, Symon vander Scivyr, decreed that no one shall be allowed to dig peat in Nieuwkerk along the west side of the road between the New Riedwijk Road and Church Street. Further, no one shall dig peat along the east side of the road between the great Zijdwijn [an embankment marking the edge of the village territory] and the Hunter's Wharf, and beyond this, no one shall be allowed to dig more than two day's worth of peat and then not within forty rods of the lake. Further, no one shall sell or transport peat outside the community of Nieuwkerk without violating this decision. Further, no one shall cut sods within forty rods of the lake according to the *heemraden*'s decision which carries the highest penalty.

In the year of Our Lord mccccxciiii [1394], the Friday after Saint Peter's and Saint Paul's Day, the *dijkgraaf* and the *heemraden* aforementioned decreed. No one shall burn nor shall anyone dig more than x lb. for every x rods from the boundary between Schieland and Rijnland, and this shall begin at the dunes next to the woods of The Hague where the canal lies, and from the woods of The Hague to Oevenzijwijn, and from Oevenzijwijn to Zoetermeer, and from Zoetermeer through to Zegwaard and further from Zegwaard through to Benthuizen. And from Benthuizen through to Hazarswoude, and from Hazarswoude through to Waddinxveen, and from Waddinxveen through as far as our authority extends. And we have made this decree against Delfland and Schieland and it shall be enforced for as long as the *dijkgraaf* and the *heemraden* of Rijnland think good.

Item. All communal drainage officials shall announce next Sunday, each in his own church in the peat districts, that no one may dig peat after the following Friday. Each shall come to the *heemraden* in Leiden on a Saturday and show where he has planted, and whoever does not do this and then later digs, he shall be fined x lb. Anno xxi [i.e., 1421].

Item. No one shall be allowed to dig peat closer than fifty rods to the lakes within our boundaries under penalty of x lb., except for Bennebroek where it shall be xxv rods. Anno xxxvi [i.e., 1436].[1]

[1]Oude archieven van Rijnland, Leiden, no. 12: "Het groote register," fols. 1v–2v, 10v–11r. See chap. 7, above.

Bibliography

ARCHIVAL SOURCES

Algemeen Rijksarchief, The Hague: Archief van de graven van Holland, 889–1581, no. 676 [formerly Archief leen- en registerkamer van Holland, no. 429].

De oude archieven van Rijnland, Leiden, no. 12: "Het groote register."

De oude archieven van Rijnland, Leiden, no. 9508: "Register van inkomsten en uitgaven van den klerk van hoogheemraden (1375)."

PRINTED PRIMARY SOURCES

Alpertus Mettensis. *De diversitate temporum*. Ed. A. Hulshof. Werken van het Historisch Genootschap, 3d ser., no. 37. Amsterdam: J. Miller, 1916. [Other editions: (1) G. H. Pertz, ed. Monumenta Germaniae Historica, *Scriptorum*, vol. 4. Hannover: Hahn, 1841. (2) Hans van Rij, ed. Amsterdam: Verloren, 1980.]

Anonymous of Utrecht. *II vita S. Bonifacii auctore presbytero S. Martini Ultrajecti*. In *Acta Sanctorum quotquot toto orbe coluntur, vel a catholicis scriptoribus celebrantur et Latinis et Graecis, aliarum gentium antiquis monumentis*. Vol. 21, *Acta Sanctorum Iunii, tomus primus*. New ed. Ed. J. Carnandet. Pt. 1, pp. 469–73. Paris: Victor Palmé, 1867.

Bergh, L. Ph. C. van den, ed. *Oorkondenboek van Holland en Zeeland*. Sec. 1, *Tot het einde van het Hollandsche Huis*. 2 vols. Koninklijke Akademie van Wetenschappen. Amsterdam: Frederik Muller; The Hague: Nijhoff, 1866, 1873.

Byvanck, A. W., ed. *Excerpta Romana: de bronnen der Romeinsche geschiedenis van Nederland*. Vol. 1, *Teksten*; vol. 2, *Inscripties*. Rijks geschiedkundige publicatiën, grote series, nos. 73, 81. The Hague: Nijhoff, 1931, 1935.

Eddison, E. R., trans. and ed. *Egil's saga* [*Egils saga Skallagrímssonar*]. Cambridge: Cambridge University Press, 1930.

Franz, G., ed. *Quellen zur Geschichte des deutschen Bauernstandes im Mittelalter*. Darmstadt: Wissenschaftliche Buchgesellschaft, 1967.

Gysseling, M., and A. C. F. Koch, eds. *Diplomata Belgica ante annum millesimum centesimum scripta*. Pt. 1, *Teksten*. Bouwstoffen en studiën voor de lexicografie van het Nederlands, no. 1. Brussels: Belgisch Interuniversitair Centrum voor Neerlandistiek, 1950.

Hamaker, H. G., ed. *De rekeningen der graafelijkheid van Holland onder het Henegouwsche huis*. 3 vols. Werken van het Historisch Genootschap, n.s., nos. 21, 24, 26. Utrecht: Kemink en Zoon, 1875, 1876, 1878.

Helbig, Herbert, and Lorenz Weinrich, eds. *Mittel- und Norddeutschland, Ostseeküste*. 2d ed. Vol. 1 in *Urkunden und erzählende Quellen zur deutschen Ostsiedlung im Mittelalter*. Darmstadt: Wissenschaftliche Buchgesellschaft, 1975.

Jacob, Georg, trans. and ed. *Arabische Berichte von Gesandten an germanische Fürstenhofe aus dem 9. und 10. Jahrhundert.* Quellen zur deutschen Volkskunde, vol 1. Berlin: W. de Gruyter, 1927.

Koch, A. C. F., ed. *Oorkondenboek van Holland en Zeeland tot 1299.* Vol. 1, *Eind van de 7e eeuw tot 1222.* The Hague: Nijhoff, 1970.

Leeuwen, S. van. *Costumen, keuren ende ordinatien van het baljuschap ende lande van Rijnland.* Leiden and Rotterdam: By de Hackens, 1667.

Mieris, Frans van, ed. *Groot Charterboek der Graaven van Holland, van Zeeland en Heeren van Friesland.* 4 vols. Leiden: Pieter van der Eyk, 1753–56.

Muldoon, James, ed. *The Expansion of Europe: The First Phase.* Philadelphia: University of Pennsylvania Press, 1977.

Muller Hz., S., ed. "Oude register van graaf Florens." *Bijdragen en mededelingen van het Historisch Genootschap* 22 (1901): 90–357.

Oorkondenboek van het Sticht Utrecht tot 1301. 5 vols. Utrecht: A. Oosthoeck, 1920–59.

Oppermann, Otto, ed. *Fontes Egmundenses.* Werken van het Historisch Genootschap, 3d ser., no. 61. Utrecht: Keminck en Zoon, 1933.

Pertz, G. H., ed. *Eigilis vita S. Sturmi abbatis Fuldensis.* In Monumenta Germaniae Historica, *Scriptorum*, vol. 2. Hannover: Hahn, 1829.

Philippus de Leyden. *De cura reipublicae et sorte principantis.* Ed. R. Fruin and P. C. Molhuijsen. Werken der Vereeniging tot Uitgave der Bronnen van het Oude Vaderlandsche Recht, 2d ser., no. 1. The Hague: Nijhoff, 1900.

Pliny the Elder [Plinius Secundus]. *Natural History* [*Naturalis historia*]. Trans. H. Rackham. Bk. 16. Loeb Classical Library, vol. 4. London: William Heineman; Cambridge, Mass.: Harvard University Press, 1945.

Richthofen, Karl von, ed. *Lex Frisionum.* In Monumenta Germaniae Historica, *Legum*, sec. 1, *Legum nationem Germanicarum*, 3:656–82. Hannover: Hahn, 1863.

Tacitus [Cornelius]. *The Histories* [*Historiae*]. Trans. Clifford H. Moore. 2 vols. London: William Heineman; Cambridge, Mass.: Harvard University Press, 1968, 1969.

Unger, J. H. W., ed. *Regestenlijst voor Rotterdam en Schieland tot in 1425.* Rotterdam: P. van Waesberge en Zoon, 1907.

Vierlingh, Andreas. *Tractaet van dyckagie.* Ed. J. de Hullu and A. G. Verhoeven. Rijks geschiedkundige publicatiën, kleine serie, no. 20. The Hague: Nijhoff, 1920.

SECONDARY LITERATURE

Abel, Wilhelm. *Agricultural Fluctuations in Europe: From the Thirteenth to the Twentieth Centuries.* Trans. Olive Ordish. New York: St. Martin's Press, 1980.

———. *Geschichte der deutschen Landwirtschaft vom frühen Mittelalter bis zum 19. Jahrhundert.* Deutsche Agrargeschichte, vol. 2. Stuttgart: Eugen Ulmer, 1967.

Acket, M. N. "De Oude Rijn en zijn omgeving." *Leids jaarboekje* 45 (1953): 73–102.

Akkerman, J. B. "De vroeg-middeleeuwse emporia." *Tijdschrift voor rechtsgeschiedenis* 35 (1967): 230–83.

Alexander, John. "The Beginnings of Urban Life in Europe." In Peter J. Ucko, Ruth Tringham, and G. W. Dimbleby, eds. *Man, Settlement and Urbanism*, pp. 843–50. London: Gerald Duckworth, 1972.

Die Anfänge der Landgemeinde und Ihr Wesen. 2 vols. Konstanzer Arbeitskreis für Mittelalterliche Geschichte, Vorträge und Forschungen, nos. 7–8. Constance and Stuttgart: Jan Thorbecke Verlag, 1964.

Baars, F. "Afdelingnieuws: afdeling Noord-Holland Noord." *Westerheem* 27 (1978): 237.

Bakker, J. A. "Een grafheuvel en oud akkerland te Hoogkarspel (N.H.)." In *In het voetspoor van A. E. van Giffen*, 2d ed., pp. 103–109, 177–78. Universiteit van Amsterdam, Instituut voor Prae- en Protohistorie, 1951–61. Groningen: J. B. Wolters, 1966.

——— et al. "Hoogkarspel-Watertoren: Towards a Reconstruction of Ecology and Archaeology of an Agrarian Settlement of 1000 B.C." In B. L. van Beek, R. W. Brandt, and W. Groenman–van Waateringe, eds., *Ex horreo: I.P.P. 1951–1976*, pp. 187–225. University of Amsterdam, Albert Egges van Giffen Instituut voor Prae- en Protohistorie, Cingula 4. Amsterdam: I.P.P., 1977.

Balthasars, Floris. *Kaarten van Rijnland, 1615*. Intro. G. 't Hart. Alphen aan den Rijn: Canaletto, 1972.

Beek, B. L. van. "Pottery of the Vlaardingen Culture." In B. L. van Beek, R. W. Brandt, and W. Groenman–van Waateringe, eds. *Ex horreo: I.P.P. 1951–1976*, pp. 86–100. University of Amsterdam, Albert Egges van Giffen Instituut voor Prae- en Protohistorie, Cingula 4. Amsterdam: I.P.P., 1977.

Bennema, J. "De bewoonbaarheid van het Nederlandse kustgebied vóór de bedijkingen." *Westerheem* 5 (1956): 88–91.

———. "Het oppervlakteveen in West-Nederland." *Boor en spade* 3 (1949): 139–49.

Besteman, J. C. "Carolingian Medemblik." *Berichten van de Rijksdienst voor het Oudheidkundig Bodemonderzoek* 24 (1974): 43–106

———, and A. J. Guiran. "Het middeleeuws- archeologisch onderzoek in Assendelft, een vroege veenontginning in middeleeuws Kennemerland." *Westerheem* 32 (1983): 144–76.

———, and H. Sarfatij. "Bibliographie zur Archäologie des Mittlealters in den Niederlanden 1945 bis 1975." *Zeitschrift für Archäologie des Mittelalters* 5 (1977): 163–231.

Beunder, P. C. "De Romeinse legerweg tussen Zwammerdam en Bodegraven." *Westerheem* 23 (1974): 216–25.

———. "Tussen Laurum (Woerden) en Nigrum Pullum (Zwammerdam?) lag nog een castellum." *Westerheem* 29 (1980): 2–33.

———. "Waarnemingen langs de Romeinse Rijnoever te Alphen a/d Rijn (Z.-H.)." *Westerheem* 26 (1977): 275–78.

Birrell, Jean. "Peasant Craftsmen in the Medieval Forest." *Agricultural History Review* 17 (1969): 91–107.

Bloch, Marc. *French Rural History: An Essay on Its Basic Characteristics*. Trans. Janet Sondheimer. Berkeley and Los Angeles: University of California Press, 1966.

———. "Occupation du sol et peuplement." In *Mélanges historiques*, pp. 124–41. Bibliothèque générale de l'École Practique des Hautes Études, sec. 6. Paris: S.E.V.P.E.N., 1962.

Bloemers, J. H. F. "Rijswijk (Z.H.), 'De Bult,' een nederzetting van de Cananefaten." *Hermeneus: tijdschrift voor antieke cultuur* 52 (1980): 95- 106.

———. *Rijswijk (Z.H.), "De Bult," Eine Siedlung der Cananefaten*. Rijksdienst voor het Oudheidkundig Bodemonderzoek, Nederlandse Oudheden 8. Amersfoort: R.O.B., 1978.

Blok, D. P. *De Franken in Nederland*. 3d ed. Haarlem: Fibula-Van Dishoeck, 1979.

———. "Het goederenregister van de St.-Maartenskerk te Utrecht." *Mededelingen van de Vereniging voor Naamkunde te Leuven en de Commissie voor Naamkunde te Amsterdam* 33 (1957): 89–104.

———. "Histoire et toponymie: l'example des Pays-Bas dans le haut moyen âge." *Annales: économies, sociétés, civilisations* 24 (1969): 919–46.

———. "De Hollandse en Friese kerken van Echternach." *Naamkunde* 6 (1974): 167–84.

———. "Holland und Westfriesland." *Frühmittelalterliche Studien: Jahrbuch des Instituts für Frühmittelalterforschung der Universität Münster* 3 (1969): 347–61.

———. "Opmerkingen over het aasdom." *Tijdschrift voor rechtsgeschiedenis* 31 (1963): 243–74.

———. "Plaatsnamen in Westfriesland." *Philologia Frisica anno 1966* no. 319 (1968): 11–19.

———. "Probleme der Flussnamenforschung in den alluvialen Gebieten der Niederlande." In Rudolf Schützeichel and Matthias Zender, eds., *Namenforschung: Festschrift für Adolf Bach zum 75. Geburtstag am 31. Januar 1965*, pp. 212–27. Heidelberg: Carl Winter–Universitätsverlag, 1965.

———. "De vestigingsgeschiedenis van Holland en Utrecht in het licht van de plaatsnamen." In M. Gysseling and D. P. Blok, *Studies over de oudste plaatsnamen van Holland en Utrecht*, pp. 13–34. Bijdragen en mededelingen der Naamkunde-Commissie van de Koninklijke Nederlandse Akademie van Wetenschappen te Amsterdam, no. 17. Amsterdam: Noord-Hollandse Uitgeversmaatschappij, 1959.

———. "De Wikingen in Friesland." *Naamkunde* 10 (1978): 25–47.

———, and A. C. F. Koch. "De naam Wijk bij Duurstede in verband met de ligging der stad." *Mededelingen van de Vereniging voor Naamkunde te Leuven en de Commissie voor Naamkunde te Amsterdam* 40 (1964): 38–51, 189.

Blum, Jerome. *The End of the Old Order in Rural Europe*. Princeton, N.J.: Princeton University Press, 1978.

———, ed. *Our Forgotten Past: Seven Centuries of Life on the Land*. London: Thames and Hudson, 1982.

Bodemkaart van Nederland, schaal 1:200,000. Comp. Stichting voor Bodemkaartering. Wageningen: Stichting voor Bodemkaartering, 1961.

Boe, G. de. "De Romeinse villa te Haccourt en de landelijke bewoning." *Hermeneus: tijdschrift voor antieke cultuur* 52 (1980): 107–13.

Boeles, P. C. J. A. *Friesland tot de elfde eeuw: zijn vóór- en vroege geschiedenis*. 2d ed. The Hague: Nijhoff, 1951.

Boer, D. E. H. de. *Graaf en grafiek: sociale en economische ontwikkelingen in het middeleeuwse "Noordholland" tussen 1345 en 1415*. Leiden: New Rhine Publishers, 1978.

———. "De verhouding Leiden-Rijnland, 1365–1414: veranderingen in een relatie." *Economisch- en sociaal-historisch jaarboek* 38 (1975): 48–72.

Bogaers, J. E. *Civitas en stad van de Bataven en Canninefaten*. Inaugural Address, Catholic University at Nijmegen, 21 October 1960. Nijmegen and Utrecht: Dekker and van de Vegt, 1960.

———. "Romeinse militairen aan het Helinium." *Westerheem* 23 (1974): 70–78.

———. "Waarnemingen in Westerheem." *Westerheem* 17 (1968): 173–79, 217–23.

Bois, Guy. *Crise du féodalisme: économie rurale et démographie en Normandie Orientale au début du 14e siècle au milieu du 16e siècle*. Cahiers de la Fondation Nationale des Sciences Politiques, no. 202. Paris: Presses de la Fondation des Sciences Politiques, 1976.

Boone, W. J. de. "De Vikingen in Velsen." *Westerheem* 7 (1958): 30–38.

Boonenburg, K. "De oudste tot op heden bekende vermelding van windmolens in het graafschap Holland (1274)." *Bulletin van de Koninklijke Nederlandse Oudheidkundige Bond*, 6th ser., 5 (1950): 62.

Borger, G. J. "Ontwatering en grondgebruik in de middeleeuwse veenontginningen in Nederland." *Geografisch tijdschrift* 10 (1976): 343–53.

———. "De ontwatering van het veen: een hooflijn in de historische nederzettingsgeografie van Nederland." *Geografisch tijdschrift* 11 (1977): 377–87.

———. *De Veenhoop: een historisch-geografisch onderzoek naar het verdwijnen van het veendek in een deel van West-Friesland*. Amsterdam: Repro Holland, 1975.

Boserup, Ester. *The Conditions of Agricultural Growth: The Economics of Agrarian Change under Population Pressure*. Chicago: Aldine Publishing Co., 1965.

———. "Environnement, population et technologie dan les sociétés primitives." *Annales: économies, sociétés, civilisations* 29 (1974): 538–52.

———. *Population and Technological Change: A Study of Long-Term Trends*. Chicago: University of Chicago Press, 1981.

Bosl, Karl. *Die Grundlagen der modernen Gesellschaft im Mittelalter: Eine deutsche Gesellschaftsgeschichte des Mittelalters*. Monographien zur Geschichte des Mittelalters, vol. 4, pts. 1, 2. Stuttgart: Anton Hiersemann, 1972.

Bouwsma, William J. "The Renaissance and the Drama of Western History." *American Historical Review* 84 (1979): 1–15.

Braat, W. C. "Early Medieval Glazed Pottery in Holland." *Medieval Archaeology* 15 (1971): 112–14.

———. "Leython." *Leids jaarboekje* 44 (1952): 79–93.

Brandt, R. W. "De archeologie van de Zaanstreek." *Westerheem* 32 (1983): 120–37.

———. "De kolonisatie van West-Friesland in de bronstijd." *Westerheem* 29 (1980): 137–51.

———. "Landbouw en veeteelt in de late bronstijd van West- Friesland." *Westerheem* 25 (1976): 58–66.

Brongers, J. A. *Air Photography and Celtic Field Research in the Netherlands.* Rijksdienst voor het Oudheidkundig Bodemonderzoek, Nederlandse Oudheden, vol. 6. Amersfoort: R.O.B., 1976.

———et al. "Prehistory in the Netherlands: An Economic-Technological Approach." *Berichten van de Rijksdienst voor het Oudheidkundig Bodemonderzoek* 23 (1973): 7–47.

Buitenen, M. P. van. *Langs de heiligenweg: perspectief van enige vroeg-middeleeuwse verbindingen met Noord-Nederland.* Verhandelingen, Koninklijke Nederlandse Akademie van Wetenschappen, afdeling letterkunde, n.s., vol. 94. Amsterdam: Noord- Hollandsche Uitgeversmaatschappij, 1977.

Butler, J. J. *Bronze Age Connections across the North Sea: A Study in Prehistoric Trade and Industrial Relations between the British Isles, the Netherlands, North Germany and Scandinavia, c. 1700–700 B.C.* [comprises *Palaeohistoria: Acta et Communicationes Instituti Bio- Archaeologici Universitatis Groninganae* 9 (1963)]. Groningen: J. B. Wolters, 1963.

———. *Nederland in de bronstijd.* Fibulareeks, vol. 31. Bussum: Fibula–van Dishoeck, 1969.

———. "Vergeten schatvondsten uit de bronstijd." In J. E. Bogaers, W. Glasbergen, P. Glazema, and H. T. Waterbolk, eds., *Honderd eeuwen Nederland*, pp. 125–42 [comprises *Antiquity and Survival* 2 (1959), nos. 5–6]. The Hague: Luctor et Emergo, 1959.

Butzer, Karl W. "Accelerated Soil Erosion: A Problem of Man-Land Relationships." In Ian R. Manners and Marvin W. Mikesell, eds., *Perspectives on Environment*, pp. 57–78. Washington: Association of American Geographers, 1974.

Buurman, Janneke. "Cereals in Circles: Crop Processing Activities in Bronze Age Bovenkarspel (the Netherlands)." In Udelgard Körber-Grohne, ed., *Festschrift Maria Hopf zum 65. Geburtstag am 14. September 1979*, pp. 21–37. Rheinischen Landesmuseum Bonn, Archaeo-Physika, vol. 8. Cologne: Rheinland Verlag; Bonn: Rudolf Habelt Verlag, 1979.

Byvanck, A. W. *Nederland in den Romeinschen tijd.* 3d ed. 2 vols. Leiden: E. J. Brill, 1945.

Casparie, W. A. *Bog Development in Southeastern Drente.* The Hague: Nijhoff, 1972.

Cheyette, Frederic L. "The Invention of the State," In Bede Karl Lackner and Kenneth Roy Philp, eds., *Essays on Medieval Civilization*, pp. 143–78. Walter Prescott Webb Memorial Lectures, no. 12. Austin and London: University of Texas Press, 1978.

Chisholm, Michael. *Rural Settlement and Land Use: An Essay in Location*. 2d ed. London: Hutchinson, 1968.

Clark, J. G. D. *Prehistoric Europe: The Economic Basis*. London: Methuen and Co., 1952.

Clason, A. T. *Animal and Man in Holland's Past: An Investigation of the Animal World surrounding Man in Prehistoric and Early Historical Times in the Provinces of North and South Holland* [comprises *Palaeohistoria: Acta et Communicationes Instituti Bio-Archaeologici Universitatis Groninganae*, vol. 13, pts. A, B]. 2 vols. Groningen: J. B. Wolters, 1967.

———. "The Antler, Bone, and Tooth Objects from Velsen: A Short Description." *Berichten van de Rijksdienst voor het Oudheidkundig Bodemonderzoek* 24 (1974): 119–31.

Cock, J. K. de. *Bijdrage tot de historische geografie van Kennemerland in de middeleeuwen op fysisch-geografische grondslag*. Groningen: J. B. Wolters, 1965.

———. "Die Grafschaft Masalant." In *Miscellanea mediaevalia in memoriam Jan Frederik Niermeyer*. Groningen: J. B. Wolters, 1967.

———. "Veenontginningen in West-Friesland." *West-Frieslands oud en nieuw* 36 (1969): 154–71.

Cohen, Mark Nathan. *The Food Crisis in Prehistory: Overpopulation and the Origins of Agriculture*. New Haven, Conn.: Yale University Press, 1976.

Cole, John W., and Eric R. Wolf. *The Hidden Frontier: Ecology and Ethnicity in an Alpine Valley*. Studies in Social Discontinuity. New York and London: Academic Press, 1974.

Cooter, William S. "Ecological Dimensions of Medieval Agrarian Systems." *Agricultural History* 52 (1978): 458–77.

———. "Preindustrial Frontiers and Interaction Spheres: Aspects of the Human Ecology of Roman Frontier Regions in Northwest Europe." Ph.D. diss., University of Oklahoma, 1976.

Darby, H. C. "The Clearing of the Woodland in Europe." In William L. Thomas et al., eds., *Man's Role in Changing the Face of the Earth*, pp. 183–216. Chicago: University of Chicago Press, 1956.

———. "The Fenland Frontier in Anglo-Saxon England." *Antiquity* 8 (1934): 185–201.

Datoo, B. A. "Towards a Reformulation of Boserup's Theory of Agricultural Change." *Economic Geography* 54 (1978): 135–44.

Deelen, D. van, and A. Schermer. "Middeleeuws akkerland onder de Castricummer duinen." *Westerheem* 12 (1963): 136–44.

Dekker, C. "The Representation of the Freeholders in the Drainage Districts of Zeeland West of the Scheldt during the Middle Ages." *Acta Historiae Neerlandicae* 8 (1975): 1–30.

———. "De vorming van aartsdiakonaten in het diocees Utrecht in de tweede helft van de 11e en het eerste kwart van de 12e eeuw." *Geografisch tijdschrift* 11 (1977): 339–60.

———. *Zuid-Beveland: de historische geografie en de instellingen van een Zeeuws eiland in de middeleeuwen.* Van Gorcum's Historische Bibliotheek, no. 87. Assen: Van Gorcum, 1971.

Des Marez, G. *Le problème de la colonisation franque et du régime agraire en Basse-Belgique.* Académie Royale de Belgique, classe des lettres, mémoires, collection en quarto, 2d ser., vol. 9, fasc. 4. Brussels: Maurice Lamertin, 1926.

Devèze, M. "Forêts françaises et forêts allemandes: étude historique comparée." *Revue historique* 235 (1966): 347–80; 236 (1966): 47–68.

Diepeveen, W. J. *De verveening in Delfland en Schieland tot het einde der zestiende eeuw.* Leiden: Eduard IJdo, 1950.

Doorn, C. J. van. *Het oude Miland en zijn waterstaatkundig ontwikkeling.* Utrecht: Kemink en Zoon, 1940.

Doorn, Z. van. "Enkele waarnemingen van oorspronkelijke Indonesische veenmoerassen ter vergelijking met de Hollands-Utrechtse venen." *Boor en spade* 10 (1959): 156–70.

Doorselaer, A. van. "De Romeinen in België en Nederland." *Hermeneus: tijdschrift voor antieke cultuur* 52 (1980): 74–86.

Dou, Jan Jansz., and Steven van Broekhuysen. *Kaartboek van Rijnland.* Intro. G. 't Hart. Reproduction of 3d ed. of 1746, reworked by Melchior Bolstra from eds. of 1647 and 1687. Alphen aan den Rijn: Canaletto, 1969.

Draye, H. *Landelijke cultuurvormen en kolonisatiegeschiedenis.* Toponymica: bijdragen en bouwstoffen uitgegeven door de Vlaamsche Toponymische Vereeniging te Leuven, no. 7. Leuven: Instituut voor Vlaamsche Toponymie; Brussels: Standaard Boekhkandel, 1941.

Duby, Georges. *The Early Growth of the European Economy: Warriors and Peasants from the Seventh to the Twelfth Century.* Trans. Howard B. Clarke. Ithaca, N.Y.: Cornell University Press, 1974.

———. "La révolution agricole médiévale." *Revue de géografie Lyon* 29 (1954): 361–66.

———. *Rural Economy and Country Life in the Medieval West.* Trans. Cynthia Postan. London: Edward Arnold, 1968.

Dumon Tak, A. M., and J. van den Berg. "Een pottenbakkersoven uit de IJzertijd te Serooskerke (Walcheren)." *Westerheem* 22 (1973): 242–47.

East, W. Gordon. *The Geography behind History.* Rev. ed. London: Nelson, 1965.

Edelman, C. H. *Over de Plaatsnamen met het bestanddeel woud en hun betrekking tot de bodemgesteldheid.* Bijdragen en mededelingen der Naamkunde-Commissie van de Koninklijke Nederlandse Akademie van Wetenschappen te Amsterdam, vol. 7. Amsterdam: Noord-Hollandsche Uitgeversmaatschappij, 1955.

———. *Soils of the Netherlands.* Amsterdam: Noord-Hollandsche Uitgeversmaatschappij, 1950.

Edelman, T. *Bijdrage tot de historische geografie van de Nederlandse kuststreek.* Rijkswaterstaat, Directie Waterhuishouding en Waterbeweging, Publicatie no. 14. The Hague: Rijkswaterstaat, 1974.

———. "Oude ontginningen van de veengebieden in de Nederlandse kuststrook." *Tijdschrift voor economische en sociale geografie* 49 (1958): 239–45.

Edgerton, Robert B. " 'Cultural' vs. 'Ecological' Factors in the Expression of Values, Attitudes, and Personality Characteristics." *American Anthropologist* 67 (1965): 442–47.

Ennen, Edith. *The Medieval Town.* Trans. Natalie Fryde. Amsterdam: North Holland Publishing Co., 1979.

———, and Walter Janssen. *Deutsche Agrargeschichte: Vom Neolithikum bis zur Schwelle des Industriezeitaltrs.* Wissenschaftliche Paperbacks, no. 12, Sozial und Wirtschaftsgeschichte. Wiesbaden: Franz Steiner Verlag, 1979.

Es, W. A. van. "Early Medieval Settlements." *Berichten van de Rijksdienst voor het Oudheidkundig Bodemonderzoek* 23 (1973): 281–87.

———. "Excavations at Dorestad: A Preliminary Report: 1967–1968." *Berichten van de Rijksdienst voor het Oudheidkundig Bodemonderzoek* 19 (1969): 183–209.

———. "Friesland in Roman Times." *Berichten van de Rijksdienst voor het Oudheidkundig Bodemonderzoek* 15–16 (1965–66): 37–68.

———. "De Nederlandse archeologie na 1945." *Westerheem* 25 (1976): 279–305.

———. "Die neuen Dorestad-Grabungen 1967-1972." In *Vor- und Frühformen der europäischen Stadt im Mittelalter: Bericht über ein Symposium in Reinhausen bei Göttingen vom 18. bis 24. April 1972*, 1:202–17. Abhandlungen der Akademie der Wissenschaften in Göttingen, Philologisch-Historische Klasse, 3d ser., nos. 83–84. Göttingen: Vandenhoeck and Ruprecht, 1973.

———. *De Romeinen in Nederland.* Grote Fibula serie. Bussum: Fibula–van Dishoeck, 1972.

———. "Vis uit Dorestad voor mijnheer Calkoen." *Westerheem* 23 (1974): 89–94.

———, et al. *Excavations at Dorestad 1—The Harbour: Hoogstraat I.* 2 vols. Kromme Rijn Projekt 1, Rijksdienst voor het Oudheidkundig Bodemonderzoek, Nederlandse Oudheden, no. 9. Amersfoort: R.O.B., 1980.

Faber, F. J. *Nederlandsche landschappen*, 2d ed. Vol. 3 in *Geologie van Nederland.* Gorinchem: J. Noorduijn, 1947.

Flannery, Kent V. "The Cultural Evolution of Civilizations." *Annual Review of Ecology and Systematics* 3 (1972): 399–426.

Fockema Andreae, S. J. " 'Aen't ende van den lande': de Hollands-Utrechtse grensstreek bij Woerden." *Zuid-Hollandse Studiën* 1 (1950): 83–94.

———. "Friesland van de vijfde tot de tiende eeuw." In *Algemene geschiedenis der Nederlanden*, 1:386–406. Utrecht: De Haan and Antwerpen: Standaard Boekhandel, 1949.

———. *Het Hoogheemraadschap van Rijnland: zijn recht en zijn bestuur van den vroegsten tijd tot 1857.* Leiden: Eduard IJdo, 1934.

———. "Middeleeuwsch Oegstgeest." *Tijdschrift voor geschiedenis* 50 (1935): 256–75.

———. *De oude archieven van het Hoogheemraadschap van Rijnland.* Leiden: Eduard IJdo, 1933.

———. "De Oude Rijn: eigendom van openbaar water in Nederland." In *Rechtskundige opstellen op 2 november 1935 door oud- leerlingen aangeboden aan prof. mr. E. M. Meijers.* pp. 699–715. Zwolle: Tjeenk Willink, 1935.

———. *Poldernamen in Rijnland.* Bijdragen en mededelingen der Naamkunde-Commissie van de Koninklijke Nederlandse Akademie van Wetenschappen, no. 4. Amsterdam: Noord-Hollandsche Uitgeversmaatschappij, 1952.

———. "De Rijnlanden." *Leids jaarboekje* 48 (1956): 45–56.

———. "De Rijnlandse kastelen en landhuizen in hun maatschappelijk verband." In S. J. Fockema Andreae et al., *Kastelen, ridderhofsteden en buitenplaatsen in Rijnland*, pp. 1–20. Leiden: Vereniging "Oud-Leiden," 1952.

———. "Stein, het ontstaan van een vrije hooge heerlijkheid op de grenzen van Holland en van hare bestuursorganen." *Tijdschrift voor geschiedenis* 47 (1932): 396–431.

———. *Studiën over waterschapsgeschiedenis.* Vol. 4, *Het Nedersticht.* Leiden: E. J. Brill, 1950.

———. "Een verdwenen dorp? Zwieten bij Leiden." In *Varia historica aangeboden aan professor doctor A. W. Byvanck ter gelegenheid van zijn zeventigste verjaardag door de Historische Kring te Leiden*, pp. 121–28. Assen: Van Gorcum, 1954.

———. "De Visbrug te Leiden." *Leids jaarboekje* 43 (1951): 56–65.

———. "Warmond." *Leids jaarboekje* 41 (1949): 69–84.

———. "Waterschapsorganisatie in Nederland en in den vreemde." *Mededelingen der Koninklijke Nederlandse Akademie van Wetenschappen* 14, no. 9 (1951): 1–22.

———. *Willem I graaf van Holland, 1203–1222, en de Hollandse hoogheemraadschappen.* Wormerveer: Iris Pers, 1954.

———, ed. Introduction. *Rechtsbronnen der vier hoofdwaterschappen van het vasteland van Zuid-Holland (Rijnland; Delfland; Schieland; Woerden).* Werken der Vereeniging tot Uitgaaf der Bronnen van het Oud-Vaderlandsche Recht, 3d ser., no. 15. Utrecht: Kemink en Zoon, 1951.

Fourquin, Guy. *Le paysan d'Occident au moyen âge.* Paris: Fernand Nathan, 1972.

Ganshof, F. L. "Het laat-Karolingische tijdperk: het ontstaan van het graafschap Vlaanderen en van de Lotharingse gravenhuizen." In *Algemene geschiedenis der Nederlanden*, 1:367–85. Utrecht: De Haan and Antwerpen: Standard Boekhandel, 1949.

———. "Het tijdperk van de Karolingen tot de grote Noormanneninval, 751–879." In *Algemene geschiedenis der Nederlanden*, 1: 306–66. Utrecht: De Haan and Antwerpen: Standaard Boekhandel, 1949.

———. "Het tijdperk van de Merowingen." In *Algemene geschiedenis der Nederlanden*, 1:252- 305. Utrecht: De Haan and Antwerpen: Standaard Boekhandel, 1949.

———, and Adriaan Verhulst. "Medieval Agrarian Society in Its Prime: France, the Low Countries and Western Germany." In M. M. Postan, ed., *The Agrarian Life of the Middle Ages*, 2d ed., pp. 290–339. Vol. 1 in *The Cambridge Economic History of Europe*. Cambridge: Cambridge University Press, 1966.

Genicot, Léopold. "Crisis: From the Middle Ages to Modern Times." In M. M. Postan, ed., *The Agrarian Life of the Middle Ages*, 2d ed., pp. 660–741. Vol. 1 in *The Cambridge Economic History of Europe*. Cambridge: Cambridge University Press, 1966.

Giffen, A. E. van. "Nederzettingssporen van de vroege Klokbekercultuur bij Oostwoud (N.H.)." In *In het voetspoor van A. E. van Giffen*, 2d ed., pp. 66–71, 158, 174–75. Universiteit van Amsterdam, Instituut voor Prae- en Protohistorie, 1951–61. Groningen: J. B. Wolters, 1966.

———. "De ouderdom onzer dijken." *Tijdschrift van het Koninklijk Nederlands Aardrijkskundig Genootschap*, 2d ser., 81 (1964): 273–86.

———. "Three Roman Frontier Forts in Holland, at Utrecht, Valkenburg and Vechten." In Eric Birley, ed., *The Congress of Roman Frontier Studies 1949*, Proceedings of a Congress held at Newcastle upon Tyne, 11–14 July 1949, pp. 31–40. Durham: University of Durham, 1952.

Glacken, Clarence J. "Man and the Earth." *Landscape* 5 (1956): 27–29.

———. *Traces on the Rhodian Shore: Nature and Culture in Western Thought from Ancient Times to the End of the Eighteenth Century*. Berkeley and Los Angeles: University of California Press, 1967.

Glasbergen, W. "De abdijkerk van Rijnsburg: opgravingen in 1949." *Leids jaarboekje* 42 (1954): 46–49.

———. "Sporen van Rothulfuashem (het vroeg-middeleeuwsche Rijnsburg)." *Leids jaarboekje* 36 (1944): 101–109.

———, and W. Groenman–van Waateringe. *The Pre-Flavian Garrisons of Valkenburg, Z.H.: Fabriculae and Bipartite Barracks*. Verhandelingen der Koninklijke Nederlandse Akademie van Wetenschappen, afdeling letterkunde, n,s., no. 85. Amsterdam and London: North-Holland Publishing Co., 1974.

———, et al. "De neolithische nederzettingen te Vlaardingen (Z.H.)." In *In het voetspoor van A. E. van Giffen*, 2d ed., pp. 41–65, 157–58, 173. Universiteit van Amsterdam, Instituut voor Prae- en Protohistorie, 1951–61. Groningen: J. B. Wolters, 1966.

———, et al. *De Romeinse castella te Valkenburg Z.H.: de opgravingen in de dorpsheuvel in 1962*. Universiteit van Amsterdam, Albert Egges van Giffen Instituut voor Prae- en Protohistorie, Cingula 1. Groningen: Wolters-Noordhoff, 1972.

———, et al. "Settlements of the Vlaardingen Culture at Voorschoten and Leidschendam." *Helinium* 7 (1967): 3–31, 97–120.

Godwin, H. "The Beginnings of Agriculture in North West Europe." In Joseph Hutchinson, ed., *Essays in Crop Plant Evolution*, pp. 1–22. Cambridge: Cambridge University Press, 1965.

Gosses, I. H. "Deensche heerschappijen in Friesland gedurende den Noormanentijd." In his *Verspreide geschriften*, ed. F. Gosses and J. F. Niermeyer, pp. 130–51. Groningen and Batavia: J. B. Wolters, 1946.

———. *Handboek tot de staatkundige geschiedenis der Nederlanden*. Vol. 1, *De middeleeuwen*, rev. ed., ed. R. R. Post. The Hague: Nijhoff, 1959.

———. "De vorming van het graafschap Holland." In his *Verspreide geschriften*, ed. F. Gosses and J. F. Niermeyer, pp. 239–344. Groningen and Batavia: J. B. Wolters, 1946.

Gottschalk, M. K. E. "De ontginning der Stichtse venen ten oosten van de Vecht." *Tijdschrift van het Koninklijk Nederlandsch Aardrijkskundig Genootschap*, 2d ser., 73 (1956): 207–22.

———. *Stormvloeden en rivieroverstromingen in Nederland*. Vol. 1, *De periode voor 1400*; vol. 2, *De periode 1400–1600*. Sociaal geografische studies, nos. 10, 13. Assen: Van Gorcum, 1971, 1975.

Goudappel, H. "Uit de kranten." *Westerheem* 29 (1980): 311.

Grand, Roger, and Raymond Delatouche. *L'Agriculture au moyen âge de la fin de l'Empire Romain au XVIe siècle*. L'Agriculture à travers les âges, vol. 3. Paris: E. de Bocard, 1950.

Green, Stanton W. "The Agricultural Colonization of Temperate Forest Habitats: An Ecological Model." In William W. Savage, Jr., and Stephen I. Thompson, eds., *The Frontier: Comparative Studies*, 2:69–103. Norman: University of Oklahoma Press, 1979.

Grigg, David B. *The Agricultural Revolution in South Lincolnshire*. Cambridge Studies in Economic History. Cambridge: Cambridge University Press, 1966.

———. *The Agricultural Systems of the World: An Evolutionary Approach*. Cambridge Geographical Studies, vol. 5. Cambridge: Cambridge University Press, 1974.

———. "Ester Boserup's Theory of Agrarian Change." *Progress in Human Geography* 3 (1979): 64–84.

———. *Population Growth and Agrarian Change: An Historical Perspective*. Cambridge Geographical Studies, vol. 13. Cambridge: Cambridge University Press, 1980.

Groenman–van Waateringe, W. "Grain Storage and Supply in the Valkenburg Castella and Pretorium Aggripinae." In B. L. van Beek, R. W. Brandt, and W. Groenman–van Waateringe, eds., *Ex horreo: I.P.P. 1951–1976*, pp. 226–40. University of Amsterdam, Albert Egges van Giffen Instituut voor Prae- en Protohistorie, Cingula 4. Amsterdam: I.P.P., 1977.

———. "Nederzettingen van de Hilversumcultuur te Vogelenzang (N.H.) en Den Haag (Z.H.)." In *In het voetspoor van A. E. van Giffen*, 2d ed., pp. 81–92, 158, 176–77. Universiteit van Amsterdam, Instituut voor Prae- en Protohistorie, 1951–61. Groningen: J. B. Wolters, 1966.

———, et al. "Een boerderij uit de eerste eeuw na Chr. te Krommenie (N.H.)." In *In het voetspoor van A. E. van Giffen*, 2d ed., pp. 110–28, 158, 178–79. Universiteit van Amsterdam, Instituut voor Prae- en Protohistorie, 1951–61. Groningen: J. B. Wolters, 1966.

———, et al. "Settlements of the Vlaardingen Culture at Voorschoten and Leidschendam (Ecology)." *Helinium* 8 (1968): 105–30.

Güray, A. R. "De bodemgesteldheid van de IJpolders." *Boor en spade* 5 (1952): 1–28.

Gysseling, Maurits. *Toponymisch woordenboek van België, Nederland, Luxemburg, Noord-Frankrijk en West-Duitsland (vóór 1226)*. 2 vols. Bouwstoffen en studiën voor de geschiedenis en de lexicografie van het Nederlands, no. 6. Brussels: Belgisch Interuniversitair Centrum voor Neerlandistiek, 1960.

Haalebos, J. K. "Het einde van de weg." *Westerheem* 25 (1976): 24–29.

———. *Zwammerdam-Nigrum Pullum: Ein Auxiliarskastell am niedergermanischen Limes*. University of Amsterdam, Albert Egges van Giffen Instituut voor Prae- en Protohistorie, Cingula 3. Amsterdam: I.P.P., 1976.

Haans, J. C. R. M., and G. C. Maarleveld. "De zandgronden." In *De Bodem van Nederland: toelichting bij de bodemkaart van Nederland*, comp. Stichting voor Bodemkaartering, pp. 178–206. Wageningen: Stichting voor Bodemkaartering, 1965.

Haarnagel, W. "Die frühgeschichtliche Handels-Siedlung Emden und ihre Entwicklung bis ins Mittelalter." *Friesisches Jahrbuch*, 1956, pp. 9–78.

———. "Die prähistorischen Siedlungsformen im Küstengebiet der Nordsee." In *Beiträge zur Genese der Siedlungs- und Agrarlandschaft in Europa*, pp. 67–84. Geographische Zeitschrift: Beihefte. Wiesbaden: Franz Steiner Verlag, 1968.

Haenens, A. d'. *Les invasions normandes en Belgique au IXe siècle: le phénomène et sa répercussions dans l'historiographie médiévale*. Recueil de travaux d'histoire et philologie, 4th ser., fasc. 38. Louvain: Bureaux du Recueil, Bibliothèque & Publications Universitaires, 1967.

———. "De post-Karolingische periode." In R. C. van Caenegem and H. P. H. Jansen, eds., *De Lage Landen van prehistorie tot 1500*, pp. 61–173. Amsterdam: Elsevier, 1978.

Hageman, B. P. "Development of the Western Part of the Netherlands during the Holocene." *Geologie en mijnbouw* 48 (1969): 373–88.

———, and H. Kliewe. "Neue Forschungen zur Stratigraphie mariner und perimariner Holozänsedimente in den Niederlanden." *Petermanns Geographische Mitteilungen* 113 (1969): 125–29.

Halbertsma, H. "De cultuur van het noordelijk kustgebied." In J. E. Bogaers, W. Glasbergen, P. Glazema, and H. T. Waterbolk, eds., *Honderd eeuwen Nederland*, pp. 178–97 [comprises *Antiquity and Survival*, vol. 2, nos. 5–6]. The Hague: Luctor et Emergo, 1959.

———. "Enkele aantekeningen bij een verzameling oudheden, afkomstig uit een terpje bij Deinum." *Jaarverslag van de Vereeniging voor Terpenonderzoek*, nos. 33–37 (1949–53): 239–56.

———. "The Frisian Kingdom." *Berichten van de Rijksdienst voor het Oudheidkundig Bodemonderzoek* 15–16 (1965–66): 69–108.

———. *Terpen tussen Vlie en Eems: een geografisch-historische benadering*. 2 vols. Vereniging voor Terpenonderzoek. Groningen: J. B. Wolters, 1963.

Hallewas, D. P. "Een huis uit de vroege ijzertijd te Assendelft (N.H.)." *Westerheem* 20 (1971): 19–35.

———, and J. F. van Regteren Altena. "Bewoningsgeschiedenis en landschapsontwikkeling rond de Maasmond." In A. Verhulst and M. K. E. Gottschalk, eds., *Transgressies en occupatiegeschiedenis in de kustgebieden van Nederland en België*, pp. 155- 207. Belgisch Centrum voor Landelijke Geschiedenis, no. 66. Gent: Belgisch Centrum voor Landelijke Geschiedenis, 1980.

Harris, David R. "Swidden Systems and Settlement." In Peter Ucko, Ruth Tringham, and G. W. Dimbleby, eds., *Man, Settlement and Urbanism*, pp. 245–62. London: Duckworth, 1972.

Havelaar, L. "Een huisplattegrond uit de vroege-ijzertijd te Vlaardingen." *Westerheem* 19 (1970): 120–27.

Helderman, E. J. "Enige resultaten van vijftien jaar archeologisch onderzoek in de Zaanstreek." *Westerheem* 20 (1971): 36–83.

———. "Een nederzetting van de Zeijenercultuur te Assendelft." *Westerheem* 16 (1967): 183–90.

Henderikx, P. A. "De oprichting van het hoogheemraadschap van de Alblasserwaard in 1277." *Holland: regionaal-historisch tijdschrift* 9 (1977): 212–22.

———. "De zorg voor de dijken in het Baljuwschap Zuid-Holland en de grensgebieden ten oosten daarvan tot het einde van de 13e eeuw." *Geografisch tijdschrift* 11 (1977): 407–27.

Herlihy, David. "Ecological Conditions and Demographic Change." In Richard L. De Molen, ed., *One Thousand Years: Western Europe in the Middle Ages*, pp. 3–43. Boston: Houghton Mifflin Co., 1974.

Hodges, Richard. "Trade and Urban Origins in Dark Age England: An Archaeological Critique of the Evidence." *Berichten van de Rijksdienst voor het Oudheidkundig Bodemonderzoek* 27 (1977): 191–215.

Hoek, C. "Heer Floris van Alkemade als oprichter van windwatermolens." *Holland: regionaal-historisch tijdschrift* 5 (1973): 142–44.

———. "Schiedam: een historisch-archeologisch onderzoek." *Holland: regionaal-historisch tijdschrift* 7 (1975): 89–95.

Hoffmann, Richard C. "Medieval Origins of the Common Fields." In William N. Parker and Eric L. Jones, eds., *European Peasants and Their Markets: Essays in Agrarian Economic History*, pp. 23–71. Princeton, N.J.: Princeton University Press, 1975.

Hulkenberg, A. M. *Keukenhof*. Hollandse studiën, vol. 7. Dordrecht: Historische Vereniging Holland, 1975.

Hulst, R. S. "Bewoning op het oostelijke rivierengebied in de Romeinse tijd." *Westerheem* 23 (1974): 233–36.

Immink, P. W. A. "Recht en historie." In his *Verspreide geschriften*, ed. N. E. Algra, pp. 23–41. Groningen: Noordhoff, 1967.

Jäger, Helmut. "Zur Geschichte der deutschen Kulturlandschaften." *Geographische Zeitschrift* 51 (1963): 90–143.

Jankuhn, H. "Der fränkisch-friesische Handel zur Ostsee im frühen Mittelalter." *Vierteljahrschrift für Sozial- und Wirtschaftsgeschichte* 40 (1953): 193–243.

———. "Rodung und Wüstung in vor- und frühgeschichtlicher Zeit." In *Die deutsche Ostsiedlung des Mittelalters als Problem der europäischen Geschichte*, pp. 79–129. Konstanzer Arbeitskreis für Mittelalterliche Geschichte, Reichenau-Vorträge, Vorträge und Forschungen, vol. 18. Sigmaringen: Jan Thorbecke Verlag, 1975.

———. *Vor- und Frühgeschichte: vom Neolithikum bis zur Völkerwanderungszeit.* Deutsche Agrargeschichte, vol. 1. Stuttgart: Eugen Ulmer, 1969.

Jansen, H. P. H. "Een economisch contrast in de Nederlanden; Noord en Zuid in de twaalfde eeuw." *Bijdragen en mededelingen betreffende de geschiedenis der Nederlanden* 98 (1983): 3–18.

———. "Holland's Advance." *Acta Historiae Neerlandicae* 10 (1978): 1–19.

———, and L. Milis. "De middeleeuwen." In R. C. van Caenegem and H. P. H. Jansen, eds., *De Lage Landen van prehistorie tot 1500*, pp. 175–342. Amsterdam: Elsevier, 1978.

Janssen, Walter. "Dorf und Dorfformen des 7. bis 12. Jahrhunderts im Lichte neuer Ausgrabungen in Mittel- und Nordeuropa." In *Das Dorf der Eisenzeit und des frühen Mittelalters: Siedlungsform-wirtschaftliche Funktion-soziale Struktur*, pp. 285–356. Bericht über die Kolloquien der Kommission für die Altertumskunde Mittel- und Nordeuropas in den Jahren 1973 und 1974, ed. Herbert Jankuhn, Rudolf Schützeichel, and Fred Schwind. Abhandlungen der Akademie der Wissenschaften in Göttingen, Philologisch-Historische Klasse, 3d ser., no. 101. Göttingen: Vandenhoeck und Ruprecht, 1977.

———. "Some Major Aspects of Frankish and Medieval Settlement in the Rhineland." In P. H. Sawyer, ed., *Medieval Settlement: Continuity and Change*, pp. 41–60. London: Edward Arnold, 1976.

Jappe Alberts, W., and H. P. H. Jansen. *Welvaart in wording: sociaal-economische geschiedenis van Nederland van de vroegste tijden tot het einde van de middeleeuwen.* The Hague: Nijhoff, 1964.

Jelgersma, S. *Holocene Sea Level Changes in the Netherlands.* Mededelingen van de Geologische Stichting, serie C-6, monografiën—reeks behandelende de uitkomsten van nieuwe geologisch-palaeontologische onderzoekingen van de ondergrond van Nederland, no. 7. Maastricht: Ernest van Aelst, 1961.

———, and J. F. van Regteren Altena. "An Outline of the Geological History of the Coastal Dunes in the Western Netherlands." *Geologie en mijnbouw* 48 (1969): 335–42.

———, et al. "The Coastal Dunes of the Western Netherlands: Geology, Vegetational History and Archaeology." *Mededelingen van de Rijks Geologische Dienst*, n.s., no. 21 (1970): 93–167.

Jellema, Dirk. "Frisian Trade in the Dark Ages." *Speculum* 30 (1955): 15–36.

Jelles, J. G. G. *Geschiedenis van beheer en gebruik van het Noordhollands duinreservaat.* Instituut voor Toegepast Biologisch Onderzoek in de Natuur, Mededeling no. 87. Arnhem, 1968.

Kieft, C. van de. "Les comtes de Hollande et les villes de la principauté au XIIIe siècle." *Revue de l'Université libre de Bruxelles* 21 (1970): 479–90.

Koch, A. C. F. "Die Datierung des Vertrags Friedrichs I, Erzbischofs von Hamburg, mit den holländischen Ansiedlern bei Bremen." In *Miscellanea mediaevalia in memoriam Jan Frederik Niermeyer*, pp. 211–15. Groningen: J. B. Wolters, 1967.

———. "Phasen in der Entstehung von Kaufmannsniederlassungen zwischen Maas und Nordsee in der Karolingerzeit." In Georg Droege, Peter Schöller, Rudolf Schützeichel, and Matthias Zender, eds., *Landschaft und Geschichte: Festschrift für Franz Petri zu seinem 65. Geburtstag am 22. Februar 1968*, pp. 312–22. Bonn: Ludwig Rohrscheid Verlag, 1970.

Koebner, Richard. "The Settlement and Colonization of Europe." In M. M. Postan, ed., *The Agrarian Life of the Middle Ages*, 2d ed., pp. 1–91. Vol. 1 in *The Cambridge Economic History of Europe*. Cambridge: Cambridge University Press, 1966.

Kruit, C. "Is the Rhine Delta a Delta?" In *Transactions of the Jubilee Convention*, pp. 257–66. Verhandelingen van het Koninklijk Nederlands Geologisch Mijnbouwkundig Genootschap, geologisch ser., no. 21. Maastricht, 1963.

Laet, S. J. de. "Les limites des cités des Ménapiens et des Morins." *Helinium* 1 (1961): 20–34.

———. *The Low Countries*. Ancient Peoples and Places. New York: Praeger, 1958.

Lamb, H. H. *Climate History and the Future*. Vol. 2 in *Climate: Present, Past and Future*. London: Methuen; New York: Barnes and Noble, 1977.

Lamprecht, Karl. *Deutsches Wirtschaftsleben im Mittelalter: Untersuchungen über die Entwicklung der materielen Kultur des platten Landes auf Grund der Quellen zunächst des Mosellandes*. Vol. 1, pt. 1. Leipzig, 1886.

Latouche, Robert. *The Birth of Western Economy: Economic Aspects of the Dark Ages*. Trans. E. M. Wilkinson. London: Methuen, 1967.

Le Goff, Jacques. "The Town as an Agent of Civilisation, 1200–1500." In Carlo M. Cipolla, ed., *The Middle Ages*, pp. 71–106. Vol. 1 in *The Fontana Economic History of Europe*. London: Collins/Fontana, 1972.

Leighton, Albert C. *Transport and Communication in Early Medieval Europe, A.D.500–1100*. Newton Abbot: David and Charles, 1972.

Leupen, P. *Philip of Leyden, a Fourteenth Century Jurist: A Study of His Life and Treatise "De Cura reipublicae et sorte principantis*." 2 vols. The Hague: Leiden University Press; Zwolle: W. E. J. Tjeenk Willink, 1981.

Levison, Wilhelm. *England and the Continent in the Eighth Century*. Ford Lectures, University of Oxford, Hilary Term 1943. Oxford: Clarendon Press, 1946.

Lewis, Archibald R. "The Closing of the Medieval Frontier." *Speculum* 33 (1958): 475–83.

———. *The Northern Seas: Shipping and Commerce in Northern Europe A.D. 300–1100*. Princeton, N.J.: Princeton University Press, 1958.

Linden, H. van der. *De Cope: bijdrage tot de rechtsgeschiedenis van de openlegging der Hollands-Utrechtse laagvlakte*. Bijdragen van het Instituut voor Rechtsgeschiedenis der Rijksuniversiteit te Utrecht, vol. 1. Assen: Van Gorcum, 1956.

———. "Iets over wording, ontwikkeling en landschappelijk spoor van de Hollandse waterschappen." *Holland: regionaal-historisch tijdschrift* 10 (1978): 101–13.

———. *Recht en territoir: een rechtshistorisch-sociografische verkenning*. Inaugural address, Vrije Universiteit van Amsterdam, 18 February 1972. Assen: Van Gorcum, 1972.

———. *De Zwammerdam*. Hoogheemraadschap van Rijnland. Leiden: Visdruk, 1971.

Linke, Gerhard. "Der Ablauf der holozänen Transgression der Nordsee aufgrund von Ergebnissen aus dem Gebiet Neuwerk/Scharhörn." *Probleme der Küstenforschung im südlichen Nordseegebiet* 14 (1982): 123–57.

Lopez, Robert S. "The Evolution of Land Transport in the Middle Ages." *Past and Present*, no. 9 (1956): 17–29.

Louwe Kooijmans, L. P. "Archeologische ontdekkingen in het Rijnmondgebied." *Holland: regionaal- historisch tijdschrift* 5 (1973): 25–32, 146–57.

———. "Mesolithic Bone and Antler Implements from the North Sea and from the Netherlands." *Berichten van de Rijksdienst voor het Oudheidkundig Bodemonderzoek* 20–21 (1970–71): 27–73.

———. "The Neolithic at the Lower Rhine: Its Structure in Chronological and Geographical Respect." In Sigfried J. de Laet, ed., *Acculturation and Continuity in Atlantic Europe Mainly during the Neolithic Period and the Bronze Age: Papers Presented at the IVth Atlantic Colloquium, Ghent 1975*, pp. 150–73. Dissertationes Archaeologicae Gandenses, vol. 16. Brugge: De Tempel, 1976.

———. "Het onderzoek van neolithische nederzettingsterreinen in Nederland anno 1979." *Westerheem* 29 (1980): 93–136.

———. "Oudheidkundige boomkorvisserij op de Oosterschelde." *Westerheem* 20 (1971): 151–88.

———. *The Rhine/Meuse Delta: Four Studies on Its Prehistoric Occupation and Holocene Geology*. Analecta Praehistorica Leidensia, vol. 7. Leiden: Leiden University Press, 1974.

Lyon, Bryce. "Medieval Real Estate Developments and Freedom." *American Historical Review* 63 (1957): 47–61.

Madsen, Per Kristian. "Medieval Ploughing Marks in Ribe." *Tools and Tillage* 4 (1980): 36–45.

Mayhew, Alan. *Rural Settlement and Farming in Germany*. New York: Barnes and Noble, 1973.

Mensch, P. J. A. "Dierresten uit de Polder Achthoven (Gem. Leiderdorp)." *Westerheem* 24 (1975): 111–16.

Miller, David Harry, and Jerome O. Steffen, eds. *The Frontier: Comparative Studies*. [Vol. 1.] Norman: University of Oklahoma Press, 1977.

Miskimin, Harry A. *The Economy of Early Renaissance Europe, 1300–1460*. Englewood Cliffs, N.J.: Prentice-Hall, 1969.

Modderman, P. J. R. "A Native Farmstead from the Roman Period near Kethel, Municipality of Schiedam, Province of South Holland." *Berichten van de Rijksdienst voor het Oudheidkundig Bodemonderzoek* 23 (1973): 149–58.

Moerman, H. J. *Nederlandse plaatsnamen: een overzicht*. Nomina geographica Flandrica, Studiën 7. Brussels: Standaard Boekhandel, 1956.

Mortensen, H. "Probleme der mittelalterlichen Siedlungs- und Kulturlandschaft." *Berichte zur deutschen Landeskunde* 20 (1958): 98–104.

Musset, Lucien. *The Germanic Invasions: The Making of Europe A.D. 400–600*. Trans. Edward James and Columba James. London: Paul Elek, 1975.

Nell, Edward J. "The Technology of Intimidation." *Peasant Studies Newsletter* 1 (1972): 39–44.

Niermeyer, J. F. "Dammen en dijken in Frankisch Nederland" In *Weerklank op het werk van Jan Romein: Liber Amicorum*, pp. 109–15. Amsterdam and Antwerpen: Weerldbibliotheek, 1953.

———. "Het midden-Nederlands rivierengebied in de Frankische tijd op grond van de *Ewa quae se ad Amorem habet*." *Tijdschrift voor geschiedenis* 66 (1953): 145–69.

———. "De vroegste berichten omtrent bedijking in Nederland." *Tijdschrift voor economische en sociale geografie* 19 (1958): 226–31.

———. *De wording van onze volkshuishouding: hoofdlijnen uit de economische geschiedenis der noordelijke Nederlanden in de middeleeuwen*. Sevire's encyclopaedie, afdeeling: Geschiedenis, B 3/2. The Hague: Sevire, 1946.

Nitz, Hans-Jürgen. "The Church as Colonist: The Benedictine Abbey of Lorsch and Planned Waldhufen Colonization in the Odenwald." *Journal of Historical Geography* 9 (1983): 105–26.

De "Noordzeecultuur": een onderzoek naar de cuturele relaties van de landen rond de Nordzee in de vroege middeleeuwen. University of Amsterdam, Albert Egges van Giffen Instituut voor Prae- en Protohistorie, working paper 2: Project middeleeuwse archeologie, 1972–74. Amsterdam: I.P.P., 1975.

North, Douglass C., and Robert Paul Thomas. *The Rise of the Western World: A New Economic History*. Cambridge: Cambridge University Press, 1973.

Oppermann, Otto. *Die Egmonder Falschungen*. Vol. 1 in *Untersuchungen zur nordniederländischen Geschichte des 10. bis 13. Jahrhunderts*. Bijdragen van het Instituut voor Middeleeuwse Geschiedenis der Rijksuniversiteit te Utrecht, no. 3. Utrecht: Oosthoek, 1920.

———. *Die Grafschaft Holland und das Reich bis 1256*. Vol. 2 in *Untersuchungen zur nordniederländischen Geschichte des 10. bis 13. Jahrhunderts*. Bijdragen van het Instituut voor Middeleeuwse Geschiedenis der Rijksuniversiteit te Utrecht, no. 4. Utrecht: Oosthoek, 1921.

Otto, J. S., and N. E. Anderson. "Slash-and-Burn Cultivation in the Highlands South: A Problem in Comparative Agricultural History." *Comparative Studies in Society and History* 24 (1982): 131–47.

Paffen, Karlheinz. "Natur- und Kulturlandschaft am deutschen Niederrhein." *Berichte zur deutschen Landeskunde* 20 (1958): 177–288.

Pannekoek, A. J., et al. *Geological History of the Netherlands: Explanation to the General Geological Map of the Netherlands on the Scale 1:200,000*. The Hague, 1956.

Parain, Charles. "The Evolution of Agricultural Technique." In M. M. Postan, ed.,

The Agrarian Life of the Middle Ages, 2d ed., pp. 125–79. Vol. 1 in *The Cambridge Economic History of Europe*. Cambridge: Cambridge University Press, 1966.

Piggott, Stuart. *Ancient Europe from the Beginnings of Agriculture to Classical Antiquity*. Edinburgh: Edinburgh University Press, 1965.

Poel, J. M. G. van der. "De landbouw in het verste verleden." *Berichten van de Rijksdienst voor het Oudheidkundig Bodemonderzoek* 10–11 (1960–61): 125–94.

Pons, L. J. "De veengronden." In *De bodem van Nederland: toelichting bij de bodemkaart van Nederland schaal 1:200,000*, comp. Stichting voor Bodemkaartering, pp. 144–77. Wageningen: Stichting voor Bodemkaartering, 1965.

———. "De zeekleigronden." In *De bodem van Nederland: toelichting bij de bodemkaart van Nederland schaal 1:200,000*, comp. Stichting voor Bodemkaartering, pp. 22–110. Wageningen: Stichting voor Bodemkaartering, 1965.

———, et al. "Evolution of the Netherlands Coastal Area during the Holocene." In *Transactions of the Jubilee Convention*, pp. 197–208. Verhandelingen van het Koninklijk Nederlands Geologisch Mijnbouwkundig Genootschap, geologische ser., vol. 21. Maastricht, 1963.

Pounds, Norman J. G. *An Historical Geography of Europe, 450 B.C.–A.D. 1330*. Cambridge: Cambridge University Press, 1973.

Prummel, Wietske. "Vlees, gevogelte en vis." *Spieghel historiael* 13 (1978): 282–93.

Rackham, Oliver. *Ancient Woodland: Its History, Vegetation and Uses in England*. London: Edward Arnold, 1980.

Regteren Altena, H. H. van, and H. A. Heidinga. "The North Sea Region in the Early Medieval Period (400–950)." In B. L. van Beek, R. W. Brandt, and W. Groenman–van Waateringe, eds., *Ex horreo: I.P.P. 1951–1976*, pp. 47–67. Univesity of Amsterdam, Albert Egges van Giffen Instituut voor Prae- en Protohistorie, Cingula 4. Amsterdam: I.P.P., 1977.

Regteren Altena, J. F. van, and J. A. Bakker. "De neolithische woonplaats te Zandwerven." In *In het voetspoor van A. E. van Giffen*, 2d ed., pp. 33–40, 157, 171–72. Universiteit van Amsterdam, Instituut voor Prae- en Protohistorie, 1951–61. Groningen: J. B. Wolters, 1966.

Renfrew, Colin. *Before Civilization: The Radiocarbon Revolution and Prehistoric Europe*. New York: Alfred A. Knopf, 1974.

Rentenaar, Rob. "De Nederlandse duinen in de middeleeuwse bronnen tot omstreeks 1300." *Geografisch tijdschrift* 11 (1977): 361–76.

Riché, Pierre. *Daily Life in the World of Charlemagne*. Trans. Jo Ann McNamara. Philadelphia: University of Pennsylvania Press, 1978.

Roelandts, Karl. "*Sēle* und *Heim*." In Rudolf Schützeichel and Matthias Zender, eds., *Namenforschung: Festschrift für Adolf Bach zum 75. Geburtstag am 31. Januar 1965*, pp. 273–99. Heidelberg: Carl Winter–Universitätsverlag, 1965.

Roeleveld, W. "De bijdrage van de aardwetenschappen tot de studie van de transgressieve activiteit langs de zuidelijke kusten van de Nordzee." In A. Verhulst and M. K. E. Gottschalk, eds., *Transgressies en occupatiegeschiedenis in de kustgebieden van Nederland en België*, pp. 292–99. Belgisch Centrum voor Landelijke Geschiedenis, vol. 66. Gent: Belgisch Centrum voor Landelijke Geschiedenis, 1980.

———. *The Holocene Evolution of the Groningen Marine-Clay District* [supplement to *Berichten van de Rijksdienst voor het Oudheidkundig Bodemonderzoek* 24 (1974)]. The Hague: Staatsdrukkerij, 1976.

Romein, J. *Geschiedenis van de Noord-Nederlandse geschiedschrijving in de middeleeuwen: bijdrage tot de beschavingsgeschiedenis*. Haarlem, 1932.

Rouche, Michel. "La faim à l'époque carolingienne: essai sur quelques types de rations alimentaires." *Revue historique*, no. 508 (1973): 295–320.

Russchen, A. "Jutes and Frisians." *It Beaken: tydskrift fan de Fryske Akademy* 26 (1964): 26–37.

———. "Keramiek en ritueel in de vijfde eeuw." *It Beaken: tydskrift fan de Fryske Akademy* 32 (1970): 121–32.

———. *New Light on Dark-Age Frisia*. Fryske Akademy, Utjeften no. 311. Drachten: Laverman, 1967.

———. "Origin and Earliest Expansion of the Frisian Tribe." *It Beaken: tydskrift fan de Fryske Akademy* 30 (1968): 127–49.

———. "Tussen Aller en Somme." *It Beaken: tydskrift fan de Fryske Akademy* 29 (1967): 90–96.

Russell, Josiah Cox. *Late Ancient and Medieval Population*. Transactions of the American Philosophical Society, n.s., vol. 48, pt. 3. Philadelphia: American Philosophical Society, 1958.

———. "Population in Europe, 500–1500." In Carlo M. Cipolla, ed., *The Middle Ages*, pp. 25- 70. Vol. 1 in *The Fontana Economic History of Europe*. London: Collins/Fontana, 1972.

Sarfatij, H. "Archeologische kroniek van Zuid-Holland over 1975." *Holland: regionaal-historisch tijdschrift* 8 (1976): 262–80.

———. "Archeologische kroniek van Zuid-Holland over 1976." *Holland: regionaal-historisch tijdschrift* 9 (1977): 245–67.

———. "Archeologische kroniek van Zuid-Holland over 1977." *Holland: regionaal-historisch tijdschrift* 10 (1978): 297–312.

———. "Archeologische kroniek van Zuid-Holland over 1978." *Holland: regionaal-historisch tijdschrift* 11 (1979): 313–39.

———. "Friezen-Romeinen-Cananefaten." *Holland: regionaal-historisch tijdschrift* 3 (1971): 33–47, 89–105, 153–79.

———. "Die Frühgeschichte von Rijnsburg (8.–12. Jahrhundert): Ein historisch-archäologischer Bericht." In B. L. van Beek, R. W. Brandt, and W. Groenman-van Waateringe, eds., *Ex Horreo, I.P.P. 1951–76*, pp. 290–302. University of Amsterdam, Albert Egges van Giffen Instituut voor Prae- en Protohistorie, Cingula 4. Amsterdam: I.P.P., 1977.

———. "Middeleeuwse mens en eeuwig water: veranderingen in landschap en bewoning aan de monden van Rijn en Maas gedurende de middeleeuwen." *Zuid-Holland: orgaan van de Historische Vereniging voor Zuid-Holland onder de zinspreuk "Vigilate Deo Confidentes"* 14 (1968): 19–27.

Savage, William W., Jr., and Stephen I. Thompson. "The Comparative Study of the Frontier: An Introduction." In William W. Savage, Jr., and Stephen I. Thompson, eds., *The Frontier: Comparative Studies*, 2:3–24. Norman: University of Oklahoma Press, 1979.

———, and ———, eds. *The Frontier: Comparative Studies*. Vol. 2. Norman: University of Oklahoma Press, 1979.

Sawyer, P. H. "Baldersby, Borup and Bruges: The Rise of Northern Europe." *University of Leeds Review* 16 (1973): 75–96.

Schaik, P. van. "De economische betekenis van de turfwinning in Nederland." *Economisch-historisch jaarboek* 32 (1967–68): 186–235.

Scholte, M. C. P. "Polders en polders en polders." *Holland: regionaal-historisch tijdschrift* 13 (1981): 114–15.

———. "Wat was er eerder: de dijk, de veenontginning of de polder?" *Holland: regionaal-historisch tijdschrift* 12 (1980): 1–9.

Schönfeld, M. *Nederlandse waternamen*. Nomina geographica Flandrica, Studiën 6. Brussels: Standaard Boekhandel, 1955.

Schothorst, C. J. "Subsidence of Low Moor Peat Soils in the Western Netherlands." *Geoderma* 17 (1977): 265–91.

Shard, Diana. "The Neolithic Revolution: An Analogical Overview." *Journal of Social History* 7 (1973–74): 165–70.

Slaski, Kasimiers. "North-Western Slavs in Baltic Sea Trade from the VIIIth to the XIIIth Century." *Journal of European Economic History* 8 (1979): 83–107.

Slicher van Bath, B. H. *Agrarian History of Western Europe, A.D. 500–1850*. Trans. Olive Ordish. London: Edward Arnold, 1963.

———. "Agrarische produktiviteit in het pre-industriëlle Europa." In his *Bijdragen tot de agrarische geschiedenis*, pp. 152–97. Utrecht and Antwerpen: Het Spectrum, 1978.

———. "Le climate et les récoltes en haut moyen âge." In *Agricoltura e mondo rurale in Occidente nell'alto medioevo*, pp. 399–425. Settimane di studio del Centro Italiana di Studi sull'alto Medioevo, vol. 13. Spoleto: Presso la Sede del Centro, 1966.

———. "The Economic and Social Conditions in the Frisian Districts from 900 to 1500." *A.A.G. Bijdragen* 13 (1965): 97–133.

———. "L'Histoire des forêts dans les Pays-Bas septentrionaux." *A.A.G. Bijdragen* 14 (1967): 91–104.

———. "De paleodemografie." *A.A.G. Bijdragen* 15 (1970): 134–201.

———. "Problemen rond de Friese middeleeuwse geschiedenis." In his *Herschreven historie: schetsen en studiën op het gebied der middeleeuwse geschiedenis*, pp. 259–80. Leiden: E. J. Brill, 1949.

———. "The Rise of Intensive Husbandry in the Low Countries." In J. S. Bromley and E. H. Kossman, eds., *Papers Delivered to the Oxford-Netherlands Historical Conference, 1959*, pp. 130–53. Vol. 1 in *Britain and the Netherlands*. London: Chatto and Windus, 1960.

———. "Volksvrijheid en democratie." In his *Herschreven historie: schetsen en studiën op het gebied der middeleeuwse geschiedenis*, pp. 305–15. Leiden: E. J. Brill, 1949.

Soesbergen, P. G. van. "The Phases of the Batavian Revolt." *Helinium* 11 (1971): 238–56.

Spooner, Brian, ed. *Population Growth: Anthropological Perspectives*. Cambridge, Mass.: MIT Press, 1972.

Stracke, D. A. "Bernlef." *Historisch tijdschrift* 4 (1925): 59–70, 150–69.

Sturdy, David. "Correlation of Evidence of Medieval Urban Communities." In Peter J. Ucko, Ruth Tringham, and D. W. Dimbleby, eds., *Man, Settlement and Urbanism*, pp. 863–65. London: Duckworth, 1972.

Stuurman, P. "Archeologie van het jaar nul." *Westerheem* 18 (1969): 62–79.

Sullivan, Richard E. "The Medieval Monk as Frontiersman." In William W. Savage, Jr., and Stephen I. Thompson, eds., *The Frontier: Comparative Studies*. 2:25–49. Norman: University of Oklahoma Press, 1979.

TeBrake, William H. "Air Pollution and Fuel Crises in Preindustrial London, 1250–1650." *Technology and Culture* 16 (1975): 337–59.

———. "Ecology and Economy in Early Medieval Frisia." *Viator: Medieval and Renaissance Studies* 9 (1978): 1–29.

Tegenwoordige staat der Vereenigde Nederlanden. Vol. 4, *De beschryving van Holland*. Amsterdam: Isaak Trion, 1742.

Tent, W. J. van. "De landschappelijke actergronden." *Spieghel historiael* 13 (1978): 205–14.

———, and P. J. Woltering. "The Distribution of Archaeological Finds on the Island of Texel, Province of North-Holland." *Berichten van de Rijksdienst voor het Oudheidkundig Bodemonderzoek* 23 (1973): 49–63.

Thomas, Charles. "Towards the Definition of the Term 'Field' in the Light of Prehistory." In P. H. Sawyer, ed., *Medieval Settlement: Continuity and Change*, pp. 145–51. London: Edward Arnold, 1976.

Uhlig, H. "Old Hamlets with Infield and Outfield Systems in Western and Central Europe." *Geografiska Annaler* 43 (1961): 285–312.

Ulbert, Günter. "Die römischen Funde von Bentumersiel." *Probleme der Küstenforschung im südlichen Nordseegebiet* 12 (1977): 33–65.

Uyl, R. G. den. "Dorpen in het rivierkleigebied." *Bulletin van de Koninklijke Nederlandse Oudheidkundig Bond*, 6th ser., 11 (1958), cols. 97–114.

Vader, J. "Een en ander omtrent de eerste bemaling met windwatermolens in het Hollandsche laagveengebied." *Tijdschrift voor geschiedenis* 28 (1913): 86–102.

Verhulst, A. E. "Historische geografie van de Vlaamse kustvlakte tot omstreeks 1200." *Bijdragen voor de geschiedenis der Nederlanden* 14 (1959–60): 1–37.

———. "Karolingische Agrarpolitik: Das *Capitulare de Villis* und die Hungersnöte von 792/93 und 805/06." *Zeitschrift für Agrargeschichte und Agrarsoziologie* 13 (1965): 175–89.

———. *De Sint-Baafsabdij te Gent en haar grondbezit (VIIe-XIVe eeuw): bijdrage tot de kennis van de structuur en de uitbating van het grootgrondbezit in Vlaanderen tijdens de middeleeuwen*. Verhandelingen van de Koninklijke

Vlaamse Academie voor Wetenschappen, Letteren en Schone Kunsten van België, klasse der letteren, no. 30. Brussels: Paleis der Academiën, 1958.

———, and M. K. E. Gottschalk, eds. *Transgressies en occupatiegeschiedenis in de kustgebieden van Nederland en België*. Belgisch Centrum voor Landelijke Geschiedenis, vol. 66. Gent: Belgisch Centrum voor Landelijke Geschiedenis, 1980.

Vollgraff, C. W. "Eene Romeinsche koopacte uit Tolsum." *De Vrije Fries: tijdschrift uitgegeven door het Friesch Genootschap van Geschied-, Oudheid- en Taalkunde* 25 (1917): 71–101.

Vons, P. "The Identification of Heavily Corroded Roman Coins Found at Velsen." *Berichten van de Rijksdienst voor het Oudheidkundig Bodemonderzoek* 27 (1977): 139–63.

———. "Op zoek naar een castellum." *Westerheem* 23 (1974): 59–69.

———. "De vervardiging van barnsteen-kralen te Velsen in de vroege bronstijd." *Westerheem* 19 (1970): 34–35.

Vos, P. C. "De relatie tussen de geologische ontwikkeling en de bewoningsgeschiedenis in de Assendelver Polders vanaf 1000 v. Chr." *Westerheem* 32 (1983): 54–80.

Vries, Jan de. *The Dutch Rural Economy in the Golden Age, 1500–1700*. Yale Series in Economic History. New Haven, Conn., and London: Yale University Press, 1974.

Vries, J. M. P. L. de. "De laatste Wikingstochten in de gewesten van den Nederrijn, XIe eeuw." *Bijdragen voor vaderlandsche geschiedenis en oudheidkunde*, 5th ser., 10 (1923): 249–56.

Waals, J. D. van der. *Prehistoric Disc Wheels in the Netherlands*. Groningen: J. B. Wolters, 1964.

Wailes, Bernard. "Plow and Population in Temperate Europe." In Brian Spooner, ed., *Population Growth: Anthropological Implications*, pp. 154–79. Cambridge, Mass.: MIT Press, 1972.

Wallace-Hadrill, J. M. "A Background to St. Boniface's Mission." In his *Early Medieval History*, pp. 138–54. Oxford: Blackwell's, 1975.

———. "Early Medieval History." In his *Early Medieval History*, pp. 1–8. Oxford: Blackwell's, 1975.

———. "The Vikings in Francia." In his *Early Medieval History*, pp. 217–36. Oxford: Blackwell's, 1975.

Wallenburg, C. van, and W. C. Markus. "Toemaakdekken in het Oude Rijngebied." *Boor en spade* 17 (1971): 64–81.

Wallerstein, Immanuel. *The Modern World-System I: Capitalist Agriculture and the Origins of the European World-Economy in the Sixteenth Century*. Studies in Social Discontinuity. New York and London: Academic Press, 1974.

Wassink, A. "Ligt de oorsprong van de stad Leiden bij het Romeinse castellum Matilo?" *Westerheem* 27 (1978): 294–98.

———. "Het Romeinse castellum te Alphen aan den Rijn." *Westerheem* 32 (1983): 296–302.

Waterbolk, H. T. "The Bronze Age Settlement of Elp." *Helinium* 4 (1964): 97–131.

———. "Evidence of Cattle Stalling in Excavated Pre- en Protohistoric Houses." In A. T. Clason, ed., *Archaeozoological Studies*, pp. 383–940. Papers of the Archaeozoological Conference, Biologisch-Archaeologisch Instituut, State University of Groningen, 1974. Amsterdam: North Holland Publishing Co.; New York: American Elsevier, 1975.

———. "Food Production in Prehistoric Europe." *Science* 162 (1968): 1093–1102.

———. "The Lower Rhine Basin." In Robert J. Braidwood and Gordon R. Willey, eds., *Courses Toward Urban Life: Archaeological Considerations of Some Cultural Alternates*, pp. 227–53. Chicago: Aldine, 1962.

———. "The Occupation of Friesland in the Prehistoric Period." *Berichten van de Rijksdienst voor het Oudheidkundig Bodemonderzoek* 15–16 (1965–66): 13–35.

———. *De praehistorische mens en zijn milieu: een palynologisch onderzoek naar de menselijke invloed op de plantengroei van de diluviale gronden in Nederland*. Assen: Van Gorcum, 1954.

———. "Siedlungskontinuität im Küstengebiet der Nordsee zwischen Rhein und Elbe." *Probleme der Küstenforschung im südlichen Nordseegebiet* 13 (1979): 1–21.

———. "Terpen, milieu en bewoning." In J. W. Boersma, ed., *Terpen—mens en milieu*, pp. 6–24. Triangelreeks. Haren, Groningen: Knoop en Niemeijer, 1970.

Watson, Andrew M. "Towards Denser and More Continuous Settlement: New Crops and Farming Techniques in the Early Middle Ages." In J. A. Raftis, ed., *Pathways to Medieval Peasants*, pp. 65–82. Papers in Mediaeval Studies, vol. 2. Toronto: Pontifical Institute of Mediaeval Studies, 1981.

Wenskus, R. *Stammesbildung und Verfassung: Das Werden der frühmittelalterlichen Gentes*. Cologne and Graz: Böhlau Verlag, 1961.

White, Lynn, Jr. "Cultural Climates and Technological Advance in the Middle Ages." *Viator: Medieval and Renaissance Studies* 2 (1971): 171–201.

———. "The Expansion of Technology, 500–1500." In Carlo M. Cipolla, ed., *The Middle Ages*, pp. 143–74. Vol. 1 in *The Fontana Economic History of Europe*. London: Collins/Fontana, 1972.

———. *Medieval Technology and Social Change*. Oxford: Clarendon Press, 1962.

Wilson, David. *The Anglo Saxons*. 2d ed. Harmondsworth: Penguin Books, 1971.

Wind, C. "Een nederzetting uit de voor-Romeinse ijzertijd te Rockanje." *Westerheem* 19 (1970): 242–61.

Winter, J. M. van. "Die Entstehung der Landgemeinde in der Holländisch-Utrechtschen Tiefbene." In *Die Anfänge der Landgemeinde und Ihr Wesen*, pt. 1, pp. 439–45. Konstanzer Arbeitskreis für Mittelalterliche Geschichte, Vorträge und Forschungen, nos. 7–8. Constance and Stuttgart: Jan Thorbecke Verlag, 1964.

———. "Vlaams en Hollands recht bij de kolonisatie van Duitsland in de 12e en 13e eeuw." *Tijdschrift voor rechtsgeschiedenis* 21 (1953): 205–24.

Wolf, Eric. *Peasants*. Foundations of Modern Anthropolgy Series. Englewood Cliffs, N.J.: Prentice-Hall, 1966.

Woltering, P. J. "Archeologische kroniek van Noord-Holland over 1975." *Holland: regionaal-historisch tijdschrift* 8 (1976): 235–61.

———. "Archeologische kroniek van Noord-Holland over 1976." *Holland: regionaal-historisch tijdschrift* 9 (1977): 187–211.

———. "Archeologische kroniek van Noord-Holland over 1977." *Holland: regionaal-historisch tijdschrift* 10 (1978): 250–76.

———. "Archeologische kroniek van Noord-Holland over 1978." *Holland: regionaal-historisch tijdschrift* 11 (1979): 242–73.

Ypey, J. "Frankisch goud in Beuningen (Gld.)." In W. A. van Es. et al., eds., *Archeologie en historie: opgedragen aan H. Brunsting bij zijn zeventigste verjaardag*, pp. 441–58. Bussum: Fibula-van Dishoeck, 1973.

———. "De verspreiding van vroeg-middeleeuwse vondsten in Nederland." *Berichten van de Rijksdienst voor het Oudheidkundig Bodemonderzoek* 9 (1959): 98–118.

Zadocks-Josephus Jitta, A. N. "Looking Back at 'Frisians, Franks, and Saxons.'" *Bulletin van de Vereeniging tot Bevordering der Kennis van de Antieke Beschaving te 's-Gravenhage* 36 (1961): 41–59.

Zagwijn, W. H. "De ontwikkeling van het 'Oer-IJ' estuarium en zijn omgeving." *Westerheem* 20 (1971): 11–18.

Zeist, W. van. "Agriculture in Early-Medieval Dorestad: A Preliminary Report." *Berichten van de Rijksdienst voor het Oudheidkundig Bodemonderzoek* 19 (1969): 209–12.

———. "The Environment of 'Het Torp' in Its Early Phases." *Berichten van de Rijksdienst voor het Oudheidkundig Bodemonderzoek* 23 (1973): 347–53.

———. "Prehistoric and Early Historic Food Plants in the Netherlands." *Palaeohistoria: Acta et Communicationes Instituti Bio-Archaeologici Universitatis Groninganae* 14 (1968): 41–173.

Index